City Visions

Pearson
Education

We work with leading authors to develop the strongest educational materials in geography, bringing cutting-edge thinking and best learning practice to a global market.

Under a range of well-known imprints, including Prentice Hall, we craft high quality print and electronic publications which help readers to understand and apply their content, whether studying or at work.

To find out about the complete range of our publishing please visit us on the World Wide Web at: http://www.pearsoneduc.com

City Visions

edited by

David Bell and

Azzedine Haddour

An imprint of **Pearson Education**

Harlow, England · London · New York · Reading, Massachusetts · San Francisco · Toronto · Don Mills, Ontario · Sydney
Tokyo · Singapore · Hong Kong · Seoul · Taipei · Cape Town · Madrid · Mexico City · Amsterdam · Munich · Paris · Milan

Pearson Education Limited
Edinburgh Gate
Harlow
Essex CM20 2JE
United Kingdom
and Associated Companies throughout the world

Visit us on the World Wide Web at
http://www.pearsoneduc.com

First published 2000

ISBN 0 582 32741 5

British Library Cataloguing-in-Publication Data
A catalogue record for this book can be obtained
from the British Library.

Library of Congress Cataloging-in-Publication Data
City visions / edited by David Bell and Azzedine Haddour.
     p. cm.
   Includes bibliographical references and index.
   ISBN 0-582-32741-5 (pbk. : alk. paper)
     1. Cities and towns--Congresses. 2. Cities and towns in literature--Congresses. 3. City
and town life--Congresses. I. Bell, David, 1965- II. Haddour, Azzedine.

HT107 .C493 2000
307.76--dc21   00-021677

Typeset by 7 in 10/12pt Garamond

Transferred to Digital Print on Demand 2009

Printed and bound in Great Britain by
CPI Antony Rowe, Chippenham and Eastbourne

Contents

Acknowledgements

This collection of essays emerges from the City Limits conference organized at Staffordshire University and sponsored by the School of Humanities and Social Sciences. Our first thanks are due to the many colleagues who supported this event. We would like to extend our gratitude to all the contributors, and to Jackie Clewlow for assistance in the preparation of the manuscript. Last but not least, our many thanks to Matthew Smith, commissioning editor at Pearson Education for his support and patience, and to Shuet-Kei Cheung for her work on the book.

Contributors

Mark Jayne is a research scholar in Sociology at Staffordshire University. He is currently completing a PhD on urban regeneration in Stoke-on-Trent.

John Hutnyk is senior lecturer in Anthropology at Goldsmith's College. He is the co-editor of *Dis-Orienting Rhythms* and *Travel Worlds*.

Azzedine Haddour is a lecturer in French at University College London. He is the author of *Colonial Myths, History and Narrative*.

Willy Maley is Professor of English Literature at the University of Glasgow. He is the author of *Salvaging Spenser: Colonialism, Culture and Identity* and *Spenser Chronology*, and co-editor of *Postcolonial Criticism*.

David Oswell is lecturer in Communication and Media Studies at the Centre for Research into Innovation, Culture and Technology, Brunel University. He is the author of *Children's Television in Context*.

Christopher Stanley is the author of *Urban Excess and the Law: Capital, Culture and Desire*.

Valerie Briginshaw teaches at University College, Chichester.

Tim Edensor is senior lecturer in Cultural Studies at Staffordshire University. He is the author of *Tourists at the Taj* and editor of a forthcoming volume, *Reclaiming Stoke-on-Trent*.

Ian Toon is a research scholar in Cultural Studies at Staffordshire University. He is currently completing a PhD on youth cultures and identities in Tamworth.

Jon Binnie is a lecturer in Human Geography at Brunel University. He is co-author of *Hard Choices: Sexual Citizenship after Queer Politics*, and is currently completing a book on globalisation and sexuality.

David Theo Goldberg is Professor of Justice Studies and Director of the School of Justice Studies at Arizona State University. He is the author of *Racist Culture* and Racial Subjects. He is also the editor of *Anatomy of Racism* and *Multiculturalism*, and co-editor of *Social Identities: A Journal for the Studies of Race, Nation and Culture*.

Iris Marion Young is Professor of Public and International Affairs at the University of Pittsburgh. She is the author of *Justice and the Politics of*

Difference, Throwing Like a Girl and Other Essays in Feminist Philosophy and Social Theory, and *Intersecting Voices: Dilemmas of Gender, Political Philosophy and Policy*.

Anthony Gorman teaches Philosophy at Staffordshire University.

Nigel Thrift is Professor of Geography at the University of Bristol. His books include *Writing the Rural* and *Spatial Formations*, and (as co-editor) *City A-Z* and *Mapping the Subject*.

David Bell teaches Cultural Studies at Staffordshire University. He is co-editor of *Mapping Desire* and *The Cybercultures Reader*, and co-author of *Consuming Geographies* and *Hard Choices*.

Chapter 1

What we talk about when we talk about the city

David Bell and Azzedine Haddour

Talking about cities has become, at the end of the millennium, a way of talking about lots of different things. 'The city' has come to be a symbol – maybe even a symptom – of almost every social and cultural process. Cities are certainly *concentrations* of those processes: the city is often read as the medium through which modernity (and then postmodernity) gets expressed, worked through, concretized. Yet, as this collection shows, there are in fact an incredible multiplicity of cities, and ways of talking about them. In addition, there are even more multiple and heterogeneous ways of experiencing cities, of living with cities (and in cities). It is not possible (nor desirable) to construct one prescribed way to approach the city – say, exclusively through urban semiotics or urban ethnography – because of this heterogeneity. Instead, what the essays in *City Visions* collectively attempt to present is a series of glimpses, sometimes of particular cities, sometimes of particular ways of thinking about cities. Each has its own approach, its own agenda, its own emphases, but all are bound by a desire to think and talk about cities and the things that go on in them. Taken together, with their polymorphous approaches, the essays in this collection intervene in the current discourses of city cultures and urban life, investigating ways of working with cities and city visions. This intervention is political, in the sense that it seeks to ask questions about what cities are and how they work (and don't work) for their citizens, as well as suggesting ways in which cities might be reshaped so that they work better. At the same time, the contributors ask that we challenge the ways that we think about cities, activating new approaches and agendas.

In this introduction, we want to suggest some connections that work between and across the essays; to trace some ways of talking and thinking about cities. Then, in the chapter that follows, the focus turns to a reading of the city from which the collection emerges: Stoke-on-Trent (otherwise known as The Potteries) in the English Midlands. There are a number of motives for placing Mark Jayne's chapter here. First, much of the work on cities to date has concerned itself either with 'The City' in an abstract sense, or with the mightiest of megalopolises, or world cities: London, Los Angeles, New York and so on. So-called 'second tier' (or third tier) cities get less of a look-in. Secondly, because the collection has been pulled together from within that city – and has its origins in a conference held here – it would seem some kind of civic disservice not to look at Stoke-on-Trent itself; cities like Stoke embody the processes of urban social and cultural transformation in distinct ways, which are revealing of the mesh of city policies and politics as they interact with everyday urban life. Thirdly, the city is currently attempting to consciously respond to the challenges of the new urban order, most notably via the development of a Cultural Quarter – and so, in sharing the tales of this city, we hope to be able to throw some suggestive light on a number of theoretical and empirical agendas for examining how cities are talked about by some of the key actors involved in reshaping and restructuring initiatives. How, in short, are city visions conjured and promoted in the context of a city like Stoke-on-Trent? To answer this question at least in part, Jayne describes the processes and outcomes of the city's drive to re-imagine itself on a post-industrial stage, setting this story in the general context of urban restructuring imperatives in the UK as well as the particular context of the Potteries conurbation. Before that, however, we propose to map the collection out. Of course, this introduces a second-order act of reading and representation: we offer our own takes on each author's take on the city. It is up to the reader to also make her or his own map.

Reading City Visions

The essays that follow in this collection take us away from Stoke, across the globe, and give us many different ways of envisioning the city. In some cases, the writers take someone else's vision – an author, a film-maker, a choreographer – and re-read it through their own eyes. From Hutnyk's coin tricks in Calcutta, talked about via *City of Joy* and with a little help from Marx, to Maley's tales of Glasgow and Edinburgh as seen in James Kelman and Irvine Welsh's fictions, or Haddour on Camus; and from Oswell on English suburbia according to Hanif Kureishi to Briginshaw's danced cityspaces or Stanley's melding of Antigone and Rachel Whiteread, we can see that one way of talking about cities is to find them in the creative imagination, and then attempt to interpret the act of representation. This has, of course, been a dominant approach to the city – to find it in film or in fiction and to 'read' it (see, for example, Clarke, 1997; Jarvis, 1998).

For John Hutnyk, then, the symbolic resonances of gift-giving by Western visitors in Third World cities settle on the coin as a condensed signifier of colonial–capitalist appropriation (or, as he puts it, as 'the icon of contemporary politics'), tying gift, exchange, market and city together into the logic of 'development' that is ushered in behind colonialism. Through a focus on Calcutta, both via experience and representation (*City of Joy*), Hutnyk begins to unpack and unpick the narratives which refract only certain versions of and responses to the Third World city. The responses articulated in *City of Joy* (as a story and as a project) centre on the 'compassionate' West's interventions: charity, development ... But, as Hutnyk concludes, '[t]here is an alternative to the extension of the market to all corners of the planet, and it is not a universal gift service. Charity contributes nothing but the maintenance of the trick.

Travellers who don't go further than the doorstep of their hotel miss the point.' So, in the end, his essay asks that we rethink our relations to cities like Calcutta; that we question the ways we are scripted into dealing with them; and that we work to find a way to unpack the politics of the coin-trick which, he argues, still structures those scripts and relations.

The spatial logics of the colonial city are at the heart of Azzedine Haddour's chapter, too. Beginning with a reading of Camus' *La Peste*, he maps out the exclusionary logic which expels the colonized body through the rhetoric of plague, setting this against the metaphorical figuring of the foreigner, the nomad or the *flâneur* in recent theorizations of subject positioning – a series of overlapping tropes that have risen to prominence in discussions of the post-colonial subject and the post-colonial city. As Haddour argues, in his critique of Stuart Hall's work on subjects-in-process and Iain Chambers' discussion of migrancy, much post-colonial theory works by side-stepping questions of history, specificity, location. The fetishization of 'cultural difference' – something we also see in the work of Iris Marion Young – and the stress on decentred identities (migrants/nomads) symptomatic of the postmodern subject, he writes, produces a paradoxical discourse of *decentred sameness*. The 'right to difference' mobilized in recent discussions of cultural citizenship (see Pakulski, 1997) is here erased under a sign of difference which lacks historicity: instead of the celebrated figure of the ever-moving nomad (a figure who encapsulates a kind of cosmopolitanism driven by a desire to experience difference – but only certain forms of difference – a bit like the tourist invoked by Hutnyk; see also Curtis and Pajaczkowska, 1994), the colonized subject is pushed into a position of homelessness, of exclusion, of pollutant. So, we can see here an overlapping concern in Hutnyk and Haddour with troubling the dominant narratives of (post-)colonial cities and citizens, and with making a political intervention that offers different narratives, rooted in historical mappings of the spaces of capitalism and colonialism.

In many ways, Willy Maley's essay on the city in James Kelman (Glasgow) and Irvine Welsh (Edinburgh) has parallels to the arguments made by Hutnyk and Haddour. Both novelists can be seen, Maley writes, as revealing a hidden side of their cities, a kind of underbelly largely excluded from dominant representations: as Maley puts it, '[w]e see parts of each city through the fiction of Welsh and Kelman that are more likely to crop up in court reports than tourist brochures' (though, of course, the notoriety of Welsh's *Trainspotting* has created a particular representation of Edinburgh that spills over into popular culture, giving it its own peculiar dominance). Both focus on displaced figures and marginal spaces that counter the boosterist images put out by the cities (Edinburgh with its Festival, Glasgow as City of Culture), yet in their championing they both run the risk of overburdening the people and places their characters stand in for with some kind of tag of authenticity. Nevertheless, their characters can serve as leitmotifs of urban transformations spotted elsewhere in *City Visions*: the busconductor Rab Hines, for example, is on the brink of displacement due to restructuring in the city's transport network. He must become occupationally (and therefore socially) mobile to survive, like so many caught in the tides of global change – which always has specific local impacts like those felt by Hines.

Of course, cities are more than just maps of social class; they are complex and shifting articulations of social mobility, and – as Haddour so potently illustrates – of social exclusion. The Edinburgh housing projects which Irvine Welsh describes are a particular kind of cityspace; a relic of earlier attempts at reordering which have also resulted in displacement and homelessness. The comparison made by Roy Strang, in Welsh's *Marabou Stork Nightmares*, between Muirhouse and Johannesburg, transposes the spatial logic of apartheid equally vividly, showing how the city is riven by structures of insider/outsider, sameness/difference – structures of inequality. As Chris Stanley writes in his essay, the inner city is so often paradoxically the 'outside' city, the place of the dispossessed – which makes the inner city for him the key site for 'an intervention on the relationship between justice and community practised through the performance of rupturing ethical transgressions'.

In sharp contrast (at least in some ways) to the Muirhouse projects (or to the inner city more broadly) comes the cityspace of suburbia. In its British (or, perhaps more accurately, English) context, suburbia is itself a condensation of similar structures played out in domestic space. Within popular culture, the suburbs are figured in particular ways; the most common motif, seen especially prevalently in sitcoms, has the suburbs as middle-class, white, familial space. It is also Southern – usually Home Counties: the cocktail belt. David Oswell takes us through this construction of the English suburbs in his essay on Hanif Kureishi's *The Buddha of Suburbia*.

Part of this construction is a sexualization, with suburbia mythologized as the site of illicit pleasures (wife-swapping, sex parties) always hidden behind net curtains, of sin clothed in respectability and mundanity (Hunt, 1998); the suburban always has a saucy or sleazy underbelly, but one which is particularly English, all morals and manners and restraint (exemplified, perhaps, in *Brief Encounter*; see Oswell, 1998). And, as Oswell plots out, 'popular' reactions to the television adaptation of *The Buddha of Suburbia* only confirmed the extent to which such a sexualization remains in denial – almost, it seems, because it is 'un-English'. As such, the novel – and perhaps more importantly the serial (since it was on BBC2) – works at undermining the equation of suburbia with straight white masculinity, producing what Oswell names 'suburban hybridity'. Unfixing spaces, complexifying associations between places and identities – these are the tasks which a work of fiction like Kureishi's attempts; to *queer* suburbia, perhaps.

Other kinds of representation give us access into other ways of talking about the city. This is particularly apparent in Chris Stanley's use of Rachel Whiteread's installation *House* to intervene on the questions of justice and community in urban space (for further readings of and writings on *House*, see Lingwood, 1995). More precisely, Stanley's essay works for a 'community of difference' togethered 'in-through a justice of otherness', and it attempts this by bringing Antigone to Tower Hamlets, the site of *House*. So, while the motif of togetherness-in-difference resonates with other essays in the collection, the particular strategy Stanley adopts in articulating this works in a very specific context. *House* stands (or, rather, stood prior to its demolition) as a monument – or an anti-monument, a ghost, a ruin. Read psychoanalytically, *House* rendered the *heimlich* into the *unheimlich*: *House* is not a home. As Doreen Massey (1995: 41) writes of Whiteread's piece, it subverts what could otherwise be 'an all-too-*comfortable* nostalgia of home and locality', making it the obverse of that currently dominant mode of 'preserving' the urban past, heritage. Instead of filling the house with 'authentic' Victorian artefacts and period-costumed mannequins, Whiteread filled it with concrete. And, in so doing, she threw it open to multiple acts of interpretation, rather than casting its meaning in the way heritage sites do. By ushering in Antigone, Stanley re-sites *House* as a place of mourning in the outside-city that is the inner city.

Valerie Briginshaw brings the dancing body into the city for her essay, offering readings of three dance videos in which the performers enact an embodied relationship to urban space. There are clear links to be made here to both Nigel Thrift's chapter, which draws on the dancing body in a broader context, and to Tim Edensor's, which uses dance metaphorically (in the sense of 'place ballets' and the like) to consider modes of movement in cities. With, as Thrift writes, dance as an instance of 'expressive embodiment' (which means that 'the body is not just inscribed, it is itself a source of inscription'), there are clear opportunities here for dancers and choreographers to rewrite body/city relationships. The three dance videos utilize different kinds of city space: a flat in *Step in Time Girls*, empty postmodern office buildings in *Duets with Automobiles*, and an alleyway in *Muurwerk* – domestic space, corporate space, (marginal) public space.

Briginshaw's reading of *Muurwerk* chimes remarkably, in fact, with work inspired by Michel de Certeau's famed discussions of 'pedestrian tactics' in *The Practice of Everyday Life* (1984). In both Edensor's and Toon's essays, such everyday tactics of what we might label either resistance or transgression are practices of embodied subjects in space. Just as Roxanne Huilmand reworks the alleyway, 'taking possession of the space, [and] reclaiming it for herself', so the young people on the streets of Tamworth negotiate and rein-scribe space (space which is, as Briginshaw reminds us, always invested with power – in the case of Tamworth, power literalized through CCTV). Additionally, we might make a constructive reading of *Step in Time Girls* alongside Stanley's discussion of *House*, since both deal with domestic space; both, too, are about, in some ways at least, reclamation (or at least perhaps reinvestment). The women's dancing bodies fill the space of the flat, showing that it is lived space – space crisscrossed with power, again. But here bodies memorialize the lives that turn a house into a home.

Duets with Automobiles sets its dancers in the heart of corporate urban space – in office buildings which resonate with capital's power and with the role of the city as a node in networks of that power. Here the motif of appro-priation (or reclamation) has an echo of what Hutnyk talks about, in terms of troubling the histories of colonial capital exploitation which corporate power still bears many traces of. As such, Briginshaw concludes, this piece is perhaps the most provocative as an intervention into the meanings given to spaces by buildings. Office accommodation is space ordered in accordance with a tight script; it is the space of business, of money, of power. Yet even here, resignification is possible.

Tim Edensor's walking tour finds the moving body in the city, too – a body sometimes cajoled by the ordering flows of urban planning, at other times moving against the prescribed tide. Edensor draws on the notion of performance to frame his discussion of embodied action in space, describing neatly the tension between control and freedom in the term 'regulated improvisation' in his discussion of Western city streets. The ordering of urban space, he argues, is in part an attempt at 'normative choreography'. As we saw with Briginshaw, spaces are scripted to permit (even encourage) certain kinds of performance while limiting (even prohibiting) others. That is why dancers look out-of-place in an office block, and why their movements can produce a critical reinscription that attempts to break away from regulation and regularity. In the context of the politics of so-called 'new social movements', such temporary reinscriptions become powerful strategies that bring into relief the lines of power that might otherwise be overlooked since they are so routinized and naturalized (actions like Reclaim the Streets, for instance). In fact, as Edensor shows, there are numerous and ever-shifting reinscriptions enacted by a host of urban dwellers – skateboarders, break-dancers, roadrunners – and there is a kind of politics at work in their actions, too. In the second half of his essay, Edensor focuses on movement through an Indian bazaar, in Agra. His walking tour here finds a very distinct, very mixed space where the logics of ordering and regulation operate very differently – a reminder that we must remember the particularities of place when talking about the city.

Similarly, Ian Toon shows how the surveillance technologies of contemporary cities are transgressed and reappropriated by street-savvy citizens – groups of young people. Teenagers' use of public space has for a long time been the focus of practices of ordering and regulation in Western cities, with young people being perceived as a 'problem' in need of solution. Toon's interviews with and observations of groups of young people on the streets of Tamworth (a town 15 miles from Birmingham, in the West Midlands) reveal a complex set of negotiations and performances, through which they attempt to carve out a space for themselves, a space in which social identity-work occurs. Their ability to 'deal with' new surveillance and security technologies – which are widely seen as straightforwardly, panoptically privatizing public space – shows us that the imperatives to ordering which Edensor outlines always have their limits: here we see an active, embodied engagement with close-circuit television which forms a transgressive or resistive tactic. In Thrift's essay, which closes *City Visions*, the 'city of play' marks a similarly embodied relation to urban logics. Like the skateboarders or break-dancers that Edensor mentions, the young people in Toon's study mark out a new geography of urban spaces, a hidden geography which they share, but which is always shifting. And again, the focus is on a particular place – Tamworth – which is not Los Angeles or London (or even Birmingham), and which therefore has to be read in that context; young people in Tamworth may share common traits with young people elsewhere, and Tamworth certainly has characteristics common to other cities, but the particular circumstances that combine to produce the 'place ballets' Toon reveals are located, specific.

Jon Binnie's discussion of the Bolton 7 case also finely highlights the gap between metropolitan cultures ('cosmopolitanism' or 'sophistication') and those of second-tier, ex-industrial, working-class, provincial cities. In the context of sexual politics, the progressive agendas and street radicalism of queer politics is hard to find in the case of seven men arrested for same-sex activity which they did not even know was illegal – their transgression of law was a million miles away from the queer kiss-ins taking place in other cities (on the Bolton 7, see also Moran, 1999). What Binnie's discussion shows, yet again, is the danger in taking 'the city' as a given, as a sign encapsulating a more-or-less common set of cultural, social and political orientations. Indeed, Nigel Thrift seizes upon this as one of his four myths of the city: 'one city tells all'.

In addition to this, Binnie discusses the paradoxical processes currently reshaping the sexual geographies of cities: the increased commercial visibility of gay enclaves (such as Manchester's Gay Village) which promote a certain kind of gay urban space, always at the expense of marginalizing other sexual spaces. In New York, for instance, zoning laws are effectively destroying the city's 'sex zone', while public sex practices are pushed out of cities and districts intent on rejuvenating their image (see Dangerous Bedfellows, 1996). Crucially, these imperatives have a clear class dimension to them. Just as gay gentrification – long heralded as a vital element in stimulating regeneration – has to be critically appraised from a class perspective, so the current wave of commercial redevelopment has to be read in the full context of what exclusions it produces, and the ways in which it represents a kind of desexualized gay enclosure.

David Goldberg's chapter, on the 'new segregation', is also concerned with practices which restructure urban space in particular ways. His discussion of the shifting processes of race-based segregation in US cities, which extends across housing, education and employment (to which Goldberg rightly adds representation), details the acute problems forced onto what he calls the 'racialized poor', located in inner-city areas, facing mounting and multiple forms of social exclusion. Looking back to Haddour's essay on *La Peste*, we see traces there, too, of the logic of segregation which must, in Goldberg's words, imagine a 'demonizing and demonized Other that needs accordingly to be set apart, isolated, circumscribed and constantly surveilled'. And, as Stanley made clear, one thing that has changed is that this Other is no longer expelled from the city walls, but is instead shoved into the 'outside city' that is the inner city. Goldberg also dissects the inner logics of the new segregation, which pathologizes blacks and naturalizes segregation while also critiquing antiracism and affirmative action. In a similar vein, Anna Marie Smith (1994) has pulled apart the British 'new right' discourses on race and sexuality, which also worked to demonize certain groups while simultaneously dividing those groups; in the case of sexuality, this produced the 'good homosexual' (assimilationist) versus the 'bad homosexual' (embodied, perhaps, in queer politics), a process which has had profound material effects on the shaping of sexual cultures in the UK, one concrete outcome of which is spatialized in our cities in the ways Binnie describes. Also in parallel with Binnie's argument, Goldberg suggests that globalizing forces (often commercial) exacerbate segregationist tendencies. While the rhetorics of cosmopolitanism stress the thrill of difference (itself refracted through consumerist logic), cities are being fractured into spaces which limit crosscultural mixing, which control movement and interaction; or, to use Edensor's argument, which set very particular scripts and allow little space for improvisation.

Towards the end of his essay, however, Goldberg points towards ways in which city living might turn away from infrastructural delimitation towards a pluralistic coming-together of cultures and people, a theme echoed in Iris Marion Young's engagement with urban citizenship as 'together-in-difference'. In order to facilitate such a coming-together, Goldberg suggests a couple of pragmatic moves. The first is an end to the separation of residential and commercial spaces, so that 'heterogenous people interact in their daily and nightly lives on the streets and in their social and commercial exchanges', the kind of move envisioned in the notion of the '24-hour city' currently enjoying the limelight in the UK. The second proposal returns us to themes picked up earlier, perhaps most notably by Edensor: that 'pedestrian contact' should be promoted in cityspace. Car culture is, of course, implicated in the emptying out of cities, as commercial activity relocates to radial retail complexes and housing expands suburbwards (or goes carceral). This is itself driven by the pathologizing of inner cities and their inhabitants, the marking off of 'dangerous spaces' and 'no-go zones'.

Goldberg leaves us with a set of questions, finally, which his discussion must provoke us to ponder. We might do well to keep them in mind while thinking through Iris Marion Young's discussion, which also focuses on segregation. Her critique of integration chimes with Smith's discussion of assimilationism as a response to UK new right imperatives, and shows the pitfalls of such a simplistic 'answer'. Like Goldberg, she offers instead a vision of city life based on togetherness which does not require sameness, a theme she has already written about in the context of the notions of community, citizenship and justice (see Young, 1990). She makes a similar manifesto – or, rather, writes a particular city vision – that encapsulates together-in-difference, which makes suggestions at the policy level for facilitating the kind of mixing Goldberg calls for. Planning and zoning laws, land-use policies, equality of opportunity equalized through public policy, welfare access and so on, help to conjure a city in which 'people live in an area because they want to be near a workplace or family or particular civic institutions or they like the architecture, and not because they are avoiding certain kinds of people, or because their neighbourhood has the best services, or because their level of income and discriminatory attitudes of others confine their choice to their neighbourhood' – urban space desegregated, but not in accordance with universalizing integrationism (which leaves no element of choice for people who wish to congregate locationally).

Young's theoretical agenda is picked up by Anthony Gorman, who turns to the question of urban villages and quarters in his discussion of Birmingham's Irish Quarter. It is instructive to read his account against Jayne's chapter on Stoke-on-Trent's Cultural Quarter, as a way into thinking about the 'quartering' process as a dominant technique in urban rebranding. To some extent, 'quartering' fits in with Young's formulation, since it produces a new way of marking distinct cityspaces *as distinct*, while attempting to address structural inequalities. However, as we have already seen in the context of gay villages, the role of commercial directives can make quarters deeply problematic spaces which exclude some of those who should, in theory at least, have a place there, while also to some extent commodifying difference (Irish theme pubs typify this trend). Quarters, especially perhaps those associated with marginalized or minority groups in cities (in contrast to, say, Birmingham's Jewellery Quarter or, indeed, Stoke-on-Trent's Cultural Quarter), can be seen as marking the recognition of that group's place and role in the city; but when this makes them little more than stop-offs on the tourist's itinerary, their broader political function must be called into question. Quarters do not hold all the answers, obviously: they don't in themselves end poverty or unequal access to resources or services, but they at least refuse to invisibilize the heterogeneity of cities.

We have chosen to close *City Visions* with Nigel Thrift's essay, which proposes to demolish some city myths while uncovering some 'hidden cities'. As such, it could be read as a coda to the discussions in the other chapters; indeed, it works as a useful reminder to keep in mind when reading any work on 'the city'. His four myths of the city concern globalization, homogenization, (in)authenticity and universalism. In each case, he persuasively counters the myth, urging us to think more carefully, and more locatedly, about what cities are – a move away from grand, overarching metanarratives of 'the city' thus far so prominent in urban social and cultural theory. The myths certainly provide a provocative lens with which to begin to talk about cities anew.

Following on from this, Thrift discusses three 'other cities' as a way to 'begin to provide a sense of a city that is constantly changing (even to stay the same), that does not necessarily hold together, that is both little and large'. The first of these, which we have already highlighted, is the embodied city, which Thrift writes about through dance (see also Thomas, 1997). He describes the many roles dancing provides as a playful human activity in urban space; or, as he has put it in another essay, as 'a "concentrated" example of the expressive nature of embodment' (Thrift 1997: 125). Second is the 'learning city', the city as a dissemination node for forms of knowledge. The concept of the 'creative city' prominent in boosterist literatures embodies this notion of the city as a fount of entrepreneurial activity (this is part of the project of Stoke's Cultural Quarter, too). Here, Thrift focuses on forms of management knowledge, what he calls 'knowledgeable capitalism', which circulate in organizational cultures, often promoted by so-called management gurus (epitomized, perhaps, by Peters and Waterman). Third is the unjust city: the city of social exclusions. Here Thrift offers a response to the almost exclusive focus on certain groups seen as emblematic of resistance – new social movements – by turning towards more mundane or everyday ways in which urban folk are addressing inequality. Specifically, he turns his attention towards 'different means of imagining money', through forms of alternative financial institutions (community banking, ROSCAs, LETS, etc.) Such institutions, he argues, are helping redefine the everyday life of the city by addressing *financial* exclusion.

Thrift's visions of the city are, in fact, particularly instructive for us here, since they clearly point out the need to keep an eye on context, remembering the city as 'a partially connected multiplicity which we can only ever know partially and from multiple places'; and also remembering that the myths he outlines 'downgrade everyday human practices in cities to the everyday, thereby stripping their magic from them'. If we are to fully understand any particular city vision, then, we need to approach the city in a similarly polymorphous way, piecing it together, never losing sight of the 'magical' human practices that criss-cross its form, dancing and flickering over space. We hope that the essays which together constitute *City Visions* contribute to that crucial task.

References

Clarke, D. (ed.) (1997) *The Cinematic City*, Routledge, London.

Curtis, B. and Pajaczkowska, C. (1994) 'Getting there': travel, time and narrative. In Robertson, G. *et al* (eds) *Travellers' Tales: Narratives of Home and Displacement*, Routledge, London.

Dangerous Bedfellows (eds) (1996) *Policing Public Sex: Queer Politics and the Future of Aids Activism*, South End Press, Boston.

de Certeau, M. (1984) *The Practice of Everyday Life*, University of California Press, Berkeley.

Hunt, L. (1998) *British Low Culture: From Safari Suits to Sexploitation*, Routledge, London.

Jarvis, B. (1998) *Postmodern Cartographies: The Geographical Imagination in Contemporary American Culture*, Pluto, London.

Lingwood, J. (ed.) (1995) *Rachel Whiteread House*, Phaidon, London.

Massey, D. (1995) Space-time and the politics of location. In Lingwood, J. (ed.) *Rachel Whiteread House*, Phaidon, London.

Moran, L. (1999) Law made flesh: homosexual acts. *Body & Society*, **5**(1), 39–55.

Oswell, D. (1998) True love in queer times: romance, suburbia and masculinity. In Pearce, L. and Wisker, G. (eds) *Fatal Attractions: Rescripting Romance in Contemporary Literature and Film*, Pluto, London.

Pakulski, J. (1997) Cultural citizenship. *Citizenship Studies*, **1**, 73–86.

Smith, A. M. (1994) *New Right Discourse on Race and Sexuality: Britain, 1968–1990*, Cambridge University Press, Cambridge.

Thomas, H. (ed.) (1997) *Dance in the City*, Macmillan, Basingstoke.

Thrift, N. (1997) The still point: resistance, expressive embodiment and dance. In Pile, S. and Keith, M. (eds) *Geographies of Resistance*, Routledge, London.

Young, I. M. (1990) *Justice and the Politics of Difference*, Princeton University Press, Princeton.

Chapter 2

Imag(in)ing a post-industrial Potteries

Mark Jayne

This chapter focuses on the emerging regeneration strategies of the city of Stoke-on-Trent, the centrepiece of which is the development of the Cultural Quarter at the heart of the city. My aim in this chapter is to analyse the ways in which Stoke is responding to current urban transformations and reorientations, as a way of highlighting the mesh of politics, policies and opinions which intersect around such a project. While there have been useful accounts of such processes already published, the case of Stoke-on-Trent allows us to gain a richer, more nuanced insight into the ways in which broad urban agendas get articulated in the fabric of a particular city.

Rebranding cities

An increasingly ubiquitous vision of urbanity promoted by city imagineers is the development or enhancement of distinct social and spatial areas – urban villages – such as cultural and ethnic quarters. In these spaces of consumption, now synonymous with the post-industrial economy, image becomes everything. The symbolic and cultural assets of the city are vigorously promoted, and cities are branded as attractive places to live, work and play. In urban villages, promotion of both high and low culture is undertaken as cities appeal to the consumption practices of the emerging *nouveaux riches* of the professional, managerial and service classes (see, for example, Gold and Ward, 1994; Hall and Hubbard, 1998; O'Connor and Wynne, 1996). Images of conspicuous consumption, in the shapes of art, food, music, fashion and entertainment, are promoted in these urban 'shop windows' (Hubbard, 1998: 199), created either through the enhancement of historically distinctive areas or by developing and generating signatures for previously culturally or spatially ambiguous areas. Moreover, in recognition of the complex plurality of the contemporary city, more prosaic 'low' street-culture, working-class traditions and ethnic celebrations are increasingly commodified in narratives of place (Hall, 1995; Urry, 1990; Zukin, 1998). In urban villages, then, the symbolic framing of culture becomes a powerful tool as capital and cultural symbolism intertwine (Harvey, 1989; Jessop, 1997; Knox, 1987).

Theorists have shown that such vibrant, cosmopolitan, entertaining and happening spaces are underpinned by strong regional economies with professional post-industrial business cores. The stimulus for these conditions ultimately appears to be the existence of a progressive or innovative institutional and political vision. When combined with activities of cultural intermediaries, this enables the development of a critical infrastructure of buildings, citizens, social relations and forms of sociability relevant to the post-industrial symbolic economy.[1] It is perhaps understandable that researchers have concentrated on providing generalizable accounts which have been grounded in case studies of such areas within cities which are near the top of the urban hierarchy (see, for example, Ley and Olds, 1988; Boyle and Hughes, 1991; Soja, 1996). However, the situation within cities which have been less successful in stimulating post-industrial economic activity has been relatively under-theorized. While issues concerning the management of cultural contest, reflexivity and risk have become central to dialectical understanding of localized responses to global processes, questions of scale, scope and efficacy across the spectrum of the urban hierarchy must also be considered.

For instance, the Stoke-on-Trent conurbation, still popularly referred to through its manufacturing heritage as The Potteries, has seemingly been woefully slow to respond or develop in light of global structural and institutional change. Economic and physical decay, unemployment, low wages, factory closure and shortened working weeks characterize the remaining manufacturing industry.[2] The poor image of Stoke-on-Trent is exacerbated by the comparable success of nearby neighbours Manchester and Birmingham, and the progress of Nottingham and Sheffield. With such a litany of decay and strong regional competition, the position of The Potteries remains bleak, with the conurbation apparently firmly locked into a spiral of decline. However, in spite of this apparent political, economic and cultural inertia, September 1999 saw the culmination of the nine-year process of consultation, successful funding bids for National Lottery and European money, and building renovation.[3] Set alongside other (modest) urban renewal projects, the process of developing a Cultural Quarter is easily identifiable as the flagship development in the region.[4] It is suggested here that, despite many shortcomings, an optimistic prognosis would suggest that the project is relevant and sustainable in the context of regional conditions.

The attempts of less successful cities to adopt 'me-tooist' strategies and import blueprints of economic and symbolic activity undoubtedly raise some fascinating issues (Jessop, 1998; Leitner and Sheppard, 1998). With the post-industrial city emphasizing mass consumption rather than mass production, local response to a decaying urban fabric relates to very specific situations. These not only include local opportunities, constraints and agendas, but the geographical dimensions of regional, national, supranational and global systems of space. By adopting frameworks which seek to detail localized responses to global processes, and by introducing three broad transformations which have been related to structural and socio-economic changes associated with the new global economy (Castells, 1977; Sassen, 1991, 1994; Soja, 1989, 1996), it is possible to discuss the merits of the Cultural Quarter in the political, economic, physical, social and spatial context of The Potteries.

Potted history

Firstly, development of post-industrial activity is seen to be predicated on the transformation of traditional industrial structures and landscapes.[5] If such a manifesto is overlaid onto The Potteries, we see that the industrial, economic and social structures which founded its provincial cultural affiliations and shaped the regional and urban economy continue to dominate the conurbation today. A product of the industrial revolution and comprising the six pottery towns federated in 1910 – Tunstall, Burslem, Hanley, Stoke-upon-Trent, Fenton, Longton, plus the adjacent market town of Newcastle-under-Lyme – the historical success and development of The Potteries can be seen to have been based on the evolution of the critical infrastructure which underpinned the pottery industry. Hence, a specific urban grounding was the basis for industrial innovation, an ecology of exchange and the modelling of a civil society founded on social and cultural innovation. While the wider social, technical and aesthetic changes of the emerging global consumer culture created the conditions for creativity, this was facilitated by a matrix of local people who effected value- and citizen-production, and guided development of the physical environment, social relations, lifestyles and values.

However, economic and social investment in one industry ensured not only the rapid decline of the city but also an inability to achieve economic and social entrepreneurialism and diversification. The lingering effects of these industrial social structures, the concentration of employment into just a few sectors, and the dispersal of creative energies into inter-town competition reinforce a perspective that the region is 'somehow special or isolated' (Phillips, 1993: xv–xvi). This political, economic and socio-spatial matrix continues to have tangible effects on the conurbation's ability to innovate in post-industrial symbolic and economic activities.

There have been several schemes to reclaim industrial or derelict land in the city, the 1986 National Garden Festival and subsequent use of the site as a retail and entertainment park being the most prominent success story. Meanwhile, of 427 empty industrial premises in the city, less than half are being actively marketed, and 4.7 million square feet of industrial floorspace lies vacant.[6] Further, there has been little or no progress in shifting industrial lifestyles and identities. For example, the local newspaper *The Evening Sentinel* continues to be dominated by reports of job losses and short-time working, while the traditional Potters' Holiday remains as a social relic of the industry. Parliamentary reports sent out by local MPs have a similar industrial bias.[7] While Loyd Grossman, a TV personality synonymous with post-industrial/postmodern preoccupations, ceremonially renamed Stoke's Museum and Art Gallery with the prefix 'The Potteries' rather than 'The City', this cannot be considered a determined attempt to consign The Potteries (or the potteries) to heritage, but rather represents a wider failure to decide how to develop, (re)package, market and brand the city.

Adopting another perspective also graphically highlights the entrenched nature of local representations of image, identity and lifestyle. At present there is little progressive championing of the Cultural Quarter, the potentialities of the post-industrial economy nor the activities of cultural intermediaries by the local media. Similarly, there appears to be little willingness to positively represent or support the identities, lifestyles and consumption practices of, for example, lesbian and gay, youth and ethnic social groups who have been so intimately related to the revitalization of other cities.[8]

Central to negotiating and achieving these processes of (in)visibility is the second identifiable constituent of post-industrial economic activity; the transformation of the working city through development of arts- and leisure-led regeneration projects in an attempt to attract new investment, financial and administrative institutions and services and cultural tourism. Elsewhere, these physical and symbolic attempts to improve the environment have taken many forms, such as postmodern office blocks, waterfront developments, science parks and urban villages. There is also the promotion of what Landry and Bianchini (1995) described as the 'Creative City'. This encompasses a spectrum of developments from aesthetic improvements of soft infrastructure, to the attraction, stimulation and promotion of a critical infrastructure of buildings and facilities for cultural producers such as musicians, artists, actors, craft workers and those with entrepreneurial capability.[9] When combined with cultural intermediaries who bring vitality to the galleries, cafés, bars, restaurants and other spaces of consumption, such cultural producers create the cultural ambience, social relations and forms of sociability which the gentrifiers of the professional, administration and service classes are considered to find so attractive (Zukin, 1982).

City of Quarters

The formulation of the concept of a Cultural Quarter can be seen to emerge from the 1990 report *A Cultural Strategy for Stoke-on-Trent*. This was commissioned by Stoke-on-Trent City Council, and completed by Comedia Consultancies, headed by Charles Landry and Franco Bianchini. The report approached multi-sector uneven economic development in the context of the very specific geographical, physical, economic, and social conditions and issues of image facing the Stoke-on-Trent conurbation, and produced a detailed sectoral analysis and recommendations concerning the development of venues, access, literature, media, music, dance, visual arts and crafts, arts in the community and transport.[10] The report importantly included suggestions about the ways in which each of the confederated towns could adopt specific roles to promote different identities in an attempt to break the cycle of replication of infrastructure and unproductive internal competition between them.[11] In sum, the central aim of the report was to identify ways in which The Potteries could begin to stimulate the presence and support of people and businesses who could facilitate a post-industrial critical infrastructure.

Arguably, however, the only major recommendation assimilated and most comprehensively taken forward by the Council was the need to address the lack of a major touring venue.[12] This led three years later to the study *Major*

Touring Venues for Stoke-on-Trent (1993), completed by Arts Business Ltd, which identified the need for a distinctive 'cultural quarter' in the city.[13] This was to incorporate existing facilities including a cinema, the Mitchell Memorial Theatre, the Potteries Museum and Art Gallery and the City Library as well as enhanced retail activities. However, the central feature of this Cultural Quarter was to be the redevelopment of the Regent Theatre and the Victoria Hall as touring venues for essentially middlebrow cultural activities.[14]

It could be suggested that the rationale dominating the Council's perspective on the development of the Cultural Quarter emanates from *Major Touring Venues for Stoke-on-Trent*, rather that the more progressive and extensive recommendations of Comedia. Following renovation, the two theatres are being contracted out to be operated by the Ambassador Theatre Group, a London-based organization. While recent national television advertisements have promoted events at the Victoria Hall, these made no mention of the Cultural Quarter. This suggests that there is a no-go area of cultural representation in which identities and lifestyles associated with post-industrial economy are considered yuppyish or pretentious (see Wynne and O'Connor, 1998). As such the development of this urban village appears flawed in two ways; firstly, while it could be argued that the development of the Cultural Quarter has been initiated to stimulate the growth of a critical infrastructure of cultural producers and consumers, this can be regarded as a 'shallow' attempt to stimulate post-industrial activity. While the presence of arts, entertainment, food, music and fashion-orientated development may suggest processes of gentrification, the attraction of employers and the promotion of lifestyles associated with managerial, professional and service economy is being less vigorously pursued. Secondly, it would appear that there is little will (or perhaps ability) to generate a distinct social or spatial signature not reliant on 'traditional' industrial lexicons.

Implicated in the above two processes is the third component of post-industrial activity: the *re-imagining* of the city through partnerships which attempt to gentrify and invest in the above programmes. Place promotion is an attempt to combat long-term decline – this can include marketing campaigns, transitory events such as garden, music and other 'cultural' festivals, and media and community discussion (see Gold and Ward, 1994). Place promotion is, of course, not a uniquely post-industrial strategy, and is recognizable in Stoke-on-Trent as early as the nineteenth century; however, more recently, 'Stoke-on-Trent Fires the Imagination' has been a modestly marketed campaign, and 'Do China in a Day' (taking in eight pottery museums and twenty-nine ceramic shops) celebrates the identities of local men and women who have worked in the pottery industry. In terms of contemporary technology, Stoke-on-Trent City Council is developing its own website and a CD-Rom promoting the Cultural Quarter, in order to augment the low-key campaign 'Stoke-on-Trent Deserves its Own West End'. It is the aim of such entrepreneurial urban practices to attempt to produce attractive and distinctive locations in the face of intense globalized inter-urban competition.[15]

Tangential to this institutional response of re-imagination is the precondition that local authorities, planners and managers pursue 'entrepreneurial governance' and form coalitions with private businesses to stimulate local economic and cultural advancement through speculative development.[16] The most singularly prominent public–private partnership is between Stoke-on-Trent City Council and St Modwen Development, a Birmingham-based organization. The interface of capital and culture in this case has been in off-centre family/car-oriented retail and entertainment parks. In sum, the modest and low-key place promotion of the conurbation highlights a *lack* of promotion, branding and marketing expertise. Moreover, the public–private partnership creating development in the region appears only able to produce off-centre spaces of consumption which are significant only at intra- rather than inter-urban levels of competition, and which certainly do not contribute to the re-energizing of city centres.

(Re)Locating The Potteries

These assertions can be more fully theorized and explained by suggesting that it is an increasingly common strategy for cities and regions to approach multi-sector uneven development and the constraining and enabling factors of opportunity, location and action, by pursuing (if unknowingly) 'structured complementarity' (Jessop, 1998: 93), 'comparable regionalism' (Cisneros, 1995: 115), or 'internal regionalisation' (Hall, 1997: 204). This means that regeneration projects are tailored to the particular spatial scale which the city or spaces within the city are perceived to be competing. For example, successful cities are able to undertake 'spectacular developments' which create new cityscapes reconfiguring or redefining the urban hierarchy and producing a new physical, social and cybernetic infrastructure. This in turn produces new types of spaces such as socially and spatially distinct urban villages for living, working and playing, and attracts cultural producers, services and people consuming a 'mix-and-match' lifestyle (Mort, 1998; Zukin, 1998). Other cities on the other hand must content themselves with improving city-centre market hubs or radial gateways and enhancing off-centre retailing. Such methods of space and place production thus create location-specific advantage for the consumption and production of goods and services and other urban activities which relate to specific markets and class fractions (Jessop, 1997).

In essence, economic competitiveness is pursued by promoting the presence and visibility of institutions, infrastructure, buildings, and spaces where different market segments of producers and consumers work, live, shop and relax. Thus not all cities or parts of those cities compete at the same social or spatial scale. Hence, global cities are more successful in attracting to the local milieu 'thick' economic and symbolic capital and diverse spaces of consumption and, with distinct indigenous potential, are more proactive in shaping local, regional, national, supranational and global economic competitiveness (Jessop, 1997). Bigger cities have a higher degree of functional centrality and degree of urban spatial influence. Similarly, weaker local states are forced to focus on local issues, often pursuing unclear development potential or internal reorganization (Herrschel, 1998). While this may have some success in making particular areas of the city socially, culturally, or economically distinct, such practices have little generic impact (or success in terms of specific market segments) upon the flows of capital or culture outside the city, region or country.

Any assessment of the success of The Potteries in stimulating post-industrial economic activity would have to conclude that this has been at best limited. The conurbation does not appear to have the political, economic or social will or ability to render industrial and working-class employment and lifestyles marginal, to significantly alter the physical and symbolic resonances of their past heritage, and to socialize a significant proportion of the population to desire and pursue jobs, lifestyles and places associated with the post-industrial economy. This, of course, includes a lack of institutional expertise and vision to successfully instigate public–private partnerships with the potential ability to produce 'spectacular developments', to attract cultural producers and consumers, and to achieve the infrastructure improvements which produce the spaces and atmosphere associated with post-industrial identities and lifestyles. This suggests that the best-case scenario for the Cultural Quarter is that it may *potentially* produce a new distinctive social and spatial area in the city, giving Hanley a significant competitive advantage over the other towns. This may attract consumers who presently go to nearby cities such as Manchester and Birmingham for certain events and for shopping. While this potentially will produce a more easily identifiable and dominant city centre (this may of course be problematic to the other towns unless they find their own economic specificity) which is more attractive and liveable for certain social groups, the importance of the Cultural Quarter will have little resonance or impact on flows of capital, culture or people beyond the conurbation.

Urban governance and city cultures

In order to explain such complex relationships between economic, political and cultural opportunities and constraints, theorists have identified urban governments as being pivotal instigators (or not) of change by virtue of authority. Urban regimes are able to create local coherence through legislation, promoting conditions favourable to capital accumulation and the maintenance of production and consumption for its citizens. For example, David Harvey's (1985: 140) depiction of 'regional class alliances' identifies how urban governments legitimize specific regimes of capital accumulation and embrace particular social relations and hierarchies, standards of living, identities and lifestyles. Similarly Amin and Thrift's (1995) 'institutional thickness' describes how institutional relations develop and can have a decisive influence on economic development as firms gravitate towards localities which offer the best institutional structures to support their needs. However, while such rational choice explanations are undoubtedly incisive, they do not fully explain how struggles over economic efficiency, fiscal equality, economics, the development of the physical and symbolic city and the promotion of identities and lifestyles are differentially negotiated (Raco, 1998; Ellis and Thompson, 1997a).

For example, Boyle (1997) suggests that to date the interpretation of the politics of local economic development ignores a multiplicity of 'strategic orientations' and 'institutional positions' which a wide variety of political, cultural, business, community groups and individuals adopt. Boyle identifies that the concept of audience has been undertheorized and that rejuvenation schemes are not only encountered and consumed in terms of incorporationalist or oppositional terms, but with reference to very different agendas. Such shortcomings can perhaps be at least partially rectified through the work of Aaron Wildavsky, who sought to theorize how policy preferences attempt to bolster preferred social relations, attitudes and beliefs and how this is a process of 'cultural contest ... polarised along ... cultural fault lines' (Ellis and Thompson, 1997b: 169). This represents a typology of political change allowing an understanding of how the institutional values, interests and beliefs of urban regimes are mediated and reciprocally related to citizens' allegiances and specific socio-cultural settings. In most cases, in order to initiate the shift from industrial to post-industrial economies, cities must first move from a domination by political and social groups with *hierarchical* or *fatalistic* perspectives (traditional working-class and industrial social structures) towards those with *egalitarian* characteristics (post-industrial mix-and-match lifestyles – the new *petits bourgeois* and new class fractions). Importantly, it is understood that change is achieved through a trade-off of pluralistic perspectives. As such, interpretation of the 'strategic orientations' and 'institutional positions' of cities and political and social groups allows an appreciation of the causal influences and processes involved in making groups and places (in)visible, central and/or marginal. As such the predominantly fatalistic political outlook of The Potteries' local government is ensuring a failure to support the development of any post-industrial critical infrastructure in advance of the failing industrial infrastructure.[17]

This failure is highly visible because it is ensuring that dominant urban regeneration programmes being undertaken in the region are based around the kind of collective consumption which Castells (1977) suggests as being definitive of the cities under a particular period of advanced capitalism. For example, the Victoria Hall is being marketed as the 'People's Palace'. Similarly, the popularity of and determination to develop more family-oriented retail and entertainment parks, as opposed to also promoting city-centre living, serves to highlight the city's political, economic, and cultural inertia.

Such dominating spaces of consumption and urban lifestyles contrast with the kind of spaces and tribalisms which Maffesoli (1992) describes as epitomizing the post-industrial/postmodern. The new 'carnivalesque' spaces of successful cities (such as urban villages), while being centred on private consumption, create a collective sense of *belonging* in the city centre. In these areas membership is expressed by paraphernalia (clothes, lifestyle, taste). Such post-industrial identities and lifestyles, while being distinctively grounded in local social relations, are expressed in terms of a fluidity of social relations no longer limited to the physical spaces of the city but expanded into the virtual spaces of the consumer society (Mort, 1998; Wynne and O'Connor, 1998).

Sharon Zukin (1998) also identifies how such relationships between modes of capital accumulation and urban lifestyles can be related to particular periods of (post)modernity. This allows a contrast between the domination of locally and regionally based capital and culture, and the economic and symbolic centrality of footloose mobile global capital and culture in developing spaces of consumption associated with vibrant city-centre living. Zukin identifies the early 1980s as the period when successful UK cities began shifting dominant symbolic and economic emphasis back to a rejuvenating city centre. While being a gross oversimplification of Zukin's schema, this does allow the graphic and simple measure which identifies that The Potteries' vision of renewed urbanity remains at least twenty years behind more successful cities.[18]

It is clear that culture- and consumption-led urban regeneration is pervaded by discourses and metaphors and a socio-spatial logic of disclosure/enclosure, centre/periphery, and issues of in/visibility (Jackson and Thrift, 1995; Clarke, 1997). Miles' (1998) contention that, to the individual, consumption is both enabling (in terms of personal fulfilment and construction of sovereignty) and constraining (as it plays an ideological role in constructing lives) can be equally applicable to cities. This 'consuming paradox' suggests that cities are not helpless pawns reacting to global processes. However, their reactions are both related to sedimented local and regional political, economic, physical and social opportunities and conditions, and are highly relational in terms of the activities of other cities. The Cultural Quarter is arguably the most important development in The Potteries since the 1960s. However, not only has the vision taken ten years to come to fruition, but the project will have little impact outside the conurbation because of both its limited rationale and the strong position of regional competitors. Similarly the political, economic and social dynamics within The Potteries will determine its potential in the long term. Nevertheless, historically the area has produced a distinctive industrial critical infrastructure and specialism, which brought an added-value cachet to the area. As such it is perhaps an optimistic but important viewpoint to assert that a competitive advantage or specialism in the post-industrial economy can also be achieved. As Zukin (1998: 837) suggests, the creativity of local ecologies and the 'ability to forge urban lifestyles continues to be the city's most important product'.

Visions of Stoke

What can we glean, then, from the story of Stoke's Cultural Quarter? At one level, it tells a very familiar tale. Cities in the post-industrial West are currently undergoing processes of reorientation and redefinition, with varying degrees of success. Understanding the unevenness of this demands looking at the internal identities of cities and their inhabitants, matching the re-imaginings of new cultural intermediaries with their reception by social groups with their own different mindsets and agendas. At another level, there is a particular kind of story emerging here; one of contestation and negotiation. As rebranding becomes a watchword for cultural and economic competitiveness at all levels, it is interesting to witness how processes of urban rejuvenation are translated into a tangible space such as the Cultural Quarter. Attempts to erode pre-existing identities and signatures meet with considerable resistance both institutionally and at the level of the vernacular, testifying to the very different visions of the city that exist in sectors of its population. As with earlier attempts to think about what's happening in this city – such as Martin Parker's writing on Festival Park (1996) – it is easy to fall into critique, to lambast the city's *nontrepreneurs* (if I may coin a neologism) for their lack of city vision, or for their inability to translate that vision into something concrete. The city produces this kind of ambivalence, it seems, especially perhaps in academics who are also residents; who know their theories but also know the place. Each new development – the Garden Festival, the Potteries Shopping Centre, now the Cultural Quarter – gives us a mix of cynicism and excitement, laced with a hint of disappointment. The fatalism in city politics identified by Wildavsky infects the populace, too. Civic pride in Stoke is always partly ironic, always tinged with an acceptance that the city won't ever be a contender, could never match its heftier neighbours. This is, perhaps, best summed up in the editorial of the local free listings magazine *Bomb* (Anon, 1999), which seems a fitting place to end my discussion:

> **1999 – There is something happening … The seeds are sown … Stoke's bid for the 2008 Olympic Games starts here.**
>
> The Regent will be 'Coming on Line' in the Autumn, Trentham Gardens are in for a refit and there is still talk of 'The John Street development', a festival park style Cinema/Bar/Club/Gym next to the Victoria Hall. The Potteries Shopping Centre is going to be pulled down to be replaced by a New York Central Park style, inner city, tree and grass facility for lazy summer days and breathing fresh air into the city proper. The bus station is coming down to be replaced by a state of the art bustling narrow street/Brighton Lanes style outdoors shopping facility enabling independent outlets to stamp a local feel to the city … The fountain of Fountain Square is to be replaced by a long lost sculpture found in the vaults of Henry Moore's house, one that he specifically designed for the spot shortly before his death. Sound5 become the coolest band in the world thus kick starting what the media term the Stoke-on-Trent movement, Sasha realises the importance of Hanley and moves his club from Nottingham to the Stage, Robbie … manages to successfully run as mayor of Stoke-on-Trent, Nick Hancock unfortunately fails by 2 votes to get onto the Council for the Stoke-on-Trent Indiependence Party.
>
> Newcastle furthered by the cinema complex to be opened later in the year, realises its role as an entertainment and leisure town and shelves all development plans for retail allowing Hanley the room it needs to become a city in its own right. Mike Lloyd kicks butt on the summer festival and along with the Ambassador Theatre Group brings top names (by now most showbiz names are local anyway) and mucho tourists to both Newcastle and Hanley.
>
> Stoke-on-Trent pottery is voted by a panel of experts in the Far East as the ideal Millennium present for a loved one, to meet demand Wedgwood and Doulton have

to re-employ the whole of the city on triple pay. Stoke's going down in history next to Venice, ... a city that ruled the world... The greenhouse effect is cured just in time as the sea reaches Crewe-on-Sea where the average temperature is in the 80s, Alsager goes Bohemian – San Francisco style, London is flooded and the money markets are transferred to Stafford. Warm artificial snow is invented by a Saturday lad at Hi-peak and Meir becomes the ski resort choice for the rich and famous. Hell, even City and Vale are tipped to be ok ... Hanley is going to be big. The region of Stoke-on-Trent is going to be internationally groovy. We are going to host the Olympics.[19]

References

Amin, A. and Thrift, N. (1995) Globalisation, institutional 'thickness' and the local economy. In Healy, P., Cameron, S., Davoudi, S. Graham, S. and Madani-Pour, A. (eds) *Managing Cities: The New Urban Context*, John Wiley, Chichester.

Anon (1999) 1999 – There is something happening ... The seeds are sown ... Stoke's bid for the 2008 Olympic Games starts here. *Bomb; a Guide to Stoke-on-Trent*, 45 (February), 5.

Arts Business (1993) *Major Touring Venues for Stoke-on-Trent*, Arts Business Ltd., London.

Bourdieu, P. (1984) *Distinction: A Social Critique of the Judgement of Taste*, Routledge and Paul Kegan, London.

Boyle, M. (1997) Civic boosterism in the politics of local economic development: 'institutional positions' and 'strategic orientations' in the consumption of hallmark events. *Environment and Planning D: Society and Space*, **29**, 1975–97.

Boyle, M. and Hughes, G. (1991) The politics of the representation of the 'real' discourses from the left on Glasgow's role as European City of Culture 1991. *Area*, **23**, 217–28.

Castells, M. (1977) *The Urban Question: a Marxist Approach*. Edward Arnold, London.

Chai, S. (1997) Rational choice and culture: clashing perspectives or complementary modes of analysis?. In Ellis, R. and Thompson, M. (eds) *Culture Matters: Essays in Honor of Aaron Wildavsky*, Westview, Boulder.

Cisneros, H. (1995) *Urban Entrepreneurialism and National Economic Growth*, US Department of Housing and Urban Development, Washington, DC.

Clarke, D. (1997) Consumption and the city, modern and postmodern, *International Journal of Urban and Regional Research*, **21**, 218–37.

Comedia (1990) *A Cultural Strategy For Stoke-on-Trent*, Comedia Consultancies, London.

Cox, K. (1995) Globalisation, competition and the politics of local economic development. *Urban Studies*, 32, 213–25.

Ellis, R. and Thompson, M. (1997a) Preface: remembering Aaron Wildavsky the cultural theorist. In Ellis, R. and Thompson, M. (eds) *Culture Matters: Essays in Honor of Aaron Wildavsky*, Westview, Boulder.

Ellis, R. and Thompson, M. (1997b) Seeing green: cultural bias and environmental preferences. In Ellis, R. and Thompson, M. (eds) *Culture Matters: Essays in Honor of Aaron Wildavsky*, Westview, Boulder.

Gold, J. and Ward, S. (eds) (1994) *Place Promotion: the Use of Publicity and Marketing to Sell Towns and Regions*, John Wiley, Chichester.

Hall, T. (1995) The second industrial revolution: cultural reconstructions of industrial regions. *Landscape Research*, **20**, 112–23.

Hall, T. (1997) (Re)placing the city: cultural relocation and the city centre. In Westwood, S. and Williams, J. (eds) *Imagined Cities: Scripts, Signs and Meanings*, Routledge, London.

Hall, T. and Hubbard, P. (eds) (1998) T*he Entrepreneurial City: Geographies of Politics, Regime and Representation*, John Wiley, Chichester.

Harvey, D. (1985) *The Urbanisation of Capital*, Blackwell, Oxford.

Harvey, D. (1989) *The Urban Experience*, John Hopkins University Press, Baltimore.

Herrschel, T. (1998) From socialism to post-Fordism: the local state and economic politics in East Germany. In Hall, T. and Hubbard, P. (eds) *The Entrepreneurial City: Geographies of Politics, Regime and Representation,* John Wiley, Chichester.

Hubbard, P. (1998) Introduction: representation, culture and identities. In Hall, T. and Hubbard, P. (eds) *The Entrepreneurial City: Geographies of Politics, Regime and Representation,* John Wiley, Chichester.

Jackson, P. and Thrift, N. (1995) Geographies of consumption. In Miller, D. (ed.) *Acknowledging Consumption: A Review of New Studies*, Routledge, London.

Jessop, B. (1997) The entrepreneurial city: re-imagining localities, redesigning economic governance, or restructuring capital. In Jewson, N. and McGregor, D. (eds) *Transforming Cities: Contested Governance and New Spatial Divisions*, Routledge, London.

Jessop, B. (1998) The narrative of enterprise and the enterprise of narrative: place marketing and the entrepreneurial city. In Hall, T. and Hubbard, P. (eds) *The Entrepreneurial City: Geographies of Politics, Regime and Representation*, John Wiley, Chichester.

Knox, P. (1987) The social production of the built environment: architects, architecture and the postmodern city. *Progress in Human Geography,* **11**, 154–377.

Landry, C. and Bianchini, F. (1995) *The Creative City*, Demos/Comedia, London.

Leitner, H. (1990) Cities in pursuit of economic growth: the local state as entrepreneur. *Political Geography Quarterly,* **9**, 146–70.

Leitner, H. and Sheppard, E. (1998) Economic uncertainty, inter-urban competition and the efficacy of entrepreneurialism. In Hall, T. and Hubbard, P. (eds) *The Entrepreneurial City: Geographies of Politics, Regime and Representation,* John Wiley, Chichester.

Ley, D. and Olds, K. (1988) Landscape as spectacle: world fairs and the culture of heroic consumption. *Environment and Planning D: Society and Space,* **6**, 191–212.

Logan, M. and Molotoch, H. (1987) *Urban Fortunes: The Political Economy of Place*, University of California Press, Los Angeles.

Maffesoli, M. (1992) *The Time of The Tribes*, Sage, London.

Miles, S. (1998) The consuming paradox: a new research agenda for urban consumption. *Urban Studies,* **35**, 1001–8.

Mollencopft, J. (1983) *The Contested City*, Princeton University Press, Princeton.

Mort, F. (1998) Cityscapes, consumption, masculinity and the mapping of London since 1950. *Urban Studies,* **35**, 884–901.

O'Connor, J. and Wynne, D. (eds) (1996) *From the Margin to the Centre: Cultural Production and Consumption in the Post-industrial City*, Arena, Ashgate.

Parker, M. (1996) Shopping for principles: writing about Stoke-on-Trent's Festival Park. *Transgressions,* **2/3**, 38–54.

Phillips, A. (1993) *The Potteries: Continuity and Change in a Staffordshire Conurbation*, Allan Sutton, Stroud.

Raco, M. (1998) Assessing 'institutional thickness' in the local context: a comparison of Cardiff and Sheffield. *Environment and Planning A,* **30**, 975–96.

Sassen, S. (1991) *The Global City: New York, London, Tokyo*, Princeton University Press, Princeton.

Sassen, S. (1994) *Cities in a World Economy*, Pine Forge Press, Thousand Oaks, Ca.

Savage, M., Barlow, J., Dickens, P., and Fielding, T. (1993) *Property, Bureaucracy and Culture: Middle Class Formation in Contemporary Britain*, Routledge, London.

Soja, E. (1989) *Postmodern Geographies: The Reassertion of Space in Critical Social Theory*, Verso, London.

Soja, E. (1996) *Thirdspace: Journeys to Los Angeles and Other Real and Imagined Places*, Blackwell, Oxford.

Stone, C. (1989) *Regime Politics: Governing Atlanta 1946–1988*, University of Kansas Press, Kansas.

Urry, J. (1990) *The Tourist Gaze: Leisure and Travel in Contemporary Society*, Sage, London.

Wynne, D. and O'Connor, J. (1998) Consumption and the postmodern city. *Urban Studies*, **35**, 841–64.

Zukin, S. (1982) *Loft Living: Culture and Capital in Urban Change*, Johns Hopkins University Press, Baltimore.

Zukin, S. (1998) Urban lifestyles: diversity and standardisation in spaces of Consumption. *Urban Studies*, **35**, 839–52.

Notes

1　Discussion of the activities of 'Cultural Intermediaries', initially suggested by Bourdieu (1984), has been critically developed by Savage *et al.* (1993) and more recently by Wynne and O'Connor (1998), who note that space and place make a difference to the attitudes and activities of these urban 'movers and shakers'.

2　There are potentially many indicators and much statistical evidence to exemplify this point. A small selection follows: while economic decline has been a long process, the period since the 1970s has been the most dramatic. During the 1980s The Potteries' unemployment average was consistent with national averages but since has risen. Employment itself has been reduced by 50 per cent since the 1950s and the pottery industry and its one sector dominance now provides only 30,000 jobs in Stoke-on-Trent. Only 24 per cent of new jobs created have been in the service economy, compared to an average of 65 per cent elsewhere. During the mid 1970s, Stoke-on-Trent had the highest proportion of derelict land in relation to the size of the administrative area anywhere in the country. There was a local joke that the city had escaped bombing during the Second World War because the Luftwaffe concluded that it had been done already (Phillips,1993).

3　£24.9 million from the National Lottery Arts Fund, £4.7 million from the City Council and £1 million from Europe.

4　A selection of the other more modest regeneration programmes includes (i) the Stoke Regeneration and Tourism Project – environmental improvements and tourism opportunities in Stoke-upon-Trent town centre. A town centre heritage trail will be created linking the main inter-city railway station and the Trent and Mersey canal to the town centre; (ii) improvement to streets and spaces in Burslem – new paving and landscaping, street lighting, signage and street furniture, plus promotion and publicity campaigns; (iii) the Dudson Centre – renovation of a 200-year-old Grade Two listed pottery works, for use by voluntary organizations, business support services and vocational training organizations.

5　In other words a (re)construction and (re)imagining, and the political, economic and symbolic centrality and marginality of people and places which can be summarized as 'an idealised white middle-class hegemonic notion of urbanity, often reclaiming or gentrifying marginal spaces, leading to the displacement and further exclusion of marginal populations, the re-definition of collective memory or (more rarely) commodification of minority ways of life' (Hall and Hubbard, 1998: 110).

6 These statistics are from 'Millions of Sq. Feet Lying Derelict in Our City', an article in the local newspaper *The Evening Sentinel,* 9 May 1999, based on the research of Dr Rick Ball of Staffordshire University commissioned by the Engineering and Physical Science Research Council.

7 For example, parliamentary reports sent out by Mark Fisher MP have had the titles *Interest Rates and Ceramic Industry* (October 1998) and *Black Day at Royal Doulton* (December 1998) as their focal debates.

8 The one dramatic exception to this trend is *Bomb*; *a Guide to Stoke-on-Trent,* a local free listing magazine which often features local artists, musicians, and cultural producers as well as engaging in critical interpretation of council planning decisions.

9 These can range from the building of squares and fountains to the greening of streets, the provision of benches and improved public spaces and the establishment of late-night shopping and 'happy hours' to make the city more 'liveable' and accessible to all sectors of the population.

10 Innovations suggested for Stoke-on-Trent included the building of a light rapid transport system and the provision of late-night buses and trains, training and skills-orientated programmes, cheap artist studios, an arts centre and gallery and music, film and animation facilities.

11 For example, Longton was described as having a potentially characterful town centre with the Gladstone St James Design Quarter producing a thriving economy of small-scale cultural producers. Stoke town was considered a perfect focus as a council, administrative and business centre, augmented by the presence of the factory outlet shops, the tourist attractions of the Trent and Mersey Canal, the North Stafford Hotel and Staffordshire University. Burslem was recommended to develop its architectural and historical atmosphere and highlight its literary links with Arnold Bennett.

12 This was to be facilitated by the appointment of a Cultural Development Officer.

13 *Major Touring Venues for Stoke-on-Trent* rather obviously concluded, with the aid of questionnaire surveys and focus groups, that for a city of almost 250,000 people and a catchment of a further 925,000, there had been an underinvestment and a lack of infrastructure which many similar towns take for granted, and therefore that Stoke-on-Trent is failing to satisfy the needs of its population.

14 The report also assessed the feasibility of reopening the disused Royal Theatre, which it rejected. The Royal was subsequently opened by local entrepreneur Mike Lloyd and has a programme of essentially middlebrow activities.

15 Such elements of self-imaging and place promotion have become central to the political rhetoric of urban governance and are constitutive of a '"glocalised" ideology of locality, place and community' (Harvey, 1989: 14).

16 This has also been variously described as 'New Urban Politics' (Cox, 1995), 'Growth Machines' (Logan and Molotoch, 1987), 'Urban Regimes' (Stone, 1989), or more commonly as 'Entrepreneurial Governance' (Hall and Hubbard, 1998; Harvey, 1989; Leitner, 1990; Mollencopft, 1983).

17 Wildavsky's 'Cultural Theory' advances Mary Douglas' grid-group typology, accenting not the primacy of social relations but the attitude towards power and authority as the basic determinator. This enables analysis of the dialogue of

culture, to interpret the actions of individuals in the context of social, political and city cultures: 'Cultural Theory' does not purport to explain all individual-level variations in preferences and beliefs, but how preferences and beliefs are acquired and reinforced in bilateral interactions between individuals and social contexts. This importantly links cultural biases with rational-choice perspectives. The relevant groups are (briefly): (i) *Fatalists:* tending towards self-orientation and lacking altruism; extremely pessimistic; expecting that attempts at improvement will fail; deep-rooted risk aversion and a strong tendency to discount failure. (ii) *Individualists:* more optimistic; tend to be risk-seekers and have long-term horizons; not altruistic; have optimistic beliefs about mutual co-operation, promotion, bidding and bargaining. (iii) *Egalitarians:* unlike the above two, they are group-oriented; have altruistic preferences; dislike collective decision mechanisms; demand control. Egalitarian groups can tolerate only small differences in preferences and beliefs. (iv) *Hierarchists:* positive altruism to fellow group members; not against outcomes where shares in collective activities are unequal; certain members have permanent authority (see Chai, 1997).

18 Zukin's important schema relates the presence of various spaces of consumption with capital accumulation and different phases of postmodernity. This understanding is exaggerated here, partially cloaking Zukin's ideas (that this is by no means a straightforward chronology and that different spaces of consumption do appear in different places at different times) in order to highlight The Potteries' inertia.

19 A few notes of local knowledge are necessary here: The Regent Theatre is the centrepiece of the Cultural Quarter, a fine art deco cinema/theatre able to seat 1630. Festival Park is a retail/entertainment park built on the site of the 1986 National Garden Festival. Trentham Gardens, former seat of the Duke of Sutherland, is undergoing extensive redevelopment as a retail, leisure and tourist complex, to include 150 holiday lodges, two hotels, the restoration of the Italian Gardens, restaurants, pubs, a farm and fishing centre, a vineyard, garden centre and retail factory outlet. Sound5 are a local band. Mike Lloyd is a local entrepreneur (mainly in the music business). Robbie (Williams) is the city's most famous celebrity, a pop singer. Nick Hancock is a Stoke-born TV personality. Sasha is an internationally renowned 'techno' DJ. Crewe, Alsager, Stafford and Meir are nearby towns. (Stoke) City and (Port) Vale are the local football clubs.

Chapter 3

Capital Calcutta: coins, maps, monuments, souvenirs and tourism

John Hutnyk

The King was in his counting house ...

> Since the producers do not come into social contact with each other until they exchange their products, the specific social character of each producer's labour does not show itself except in the act of exchange. (Marx, 1867/1967: 73)

This chapter begins with the charitable gift of a coin passed from the rich West to the poor East in the city of Calcutta. Specifically, with two scenes in the cinema of travel that celebrate such exchange. A traveller arrives in India. There are many arrival scenes – by train, plane and taxi. The visitor is confronted outside his hotel by begging children (the traveller in cinema is so often male). He reaches into his pocket for a trinket. Some bauble to amuse the kids. In thinking about the city in the Third World today, the arrival moments I am interested in are the first exchanges between travellers and locals. Patrick Swayze doing coin tricks outside his hotel in *City of Joy* and Bill Murray outside his in *The Razor's Edge*. Claude Lévi-Strauss could be here too, where he describes:

> Every time I emerged from my hotel in Calcutta, which was besieged by cows and had vultures perched on its window-sills, I became the central figure in a ballet which would have seemed funny to me, had it not been so pathetic. The various accomplished performers made their entries in turn: a shoeblack flung himself at my feet; a small boy rushed up to me; whining 'One anna, Papa, one anna!' [an anna is a small coin] a cripple displayed his stumps. (Lévi-Strauss, 1955/1973: 134)[1]

What I want to illustrate are the ways the city scene in the Third World is caught up in a counterfeit exchange. Derrida says of the counterfeit coin, that it must pass itself off as real. Swayze's sleight of hand trick on the kids tells us much about the deceit here, Murray's version of the same is refused – neither gift is adequate.

In the 1946 version of *The Razor's Edge*, the hero, Tyrone Power, visits a studio-based 'Indian' city high in the Himalayan hills, replete with Greco–Roman columns, peacocks, Chinese gongs and blacked-up mystics. Shangri-La, above the clouds, not in Tibet, or Bhutan or Nepal or India, but in Hollywood – just as with *City of Joy*, director Roland Joffe spent millions building a slum set south of Calcutta since no real slum was slum enough for his film about a charity outfit servicing the slums of Calcutta.[2] The Third World city is a site of many fantasies and deceits for the Western visitor. By giving a coin in charity the attempt is to recompensate an inequality: however, this gift works by way of emphasizing that inequality.

These are more recent examples of a founding scene of city-building. The first coins in India, according to Allchin (1995), appeared between the sixth and fourth centuries BC, the earliest coins being silver, but copper currency was minted soon after. Allchin links the spread of urban settlement and coinage closely, tracing the same route north-west to the Ganga Valley then to the peninsula South (Allchin, 1995). Coins, of course, are not found only in cities, since it is along trade routes across the Himalayas and along the Ganges that they travel, but the debate over location does exercise the experts. Numismatists discuss the relative merits of coin hordes, weights, markings, the rubbing of coins, scrapings of value, wear and tear, and how those most often found in the archaeological record nearby their site of minting are the heavier of the type, since thinner ones can be assumed to have passed through many hands, and so travelled far from the mint, etc. (see the excellent Marxist historian Kosambi, 1956: 173). Here it would be protocol to insert the famous quote from Nietzsche about truths being metaphors whose meanings have rubbed off like old coins, and Derrida's discussion of this in his 'White mythology' essay (Derrida, 1972/1982). But coin rubbing is also something that preoccupied poor, cold and grumpy Marx:

> The coin, which comes into contact with all sorts of hands, bags, purses, pouches, tills, chests and boxes, wears away, leaves a particle of gold here and another there, thus losing increasingly more of its intrinsic content as a result of abrasion sustained in its worldly career. While in use it is getting used up. (Marx, 1857/1987: 343)

Marx also notes the 'contradiction between gold as coin and gold as universal equivalent, which circulates not only within the boundaries of a given territory, but also on the world market' (Marx, 1857/1987: 345–6). Subsequently a whole archive of discussion of coins in philosophy has been unearthed by Caffentzis, discussing Newton, Locke, Simmel and Marx: 'a counterfeiter presupposes the existence of a civil government' (Caffentzis, 1989: 72). Whatever the case, ill-gotten gains or not, coin hordes in the record generally mark an extended productive economy with one or more urban 'centres' within a more or less integrated domain which can be read as an archive of plunder.

More recently than the old archaeologicals, but not so recent as Marx perhaps, is the coin trick perpetrated by the East India Company at the foundation of the British presence in India, where a few coins begin the process of capital formation – the capital of the Raj – the site of capitalist extraction of wealth to Europe. I want to look at the contemporary traces of this scene in the city in the experience of travellers today, as compared to those members of the East India Company, in the early days who 'converted', who went in search of riches: what are/were they up to, those heroes of the Company who became massively rich, Job Charnock, Clive, the 'robber-baron' and his suicide? So, these themes: coins and extraction, adventurism and ideology. This token hunt encompasses tricks, deceit, exploitation and appropriation. The framing tropes here are forms of violence encapsulated in the signs of the marketplace. The scenes in the cinema celebrate this extraction, the methods of the Company, and the capitalist 'gift' of wealth to Europe. The very coins that founded Calcutta were slave-trade booty. The main point is that the financial centre of Empire was never London, but Calcutta, and that understandings of the meaning, money, motivation and memory of Empire have travelled the wrong way for three hundred years. This is a violence consequent upon the formations of disciplines of knowledge and protocols of culture reportage insofar as history and travel guides perpetuate myths of Calcutta. Against the escape velocity of conventional alibis for the colonial record, some other moves are possible. The movement against imperialism begins in the city, as does the movement of products – tea, cotton, back and forth – and of a cultural politics which leads today back to the cinema, to the movement of the image and of meanings, and of the memory, or mimesis, of development and anti-development, and a transformatory project that can reconfigure the world. Contemporary traveller-visitors and their losses, the ways they get around, the purchase they carry today in what was the centre of Empire – 'built on silt, but gold', Kipling wrote – all this needs to be recast.

So if it is coins that travel, it is also coins that mark the foundations of even the oldest cities. Pieces of Eight. The travel of coins marks out the value, signifies trade, locates the market: from here the city is a predictable arrival. It would of course be too much like a Lewis Mumford type simplification to see this as anything more than a grand narrative convenience, but the coincidence of coins, travels, markets and cities is sure enough to serve as a point of confluence.

There are so many coin exchanges we really must be going to market: *The Razor's Edge, City of Joy*, Lévi-Strauss, and soon Moorhouse with blood-curdling imageries. The purchase of land for a factory (see the film *Tales from Planet Kolkata* as an antidote to *City of Joy*). Souvenirs. Monuments. Maps. Charity. It should not be thought that I am suggesting any simple equation of coins, market and urbanization, and Marx himself was careful to avoid simplifying complex relations in his discussions of the city as the centre of rural life, as the centre of warfare, of crafts and of trade (Marx, 1857/1986: 402–7).

If exchange and the international division of labour is the scene to be analysed we could do no worse than study Charlie Chaplin's film *The Immigrant* (1917 – perhaps significantly) and the scene where the tramp conjures with a coin to purchase – trick, con, contrive – a meal in a city restaurant. He finds a coin on the street and goes in to eat, but the coin has fallen through a hole in his pocket. Another tramp enters having found the coin, and pays the waiter for his meal. The waiter loses the coin through a hole in his pocket, Charlie tries to pick it up (vaudeville here that need not be detailed) but the waiter keeps on looking around at the wrong time. Eventually he pays, but the waiter bites the coin and finds it is a counterfeit. Then a to and fro about who will pay the bill, with Charlie eventually appropriating someone else's change so as to pay. This parable on the ingenuity of the beggar is a useful parallel text to Baudelaire's one on the counterfeit coin as discussed by Derrida in *Given Time* (1991/1992: endpiece). There the coin need not be real to serve as charity, the confession by the giver that the coin was counterfeit may itself be false, tricks upon tricks. The marketplace is full of them and you need to be sharp to prevail.

The Queen was in the parlour ...

> Thus were the agricultural people, first forcibly expropriated from the soil, driven from their homes, turned into vagabonds, and then whipped, branded, tortured by laws grotesquely terrible, into the discipline necessary for the wage system. (Marx, 1867/1967: 737)

It is necessary to explore three factors in the movement of capitalism in the last fifty, and three hundred, years. First the well established mobility of capitalist production as it crosses borders in search of profit – usually cheap labour (and sites of low labour regulation) but also super-profits, a kind of adventure speculation; secondly the distribution of products, or rather the increasing reach of commerce, of consumption, of everybody doing flips and twists to get into a pair of American blue jeans and drink Coca-Cola; and thirdly, a countervailing factor, or rather reneging, that closes borders and expels, that rejects those workers brought to the metropolitan centres of Europe in a time when capital did not seem to move so fast. This third factor is closely linked to the first, just as in an earlier period migration was linked to the need for workers inside the 'West'. In both scenarios the extension of money-mediated exchange and consumption to all corners continued unabated, and intricately bound up with representation and repression.

It is the first factor, movement motivated by profit motives, that brings the East India Co. to Calcutta. Not today, perhaps (but witness the abundance of trade delegations visiting the newly deregulating India, including communist West Bengal), but certainly as the motive for the officials and buccaneers in search of fortune (no doubt there were some philanthropes too, but few) was no subtle tourism. Just over three hundred years ago Job Charnock and the English came and built a factory on the site of three villages. The British traders had entered India in 1612 and purchased land from local rulers with silver coin earned through the brutal slave trade markets of the West Indies. These first tainted coins begin a city by a process on the backs of people becoming commodities (Marx, 1857/1986: 419–20). This twisted social relation must be examined:

> The mysterious character of the commodity form consists therefore simply in the fact that the commodity reflects the social characteristics of human labour as objective characteristics of the products of labour themselves ... Through this substitution, the products of labour become commodities, sensuous things which are at the same time suprasensual or social. In the same way the impression made by a thing on the optic nerve is perceived not as a subjective excitation of that nerve but as the objective form of a thing outside the eye ... I call this fetishism. (Marx, 1867/1967: 164–5)

There is a whole range of images, narratives and meanings that cause us to accept this routine. The discipline required of the market interests me. I would agree with Andrew Lattas that 'an analysis of gifts has also to ground gifts in those structures of reification, self-mystification and legitimation which gifts make available. It is not a question of denying or dismissing as false the gift's ideology, but of exploring its constitutive power' (Lattas, 1993: 108). Just what is Swayze up to with those children? Offering a coin to teach kids to love capitalism – something like the role of Santa in the West? A commodified social contract? Is there no alternative to the market here? Cities become sites for the deal – just a glorified employment exchange. Marx shows that it is capital which travels (it circulates) while labour, even where it moves a little – as in vagabondage, urban migration, labour migrates to sell itself – is soon disciplined. Of course, without labour to make things move in the first place there could be no capital. Who, after all, built Calcutta? Still, the mystification of the social foundation of capital begins with a trick. In the 'Economic Notebooks' of 1857-8 (The *Grundrisse*), Marx sets out this moment in a vivid, if abstracted, 'sketch':

> when the great English landowners dismissed their retainers, who had consumed with them the surplus produce of their land; when their tenant farmers drove out the small cottagers, etc., then a mass of living labour power was thrown on to the labour market, a mass which was free in a double sense [*eine Masse, die in doppeltem Sinn frei war*]: free from the old client or bondage relationships and any obligatory services, and free also from all goods and chattels, from every objective and material form of being, free of all property. It was reduced either to the sale of its labour capacity or to beggary, vagabondage or robbery as its only source of income. History records that it tried the latter first, but was driven off this road and on to the narrow path which led to the labour market, by means of gallows, pillory and whip. (Marx, 1857/1986: 431 [1857/1974: 406])

The goods that had previously been consumed by the feudal lords and their retainers, and the released produce of the land, were now thrown on to the exchange market, as were those who would be henceforth known as labourers. The basis of this trick is that the sale of labour power must be instilled by disciplinary force – the gallows, the workhouse, the prison (Michel Foucault's work on asylums, clinics, punishments etc., emerges from here) – and so becomes the only 'choice'. Even the poor-houses and their charity instil the discipline of work (only Dickens' Oliver dares ask for 'more' in literary versions; no doubt there were many 'Olivers'). That this was conceived by Marx as part and parcel of capitalist development can be confirmed from other (re)writings of almost the same paragraph.

In *Capital* Marx returns more than once[3] to this scene:

> For the conversion of his money into capital, therefore, the owner of money must meet in the market with the free labourer, free in the double sense [*frei in dem Doppelsinn*], that as a free man he can dispose of his labour-power as his own commodity, and that on the other hand he has no other commodity for sale, is short of everything necessary for the realisation of his labour-power. (Marx, 1867/1967: 169 [1867/1975: 183. Notice how the rewritten version is more elegant: it is worth keep-

ing in mind, at a time when The *Grundrisse* seems more often read and quoted – even here – that *Capital* was the text actually prepared for readers.])

That this meeting in the market is no equal exchange is the trick of all tricks. Though it would seem that in the market-place the capitalist offers a 'fair' price – money for labour, wages – and that the entire history of reformist unions has been to ensure the 'fair trade' of this exchange, this is of course the big deception of capitalist appropriation, since the capitalist does not pay for every hour that the labourer works (nor for all the costs of reproducing labour power). Marx writes: 'An exchange of equivalents occurs, [but it] is merely the surface layer of a [system of] production which rests on the appropriation of alien labour without exchange, but under the guise of exchange' (Marx, 1857/1986: 433). Here, at the crucial point of the labour theory of value, the expansion of the trick of the market is played out in the coin of wages, and this trick is the foundation of the city as labour exchange, as Marx discusses in a passage that sets the whole movement out clearly:

> The other circumstances which e.g., in the sixteenth century increased the mass of circulating commodities as well as money, created new needs and therefore raised the exchange value of native products, etc., increased prices, etc., – all these fostered the dissolution of the old relations of production, accelerated the separation of the worker from the objective conditions of his own reproduction, and thus hastened the transformation of money into capital. Nothing is therefore more foolish than to conceive of the original formation of capital as having created and accumulated the original conditions of production – means of subsistence, raw materials, instruments – and then having offered them to workers stripped of them. For it was monetary wealth which had partly helped to strip off these conditions of labour power of the individuals capable of work. In part this process of separation proceeded without the intervention of monetary wealth. Once the formation of capital had reached a certain level, monetary wealth could insinuate itself as mediator between the objective conditions of life thus become free and the freed but also uprooted and dispossessed living labour powers, and buy the one with the other. As regards the formation of monetary wealth itself, prior to its transformation into capital, this belongs to the prehistory of the bourgeois economy. Usury, trade, urbanisation and the development of government finance which these made possible, play the main role here. (Marx, 1857/1986: 432)

This moment is exported universally and urbanization plays a main role. It will be no surprise to learn that the 'veiled slavery of the wage-workers in Europe needed, for its pedestal, slavery pure and simple in the new world' (Marx, 1867/1967: 759–60). Marx adds a footnote that there were 'ten slaves to one free man' in the English West Indies, 'in the French fourteen to one, in the Dutch twenty-three for one' (Marx, 1867/1967: 760n). The labourers separated from their social means of production are thus named in English legislation as the 'free labouring poor', or the 'idle poor' or the 'labouring poor', a terminology which even the 'sycophant' Edmund Burke, 'in the pay of the English oligarchy', called 'execrable political cant' (Marx, 1867/1967: 760n. Marx asks us to judge Burke's good faith here alongside his other pronouncements to the effect that the 'laws of commerce are the laws of nature and therefore the laws of god').[4] The city is the site of the natural disciplining of labour, and this is achieved on the basis of a coined counterfeit. Again in another part of Volume 1 of *Capital* labourers are 'free workers in a double sense' [*Frei Arbeiter in dem Doppelsinn*, 1867/1975: 742]:

> The capitalist system presupposes the complete separation of the labourers from all property in the means by which they can realise their labour. As soon as capitalist production is once on its own legs, it not only maintains this separation, but reproduces it on a continually extending scale. (Marx, 1867/1967: 714)

Marx's story of the development of labour for sale as a world-wide system was only a 'sketch' (though the history of this expropriation is written 'in letters of blood and fire' (Marx, 1867/1967: 715)). But in a note to the editors of the paper *Otechestvennye Zapiski* in the last years of his life, Marx warned that the chapter which set this out in the most detail, Chapter 27, should not be 'transformed' from an historical sketch of the genesis of capitalism in Western Europe to a 'theory of the general course fatally imposed upon all peoples, whatever the historical circumstances in which they find themselves placed' (Marx, 1878 in Shanin, 1983: 136). Far too often the technical abstractions necessary in setting out Marx's *Capital* which begins with commodities and expands in complexity to encompass trade, circulation of capital, rent etc., lead hasty readers to orthodox fixities and dogma. Nevertheless, the general point of the expansion of the logic of market exchange can be illustrated thus.

There is little need to go further into the hagiographic mode of repeating Marx as oracle. There are sufficient examples. In any study of the ways colonialism 'had to use force to make the indigenous populations accept the commodity form' (Cleaver, 1979: 77), the various examples would range from slavery and death to persuasion and, today, co-options of all kinds. Cleaver lists 'massacre, money taxes, or displacement to poor land' as the ways that capital dealt with resistance and refusal to be put to work. On the basis of this comes the 'civilizing' mission of the West, that would teach 'backward' peoples the values of thrift, discipline and saving.

Many heathens saved, no doubt. Mother would be proud. Patrick Swayze himself says it was a near revelation to work in Calcutta. The point is that here city building, civilizing mission, urbanization, whether through the mechanism of the sketch, or variations such as in-migration from the countryside, other strategies of 'development' (such as the weavers herded into the factories under the discipline of the coin) etc., reinforces the ramparts of the Third World city as a market with no alternative, as a glorified labour exchange, and all the Mother Teresas and coin tricks that can be marshalled should not be able to disguise such a trap.

Perhaps this is far too harsh. But then the city of Calcutta suffers from a bad press, as does development. I am resolutely not against development.[5] The problem is the fetishization of development and capital as a kind of juggernaut beyond control, and of course, the question of control by whom. The circulation and expansion of capital today is fetishes as speed. So:

There is nothing to regret, the world moves in every which way, men and women cross the planet every which way, through interposed images and sounds, or directly through the displacement of their òwn person. But let us immediately pick up the paradox. Everything circulates: the types of music, the advertising slogans, the tourists, the computer viruses, the industrial subsidiaries and, at the same time, everything seems to freeze, to be stationary, as the differences fade between things … everything has become interchangeable, equivalent within standardised spaces. (Guattari, 1992: 123)

But what actually moves? Products to the market, images, sites? The market and the city, city images and the factory – the shopping centre? Swayze tò the children (dirty dancing)? Or labour? Is revolution (the movement) more than a metaphor? Against immigration laws. Against roads as the warehouses of just-in-time delivery. Other struggles circulate here.

So what is in the traveller's suitcase? An old book. In *The Age of Revolution* Hobsbawn notes that until the industrial revolution Europe had always imported more from the East than it had sold there (Hobsbawn, 1975: 34). Balance of trade. But with the industrialization of cloth production and the rise of the Manchester mills – corresponding to the destruction of the rural and village or 'artisanal' weaving in India (sometimes by way of the amputation of weaver's thumbs by the Company, a blood curdling remembrance of *The Razor's Edge*) – weavers were forced into agriculture or into the urban centres and so into the machine shops and warehouses. This is the urban discipline of the money-wage system. The machine shops were of course first in Manchester, but soon industrialization also moved them to India and machines abstracted and multiplied the same skills that had been the – refined – preserve of the weavers' looms. Technology replaced the weaver at the very same time it brought them into the factory to work.[6]

> By ruining handicraft production in other countries, machinery forcibly converts them into fields for the supply of its raw material. In this way East India was compelled to produce cotton, wool, hemp, jute and indigo for Great Britain. (Marx, 1867/1967: 451)

And, summarising a period of English manufacturing monopoly, Marx notes:

> 1830 glutted markets, great distress; 1831 to 1833 continued depression, the monopoly of the trade with India and China withdrawn from the East India Company; 1834 great increase of factories and machinery, shortness of hands. The new poor law furthers the migration of agricultural labourers into the factory districts … 1835 great prosperity, contemporaneous starvation of the hand-loom weavers. (Marx, 1867/1967: 455)

Be it in Britain or in the 'Eastern markets' this violent conjunction of capital extraction, technological development and urbanization has transformed the world under the sign of the coin.

Thinking about weavers who now produce for the souvenir trade as technological development brings mass tourism to India adds another thread to the discussion. In this context I want to look at the ways a tourist experiences the city. The vehicles for this are the codifications that surround travel and capital. Here I mean to play on different meanings of travelling capital: the travelling capital of Europe's expansion, the physical movement of people, power, meanings and money, the exchanges of people, power, meaning and coins, and the capital of this complex. Capital as money on the move, but also the movement of the capital – a movement concerned with what happens in that old capital of Empire, Calcutta.

The Maid was in the garden ...

> The social character of activity, as also the social form of the product and the share of the individual in production, appears here as something alien to and existing outside the individuals; not as their relation to each other, but as their subordination to relationships existing independently of them and arising from the collision [*Anstoss* – also could be 'bump', echoes of rubbing coin] between different individuals. The general exchange of activities and products, which has become the condition of life for every single individual, their mutual connection, appears to the individuals themselves alien, independent, as a thing. In exchange value, the social relationship of persons is transformed into a social attitude of things; personal capacity into a capacity of things. (Marx, 1857/1986: 94 [1857/1974])

Let me leave money to one side for a moment if I can, since it is the equivalence of all things (a point to come back to: see Spivak's critique of the limits of Derrida's saying 'hello' to Marx, where she shows that Marx is exhorting the worker not to fall for the trick of the money-based explanations of the capitalist, but to remember that labour power, spent in time, is productive [Spivak, 1995]).[7] It might be worth looking to what first of all travellers get for their currency. Today, as ever, when you arrive (after the traveller's cheque swap), the first thing you need is a guide. Among the myriad cultural productions that make up the representations of Calcutta available to visitors today – films, books, photographs – by far the most explicit representational modes are those produced for the immediate consumption of tourists. Maps of the city, guide-books and postcards of monuments present the city in handy, portable, two-dimensionally convenient ways and it is these mechanisms which govern (discipline) social relationships for the traveller. Touts are available outside the hotels offering all manner of services (this in fact is noted so often in the texts of visitors it becomes a trope of the Eastern city of iniquity – again the *City of Joy* and Lévi-Strauss hotel arrival scenes).

Get a map. Finding a way through the city is a major project for all visitors. 'New' cities are easy to get lost in and so guide-books and maps are necessary and monuments become landmarks oriented more towards the city than the histories they memorialize. Such markers offer a key to the ways a city can be made and experienced. Residents and visitors alike would often be lost without reference points; but tourists 'need' maps and guide-books which calibrate with expectations and evocations of the city formed before arrival (often as a place of immanent exotic adventure) and often throughout the stay. Maps are an adjunct to the monumental vision which orients the traveller in a foreign place when the entire world is something to be seen (as an open market, or the phantasmagoria of the world fair as described by Marx in 1865). At the same time, however, for some a kind of 'alternative' traveller protocol requires a renunciation of the convenience of the guide in favour of a more individualistic, and 'authentic' exploration. Yet even this alternative is guided by a host of expectations and prior mappings – of Calcutta, India, of the Third World – and which as often as not has little correspondence with local residents' versions of their city. Calcutta's global image – a teeming poverty Ma T enhanced frightscape – gives it a bad press everywhere else, but not there. Another counterfeit. Alternative, disengaged or prefigured, the visitor's experience of the city fits pre-packed units like a code so that physical representational 'souvenirs' of maps, images of monuments and postcards become a mode, if often kitsch, of inscribing presence in, or of, a place.

In 1989 a central government edict declared illegal any map of India which did not comply with topographical Survey of India maps. In the intersections between Calcutta's political history and the seemingly more innocuous trappings of tourism, such as the paraphernalia of maps, guide-books and souvenirs, is where I would want to use the work of Henri Lefebvre, who raises questions about the relation between mapping and travel when he suggests that if 'the maps and guides are to be believed a veritable feast of authenticity awaits the tourist' (Lefebvre, 1974/1991: 84). In his book *The Production of Space*, Lefebvre argues that it is capitalism which has produced 'space' in such a way that, with the aid of the tourist map, a 'ravenous consumption' raids the landscape. He argues that 'Capitalism and neo-capitalism have produced abstract space, which includes the "world of commodities", its "logic" and its world-wide strategies, as well as the power of money and that of the political state' (Lefebvre, 1974/1991: 53). A vast network links the power of the state through a complex of financial institutions, major production centres, motorways, airports and 'information lattices' which lead to the 'disintegration' of the town as anything other than a space to be consumed (Lefebvre, 1974/1991: 53). Reading this body of work in Calcutta, particular attention here should be paid to representations of monuments and bridges, maps, guides, souvenirs – the code – particularly the Victoria Memorial, and the massive expanse of Howrah Bridge, which in postcards become equivalents.

But these maps mark out another experience more clearly. The orientation of visiting Calcutta – as metaphor of inscrutable India – is the coin trick of Patrick Swayze. This is the market moment exoticized – the urban sophisticate meets the uncomprehending (pre-market relation) others. The world-wide strategy of the commodity logic, as Lefebvre calls it, appears in the evangelical coin trick played on a few kids on the street with the same violence as the all too bloody massacres of colonial genocide.

And down came a blackbird ...

In order for there to be counterfeit money, the counterfeit money must not give itself with certainty to be counterfeit money; and this perhaps is also the intentional dimension, that is, the credit, the act of faith that structures all money, all experience or all consciousness of money, be it true or false. (Derrida, 1991/1992: 95)

Next to the map in the back-pack is the image machine. When travel is a signifier in the contemporary moment, the camera is never far. Click click. The tourist has become associated with the mechanical eye, making miniature equivalences of everything it sees. Cities, people, poverty – all become photogenic. Photogenic Calcutta, cinematic cities, the videographic construction of the subcontinent. To get into this we might take up these details presented through the artifice of the camera. I have become interested in another coin trick, a scene where a traveller takes a photograph of a destitute family living on the street in a congested part of town, and gives five rupees in exchange. You can be paid for the (perceived) misery of your condition. Poverty framed. This is not an unfamiliar or atypical moment on the 'banana-pancake trail' of Western budget tourism in the Third World. For travellers at the front-line of capitalist expansion today, photographs of the 'locals' are also monumental souvenirs. The budget travel Lonely Planet Guide Book, *India: A Survival Kit*, comments on the propriety of paying for snaps. But its editor, Tony Wheeler, millionaire, also once suggested that in situations where locals demanded they be photographed you could carry your camera without film and set it so the flash goes off but nothing else – a counterfeit photo. The scene of coin for photo exchange has a long, long history. It goes back to the early days of black and white cinema, *The Razor's Edge* (version one, starring Tyrone Power, 1946, directed by Edmund Golding), back further to the introduction of the camera into India (Dadasaheb Phalke, who introduced film technology to India, learning tricks from a magician). Back to the coin exchanged for membership of an anti-colonial organization (one rupee Congress membership). Back to the exchange of coin for the cloth of the Bengali weavers, back to the coins paid by the Company to Suraj-ud-duala and the (in)famous black hole/black box, further back to the coins paid by Job Charnock to establish a British factory on the shore of the Hooghly River.

Since the story of the Black Hole must be told here as well, it can be in a critical version: Marx calls the incident a 'sham scandal' (Marx, 1947: 81). In an extensive collection of notes made on Indian history, Marx comments that on the evening of 21 June, 1756, after the Governor of Calcutta had ignored the order of Subadar Suraj-ud-duala to 'raze all British fortifications' in the city:

> Suraj came down on Calcutta in force ... fort stormed, garrison taken prisoners, Suraj gave orders that all the captives should be kept in safety till the morning; but the 146 men (accidentally, it seems) were crushed into a room 20 feet square and with but one small window; next morning (as Holwell himself tells the story), only 23 were still alive; they were allowed to sail down the Hooghly. It was 'the Black Hole of Calcutta', over which *the English hypocrites have been making so much sham scandal to this day*. Suraj-ud-duala returned to Murshidabad; Bengal now completely and effectually cleared of the English intruders. (Marx 1947: 81, my italics)

Marx also reports on the subsequent retaliation against and defeat of Suraj-ud-duala by Lord Clive ('that Great Robber', as he calls him elsewhere; Marx, 1853/1978: 86), and Clive's 1774 suicide after his 'cruel persecution' by the directors of the East India Company (Marx, 1947: 88). There seem to be very good reasons to conclude that the black hole incident is counterfeit. The single report from a 'survivor' some months after Clive's savage response to Suraj-ud-duala's occupation of Calcutta – the famous/notorious Battle of Plassey – reads very much like a justification forged to deflect criticisms of brutality on the part of the British forces.[8] The black hole is a kind of souvenired past of imperial history faked to stand in for the theft of a city.

If only there was better documentary evidence for this tale. Where are the paparazzi when you need them? In so many different ways photographs are souvenirs of the experience of experience. They signify travel. The trick here is the incommensurability of value in the photograph that is taken on the streets of Calcutta, the coin exchanged for the privilege, the value of the image which is so difficult to calculate. The reinvestment of these images, these mimetic aids for storytelling, and the various contexts and uses for such stories should make me cautious. The images circulate again in the slower rhythms of allegedly scholarly application. At the same time that we write these histories, today, the travel guide, the cinema, the documentary film, television and the text participate in an uneven exchange between cities like Calcutta/Mumbai and those of London or Manchester. Same as it ever was?

In the midst of an excellent essay on Phalke, Ashish Rajadhyaksha quotes E. B. Havell, Principal of the Government School of Art, Calcutta, as one of those who complained of the loss of the artisan's skills in the face of technology. Havell's analysis seems to be sound, as he noted that the handloom workers were driven into powerloom factories and that the 'most skilled weavers in the world' were being 'concentrated in the great Anglo–Indian industrial cities [i.e. Calcutta] and delivered, body and soul, into the hands of Indian and European capitalists' (Havell in Rajadhyaksha, 1993: 51).

Perhaps it is no accident that the most prominent site of Mother Teresa's death cult, which provides a short-term 'home for dying destitutes' – those spat out by the machine city – is in Kalighat. The extent to which the Teresa image has blocked any other view of the city but those of the international clichés of teeming squalor, hunger, poverty and shit has not been often enough remarked. It is no coincidence that the bad press attached to this city coincides with the emergence of anti-imperialist struggle. Another mode of discipline. Also:

> It is interesting to note that the cult of the goddess Kali was practically unknown before the eighteenth century, a period when a great change was taking place in the social and political life of Bengal. The Kali cult in its present form owes its inspiration to Krishnanada Agamavagisa … it was popularised by Maharaja Krishnachandra … before whose eyes the establishment of the East India Company took place. Many of the local rebellions that took place after the establishment of Company rule – e.g. the Sanyasi rebellion, the Chuar rebellion and so on – were inspired in the name of Kali. (Battacharya in Rajadhyaksha 1993: 60)

It is reported, by Moorhouse, that assassins who were devotees of Kali knotted a coin in one end of a cloth – something like a large cotton scarf – to improve their grip when strangling someone. What then is exchanged today? Engels in Manchester, Marx in London, their texts on colonialism in India are often discussed, but always in terms of past histories. All this is interesting enough. The conjunction of rebellions against the Company, the role of technology and capital, the travels of money, cotton, wefts and weaves, and the violence of extraction, industrialization and urbanization under Empire – all this can form the subject of interesting discussion, but only insofar as to stress the continuities with extraction today, colonialism now. Coins and charity are violence. *For* a cinematic Kali cult. Photogenic Hinduism.

A Kali cult? The fear and trembling that this inspires sets tea-cups a-shaking. The fear of the demonic image of Kali for the British should be explored, but with a caution in light of both an inverted exoticism (of dark satanic imagery that fascinates the Western traveller) and in the context of rampant Hindutva and the rise of the BJP. Any danger, however, that the memory of resistance in a Kali politics leads too easily into religious chauvinism has been countered by the mobilization of the secular forces under a new democratic communist movement, of various stripes, in West Bengal. Just as the struggle against imperialism begins in Calcutta, so does a contemporary anti-capitalist politics. A struggle against extraction and exploitation which continues to be facilitated by the movement of products and profits – work conditions on the tea plantations and the warehouses, back and forth with privatization and trade delegations – and all this with little yet said of a cultural politics of the most prominent local Calcuttan figures, Netaji Subhas Chandra Bose, Charu Muzumdar and Jyoti Basu, all of whom have become iconic and even cinematic, subject to the movement of the image and of meanings, and of the memory, or mimesis. So let us remember what currency plays here.

To make sense of this overdetermined scene where coin tricks are played with photographs, I would have Patrick Swayze turn towards Marx. Could it be possible that he would learn to read the 'double' structure in Marx, that of the commodity and the money form? Instead of the strange motivations which have him forgoing his usual $7 million fee for a mere $1 million because he 'so wanted to do this film for the people of Calcutta' (why not do *Dirty Dancing 2* and donate the spare $6 million?), it is possible to force a reading lesson which would have him consider his position in relation to colonial travel and the othering experience of the Western visitor. Here our hero could reflect upon the motive of profit and the imposition of a commodity form on a non-commodity relation (if that), the disguise of this deception in the trick of the giving of charity, specifically of the coin passed to a beggar in the scene of tourism as advocated by the guide-books, and the ways this offers a currency for thinking the relations of profit and its fetish disguises. But maybe Patrick does not read this sort of literature. No matter, in the scene of the coin passed to a child, the coin trick as a sleight of hand that impresses/mystifies the locals, the whole of Marx, the deceit of money, the city and capital is displayed. The coin exchange initiates a complex web of industrialization and transition, the investment of meanings and images, the accumulation and speculation, and the inexorable spiral that leads from lost thumbs to Cola wars. (Thumbs Up to Pepsi Generations). *City of Joy* is merely the narrative condensation of so many of these developments.

The coin trick underlies the scene of deception in colonial relations as well as in the economics of tourism, gifts, knowledge and meanings. It is important here to note the similarities and differences between taking a photo, buying a souvenir, giving a coin in exchange or charity and the varied experiences these imply. As I draw examples from the travel tales of Somerset Maugham, from popular cinema such as *City of Joy* and *The Razor's Edge*, it is the example discussed by Derrida, Baudelaire's passage on deceitful gifting and that counterfeit trick of this gift, and Charlie Chaplin's more subaltern rendering of such a scene, that suggests the coin as the icon of contemporary politics.

Third World-destined movie star travellers offering charity to the poor only help, guide and serve themselves in a show of how big (hearted) they are. This is the essence of the *potlatch*, with a grand humility in charity, even as they disavow all they do as 'just a drop of water in the ocean'. What sacrifice. A transformatory project for redistributive justice does not begin with this coin. What it demands is a rethink of the front-line role of the charitable organizations, of Western NGOs, of even those progressive 'fair trade' or alternative development types who would, for example, advocate revolution in Bengal while leaving their own little backwaters – say, London – untouched by such militant fervour.

There is an alternative to the extension of the market to all corners of the planet, and it is not a universal gift service. Charity contributes nothing but the maintenance of the trick. Travellers who don't go further than the doorstep of their hotel miss the point.

There remains much more to be done to work out how to travel to Calcutta. And this has, of course, been a partial study. Let us take Derrida at his word on travel. In an essay on the gift and charity, Derrida identifies two 'risks' of travelogues in the possible meanings of the term: 'The first is that of selectivity', and he describes a *récit raisonné* as a 'narrative that, more than others, filters or sifts out the supposedly significant features – and thus begins to censor' (Derrida, 1993: 197–8); and the second, from the first; '*raisonner* also signifies, in this case, to rationalise … active overinterpretation' (Derrida, 1993: 198). These two themes of perspective and ordering selection are the themes of this work which take up Derrida's call (his is not the only call of this sort) alongside a Marxist analysis of money, for a 'systematic reflection on the relations between tourism and political analysis' at a time when tourism has become highly 'organized'. Derrida writes that such an analysis 'would have to allow a particular place to the intellectual tourist (writer or academic) who thinks he or she can, in order to make them public, translate his or her "travel impressions" into a political diagnostic' (Derrida, 1993: 215). The politics of coin tricks remains to be unpacked.

References

Ahmad, Aijaz (1992) *In Theory: Nations, Classes, Literatures*, Verso, London.

Allchin, F. R. (1995) *The Archaeology of Early Historic South Asia*, Cambridge University Press, Cambridge.

Bataille, Georges (1967/1988) *The Accursed Share*, vol. 1, Zone Books, New York.

Caffentzis, Constantine (1989) *Clipped Coins, Abused Words and Civil Government: John Locke's Philosophy of Money*, Autonomedia, New York.

Cleaver, Harry (1979) *Reading Capital Politically*, Harvester Press, Sussex.

Derrida, Jacques (1972/1982) *Margins of Philosophy*, trans. Alan Bass, University of Chicago Press, Chicago.

Derrida, Jacques (1991/1992) *Given Time: Counterfeit Money*, trans. Peggy Kamuf, University of Chicago Press, Chicago.

Derrida, Jacques (1993) Politics and friendship: an interview. In Kaplan, E. Ann and Sprinkler, Michael (1993) *The Althusserian Legacy*, Verso, New York, 183–232.

Derrida, Jacques (1995) *Spectres of Marx*, Routledge, London.

Guattari, Felix (1992) Space and corporeity: nomads, city, drawings. *Semiotext(e) Architecture*, 118–25.

Hobsbawm, Eric (1975) *The Age of Revolution 1789–1848*, Weidenfeld and Nicolson, London.

Hutnyk, John (1996) *The Rumour of Calcutta: Tourism, Charity and the Poverty of Representation*, Zed Books, London.

Hutnyk, John (1997) derrida@marx.archive. Manchester University Anthropology Working Paper.

Hutnyk, John (1998) Argonauts of Western pessimism: Jim Clifford's ethnographica. In Clarke, Steve (ed.) *Travel Writing and Empire*, Zed Books, London.

Kalra, Virinder (1997) *From Textile Mills to Taxi Ranks: Experiences of Labour among Mirpuris/(Azad) Kashmiris in Oldham*, Ph.D thesis, University of Manchester.

Kalra, Virinder and McLaughlin, Sean (1999) Wish you were(n't) here? Discrepant representations of Mirpur in narratives of migration, diaspora and tourism. In Kaur, R. and Hutnyk, J. *Travel Worlds: Journeys in Contemporary Cultural Politics*, Zed Books, London, 120–36.

Kaur, Raminder and Hutnyk, John (1999) *Travel Worlds: Journeys in Contemporary Cultural Politics*, Zed Books, London.

Kosambi, Damodar Dharmanand (1956) *An Introduction to the Study of Indian History*, Popular Prakasha, Bombay.

Lattas, Andrew (1991) Sexuality and cargo cults: the politics of gender and procreation in West New Britain. *Cultural Anthropology*, **6**(2): 230–56.

Lattas, Andrew (1993) Gifts, commodities and the problem of alienation. *Social Analysis*, **34**: 102–15.

Lefebvre, Henri (1966/1968) *The Sociology of Marx*, Penguin, Harmondsworth.

Lefebvre, Henri (1974/1991) *The Production of Space*, Basil Blackwell, Oxford.

Lévi-Strauss, Claude (1955/1973) *Tristes Tropiques*, Jonathan Cape, London.

MacFarlane, Iris (1975) *The Black Hole, or the Makings of Legend*, Allen & Unwin, London.

Marx, Karl (1853/1978) *On Colonialism*, Progress Press, Moscow.

Marx, Karl (1857/1974) *Grundrisse der Kritik der Politischen Ökonomie*, Dietz Verlag, Berlin.

Marx, Karl (1857/1986) Economic manuscripts of 1857–8. In Marx, K. and Engels, F. *Collected Works*, vol. 28, Lawrence & Wishart, London.

Marx, Karl (1857/1987) Economic manuscripts of 1857–8. In Marx, K. and Engels, F., *Collected Works,* vol. 29, Lawrence & Wishart, London.

Marx, Karl (1867/1967) *Capital: A Critique of Political Economy, Vol. 1: The Process of Capitalist Production,* International Publishers, New York.

Marx, Karl (1867/1975) *Das Kapital,* Vol. 1, Dietz Verlag, Berlin.

Marx, Karl (1898/1950) Wages, prices and profit. In *Selected Works*, vol. 1, Foreign Languages Publishing House, Moscow.

Marx, Karl (1947), *Notes on Indian History*, Foreign Languages Publishing House, Moscow.

Moorhouse, G. (1971) *Calcutta: The City Revealed*, Penguin, Harmondsworth.

Rajadhyaksha, Ashish (1993) The Phalke era: conflict of traditional form and modern technology. In Niranjana, Tejaswini, Sudhir, P. and Dhareshwar, Vivek (eds) *Interrogating Modernity: Culture and Colonialism in India,* Seagull, Calcutta.

Shanin, Teodor (ed.) (1983) *Late Marx and the Russian Road*, Monthly Review Press, New York.

Spivak, Gayatri Chakravorty (1995) Ghostwriting. *Diacritics*, 25 (Summer).

Notes

1   Half way through *Tristes Tropiques*, Lévi-Strauss describes his 1950 visit to Calcutta. He arrived at Calcutta airport, mid-century amidst a torrential downpour, and was quickly whisked away to his hotel. From there he describes the city: 'the large towns of India are slum areas … Filth, chaos, promiscuity, congestion; ruins, huts, mud, dirt; dung, urine, pus, humours, secretions and running sores: all the things against which we expect urban life to give us organised protection, all the things we hate and guard against at such great cost, all these by-products of cohabitation do not set any limitation on it in India. They are more like a natural environment which the Indian town needs to prosper. To every individual, any street, footpath or alley affords a home, where he can sit, sleep, and even pick up his food straight from the glutinous filth … the tragic intensity in the beggar's gaze as his eyes meet yours … could easily be transformed into a howling mob if, by allowing your compassion to overcome your prudence, you gave the doomed creatures some hope of charity' (Lévi-Strauss, 1955/1973: 134–5).

2   *City of Joy* originally was a book by Dominique Lapierre about a Polish priest doing charity work in Calcutta. In 1989 Rajiv Gandhi's Central Government gave British director Roland Joffe (*The Mission, The Killing Fields*) permission to make a film of the book. Despite Joffe's arguments that the film would 'project the indomitable spirit of the slum-dwellers of Calcutta' (*The Telegraph*, 24 December 1989), the CPI-M Government of Bengal withdrew Joffe's permission later in the year on the grounds that there was 'no need to show only the slum-dwellers to show the indomitable spirit of Calcuttans' (*The Telegraph*, 24 December 1989). Debates about censorship and freedom of information raged over the following months as Joffe refused to accept the decision and conscripted prominent Calcutta personalities to his cause. Coffee house discussion turned often to the merits of not only the proposed Joffe film, but also other filmed representations of the city. The stars of *City of Joy* were Patrick Swayze, Om Puri and Shabana Azmi.

3   Also: 'They were turned *en masse* into beggars, robbers, vagabonds, partly from inclination, in most cases from stress of circumstances. Hence at the end of the fifteenth and during the whole of the sixteenth century, throughout Western Europe a bloody legislation against vagabondage. The fathers of the present working-class

were chastised for their enforced transformation into vagabonds and paupers. Legislation treated them as 'voluntary' criminals, and assumed that it depended on their own good will to go on working under the old conditions that no longer existed' (Marx 1867/1967: 734). Recently it was missionaries who exported this 'necessary discipline' by chastising throughout the world; today it is the NGOs and alternative credit banks who do so.

4   See also the section in *Capital* where Marx analyzes the notion of laziness versus industriousness as a parable of sin – 'Adam bit the apple' (Marx, 1867/1967: 713).

5   'What would be the meaning of a destruction of capitalism that would be at the same time the destruction of capitalism's achievements? Obviously it would be the crudest possible denial of Marx's lucidity. The humanity that would have destroyed the work of the industrial revolution would be the poorest of all time; the memory of the recent wealth would finish the job of making that humanity unbearable' (Bataille, 1967/1988: 170). For a useful corrective to simplistic renditions of Marx's lucidity on the role of industry in India, see Ahmad, 1992, Chapter 6.

6   And back and forth: see Virinder Kalra's 1997 Ph.D. study of Kashmiri workers in Oldham near Manchester, and his subsequent discussion in Kalra and McLaughlin, 1999.

7   I have several times to a greater or lesser extent followed Spivak's arguments in my own work, on Calcutta (Hutnyk, 1996), on Derrida (Hutnyk, 1997) and on Clifford (Hutnyk, 1998).

8   For a comprehensive and readable discussion see MacFarlane, 1975.

Chapter 4

Citing difference: vagrancy, nomadism and the site of the colonial and post-colonial

Azzedine Haddour

The European City is not the prolongation of the native city. The colonizers have surrounded the native city; they have laid siege to it. Every exit from the Kasbah of Algiers opens on enemy territory. And so it is in Constantine, in Oran, in Blida, in Bone. The native cities are deliberately caught in the conqueror's vise. To get an idea of the rigor with which the immobilizing of the native city, of the autochthonous population, is organized, one must have in one's hands the plans according to which a colonial city has been laid out, and compare them with the comments of the general staff of the occupation forces. (F. Fanon (1989) *Studies in a Dying Colonialism*, Earthscan, London, 51–2)

Writing is ... a living dead, the reprieved corpse, a deferred life, a semblance of breath. The phantom, the phantasm, the simulacrum ... of living discourse is not inanimate, it is not insignificant; it simply signifies little ... This signifier of little, this discourse that does not amount to much is like all ghosts: errant ... like someone who lost his rights, an outlaw, a pervert, a bad seed, a vagrant, an adventurer, a bum. Wandering in the streets, he does not even know who he is, what his identity ... what his name is ... but he can no longer repeat his origin. Not to know where one comes from or where one is going, for a discourse with no guarantor, is not to know how to speak at all, to be in a state of infancy. Uprooted, anonymous, unattached to any house or country, this almost insignificant signifier is at everyone's disposal. (J. Derrida (1981) *Dissemination*, Athlone, London, 143–4)

Introduction

Camus' *La Peste* mediates the architecture and cartography of power which Fanon adumbrates in the above quotation. The metaphor of the plague captures the ideological containment of the colonized, their exclusion from the city. In Derrida's terms, this exclusion is the outcome of the administering of pharmacy which owns up to the endeavour of the *pharmakon* (as being at one and the same time poison and cure, evil and good, outside and inside), and to the exclusion of the *pharmakos*, the scapegoat, that representative of the outside which is perceived as a threat to the order of the city. In Derrida's view, the *pharmakon* regulates the economy of supplementarity which governs Western writing. Derrida establishes a correlation between the *pharmakon* and writing perceived in Western metaphysics as an empty signifier, a surplus, a representative of the outside to be excluded. He describes this ostracized or rather errant signifier as a wanderer in the street, as a democrat: having 'no essence, no truth, no patronym, no constitution of [its] own', this 'vagrant' signifier 'belongs to the masses'; 'it is at everyone's disposal'. James Donald perceives the city as the locus where the rights of citizenship are ratified; his notion of citizenship is akin to Derrida's vagrant writing.[1] Donald comprehends the citizen 'not as a type of person ... but as a position in the set of formal relations defined by democratic sovereignty'. He contends that 'just as "I" denotes a position in a set of linguistic relations, an empty position which makes my unique utterances possible but which can equally be occupied by anyone, so "the citizen" too denotes an empty place'.[2] He represents a tendency in cultural theory which reduces the position of the subject to the formal function of language. (As we shall see later, both Stuart Hall with his notion of the decentred subject in process and Iain Chambers with his tropology of the nomadic subject in transit subscribe to this tendency.) By voiding the category of citizenship, Donald universalizes its discourse in an abstract language which disenfranchizes subjects who do not speak it. Likewise, Derrida overcomes, by idealizing, the reality of the errant vagrant. There is a gap between the existence of the expropriated vagrant and the notion of writing conceived by Derrida as democracy at everyone's disposal.

However, and in spite of this, the Derridean *pharmakon* is akin to this notion of writing. I will appropriate the former to grasp the principle which excludes difference, the discourses which police the architecture of power in the colonial and post-colonial city. To elaborate upon these discourses, in the first part of this chapter, I will examine the racism which is bound up with the administration of pharmacy in the plague city of *La Peste*. What is obfuscated in the allegory of the doctor combating the plague symbolizing Fascist ideology is the reality of the dispossessed colonized Algerians who are turned by French colonialism into a vagrant people, disenfranchised, denied the rights which political citizenship bestows on subjects of law. Ostracized from the city and the order of its politics, denied speech and representation, the caravans of this vagrant people cannot occupy a subject position. In this part, I will also appropriate the metaphor of the plague to account for the ideological containment inherent within racism. Explicitly and implicitly, the differing concerns of 'subjects on the move' in the first and second part of the chapter establish an intersection between the discourses of colonialism and post-colonialism. In the second part, I will oppose the vagrants of colonialism with the errant nomads of post-colonialism in order to throw into relief the disjunction between these two discourses: a discourse which is shaped by French colonial history and practice on the one hand, and post-colonialism – which is arguably a British invention – a discourse which emerges from the experience of a British settler society. The disjunction in itself is interesting: it adumbrates a gap between history and theory. I intend to deconstruct the universalizing propensities of post-colonial theorizing in order to warn against the danger of dismissing the significance of history. I will argue that there is a parallel between the exclusion of the colonized in colonial French Algeria and the discourse of post-colonialism which repeats at a conceptual level the exclusion of the colonized by jettisoning the past in order to overcome history.

Policing the margins of the colonial city

During the Nazi occupation, although France was exiled from Europe she asserted her primacy in the North African exile as a symbol of free Europe. The Second World War provides one of the ironies of history: France, the colonizer, assumes the role of the colonized in her colony, Algeria. In *La Peste*, the walled city of Oran metaphorically bodies forth France's political exile. In a controversial letter to Roland Barthes, Camus argues that terror has several faces. In *La Peste*, terror appears under the 'visage' of a bubonic plague. The nature of this plague, the grossness of its symptoms, makes it an appropriate metaphor to convey the malevolence and evil of German terrorism. Camus could have represented this terrorism in the metaphor of typhus. As we shall see, this supposition is crucial to the historical and political specificities of the Camusian text: 'Supposez un typhus, une peste, cela arrive, cela s'est vu. C'est plausible en quelque sort. Eh bien, tout est transformé, c'est le désert qui vient à vous'.[3] *La Peste* ends with a final reflection to the effect that the plague will awaken the rats once more, and that these rats will return to die in a happy city.

In Camus' *La Peste*, thousands of rats invade the city of Oran to disturb its order. Through the agency of its allegory, the novel presents the German ideology in the animalistic image of the rats spreading their plague. In *Le Métier à Tisser*, Dib describes an influx of beggars who invade the streets of Tlemcen, the daily deportation of these vagrants from the precincts of Tlemcen by the officials, but despite these sanitary measures the rural dispossessed return incessantly to infiltrate the city like the rats in Camus' narrative. The official policy of rendering invisible the problem of the displaced rural persons which Dib delineates ironically replicates the impetus in Camus' works, accounting for the absence of the indigenous Algerians from the narrative of *La Peste*. Dib shows how colonialism turned the Algerian peasantry into a group of vagrants, 'an army of famished ghosts' (*une armée de fantômes affamés*).[4] Nouschi presents an account of the colonized's condition similar to that presented by Dib in *Le Métier à Tisser*: the fellahs 'ne retrouvant plus à s'employer … sont livrés à la mendicité avec familles. Un exode de mendiants presque nus, ne trouvant plus de secours dans leurs douars, viennent en ville implorer la charité publique'.[5] Famine accentuated a typhus epidemic which decimated these roving fellahs.[6]

The rats bring the desert to Oran; Rieux and his colleagues establish their nobility in combating the plague, the novel's equivalent to the elemental ahumanity of the desert. They challenge the omnipotence of the desert, thereby demonstrating in this revolt against Nature their freedom in misery. Dib demystifies this romanticized representation of the desert in his denotation of the enforced nomadic life under colonialism as vagrancy in an urban desert. His more penetrating vision of the Department of Oran as a compartmentalized space enables the reader to perceive the blanks in Camus' stylized portrait of Algeria. Dib represents a colonized Algeria suffering from different types of plagues – famine, typhus and vagrancy – which represent the true 'visage' of colonialism. He uncovers a reality which is located outside the demarcated and enclosed setting of Oran.

In *La Peste*, Camus refers to the typhus epidemics which devastated Algeria, especially the Department of Oran in the 1930s and 1940s. His explicit reference to typhus is drawn from the experience of the colonized Algerians in the douars. The figures, taken from the archives of the Institut Pasteur d'Algérie by Quilliot, indicate that the typhus which raged in Algeria killed thousands.[7] Typhus reached plague-like proportions. But the narrative silences the resonance which typhus has with the bubonic plague: French colonialism and its tyrannical rule in Algeria.

In 'Albert Camus' Algeria', R. Quilliot interprets *La Peste* as a parable for the uprising on 8 May, 1945: 'The rats [which] continued to die nevertheless, invading the gutters and the mouths of sewers' are victims of colonialism and its oppression.[8] Rambert's visit to Oran as a journalist inquiring into the dire conditions of the colonized Arabs conjures up Camus' visit to Algeria as a reporter for *Combat* in order to investigate the circumstances of the 1945 uprising. However, Camus fails to recognize this crisis as a sequela of colonial oppression. He pictures the Oran of *La Peste* as a cosmopolitan city, dramatizing Western political and religious issues. He expediently uses the colonial context to unfold his univocal views. He transforms the specificity of Fascism into a general account of the combat against evil (the rats) to universalize Western political intrigues. Hence, the carriers of the plague in *La Peste* cannot be seen as 'the rats of colonialism, an old sickness that was dragging on in Algeria'.[9]

In *Discourse on Colonialism* Césaire argues that Europe was the 'accomplice' of Nazism before it was its 'victim'; it tolerated it before it came back to roost in its land; it legitimized it when it was applied to non-Europe people.[10] What Nazism represented is not 'the crime against man' as such, but 'the humiliation of the white man' in Europe, the fact that the white man of Europe was subjected to the racist and colonialist procedures which were hitherto reserved exclusively for the colonized Algerians.[11] The plague as a symbol of Fascist ideology perceived by Camus as a universal problem must be reviewed, in the light of Césaire's remark, as a form of racism. *La Peste* reproduces this racism: Camus could not conceive of the massacres of over 45,000 Algerians, which ironically took place on VE day (8 May 1945), in Sétif, Guelma and Kharata as Fascism.

In the Camusian text, the colonial city operated as a site in which the indigenous were categorized, controlled and segregated.[12] As we shall see, the text could be read as a trope encapsulating the discourses of exclusion of otherness to be contained like the plague. What I want to highlight in the first section of this essay is the diremption between the universal discourse of *La Peste* which obfuscates its own specificities and the colonial space and history which this discourse purports to represent. By comparing and contrasting

Camus' and Dib's texts, the object of my analysis is to put into relief the tension between the universal and colonial.

Dib delineates the effects which the colonial laws had on the indigenous by dispossessing them and forcing them to seek a livelihood in metropolitan centres. It is not my intention to study the history of immigration from North Africa to France. It suffices to note in passing that the rural exodus of the vagrant peasants from the douars of French colonization to Algeria's urban centres, which Dib represents in his trilogy, anticipated the first immigratory waves to mainland France and its cities. In *Les Chemins qui montent* (Sevil, 1957), M. Feraoun describes the racism interpellating these migrants as 'bicot', 'Noraf', 'ratton', and 'chancre'. His phraseology, such as 'A strange and inhuman evil which suddenly hits your big city' [*un mal étrange, inhumain qui frappe subitement votre grande ville* (p. 126)] and 'An obstinate chancre which will insinuate itself in the nice cities of France' [*un chancre obstiné qui va se fixer dans les bonne villes de France* (p. 198)], conjures up at one and the same time Dib's depiction of the colonized fellahs as vagrant immigrants threatening French cities, and Camus' representation of the plague as a foreign agent carrying and disseminating disease, 'chancre'. In *Hospitalité française,* written almost half a century after the publication of Camus' *La Peste* and over three decades after the publication of Dib's trilogy and the independence of Algeria, Tahar Ben Jelloun maintains that the same racist language which was used to describe immigrants as 'carriers of fleas, typhus and plague' still has currency in post-colonial French cities. Migrants are seen as lazy, unemployed, or delinquents, as parasites living at the expense of the French.

Driss Chraïbi appropriates Camus' metaphor of the plague in order to account for the conditions of immigrants in post-colonial France. He asks: does racism still exist in France? The immigrants who continue to work in this highly civilized country are relegated to the margins of human society; is it still true, according to A. Camus, the plague will never disappear?[13] Chraïbi ironically ignores the colonial specificities from which the trope emerges by giving it a post-colonial configuration. However, one cannot answer his questions without dealing with the issues of racism and ethnicity. According to D.T. Goldberg, an understanding of the ghetto, as a symbol of exclusion and segregation, as a space of quarantine, is intrinsically associated with immigrants who are regarded as prime carriers of disease, of indolence and moral turpitude, the inner cities as the repositories of 'this accumulated refuse'.[14] Furthermore, '[s]tratified by race and class, the modern city becomes the testing ground of survival, of racialized power and control: the paranoia of losing power assumes the image of becoming Other, to be avoided like the plague'.[15]

La Peste delineates the administration of pharmacy, the rhetoric of drugs and the containment of the plague. Not only does the 'pharmaceutical operation' contain the plague, but it suppresses the latent conflict between the metropolis and its colonial periphery. To echo Derrida, this operation consists in keeping the representatives of the outside out in order to safeguard and guarantee the internal security of the inside store of the city.[16] Derrida draws on the mythology of Oedipus which associates the plague both with the fear that the foreigner might insinuate itself into the city to disrupt its harmony and with the taboos of incest and miscegenation which threaten to abolish the difference between the inside and outside. Racism manifests the fear of interbreeding and miscegenation: the threat that the distinction between sameness and otherness is violated. 'Incest is also a form of violence', Girard remarks, 'an extreme form, and it plays in consequence an extreme role in the destruction of differences.'[17] In using the notion of 'patricide' Girard conflates the polis with the family; he argues that incest is a fundamental threat to the taboos which safeguard the 'inviolable distinction within the group'.[18] Although both incest and miscegenation are two prohibitions which hinge on the discourse of the family, a distinction between incest and miscegenation must be made. Incest expresses the fear that difference within the family might be abolished; miscegenation, the fear that difference from without would come to corrupt family relations within the larger community. Racism could be defined as the fear of mixing (incest) and intermixing (miscegenation) which keeps the outsider out and contains difference. Human relations are rendered illicit and illegitimate by racism, to be avoided like a contagion. Racism is the conjugation of the fear of difference which must be excluded and contained like the plague.[19] To repeat, this exclusion has a pharmaceutical function: it ensures the hygiene and well-being of the collective subject, as well as the decontamination of the community from that which jeopardizes its 'integrity'. Two interrelated processes are at work here in this scapegoating of difference: (i) guilt, misery, plague: in short, the violence of racism crystallizes around difference, which becomes the very embodiment of this violence; (ii) by a perverse logic, the blame is pushed onto the victim of such violence. We witness here a reversal of roles of victimizer/victimized. As Memmi avers, attributing failure, public or private, to our enemies or opponents excuses our own shortcomings.[20] This sums up the logic of racism which endeavours to legitimate its violence by projecting itself upon its victim. The supplementary economy inherent within this logic which puts difference out manifestly expresses the fear of incest and miscegenation.

Racial violence is not a leveller. If this violence is excessive it is because it is exclusive. As *La Peste* illustrates, the protagonists involved in the colonial drama are not 'identical', 'symmetrical', 'twin doubles', one standing as a substitute for the other.[21] The colonized, as a scapegoat, is not kept at the heart of the colonial city, in its 'inside store'.[22] On the contrary, this kind of scapegoat is not allowed representation, for it is perceived as a threat to the body of the colonizer's community, its property, and propriety. The expulsion of the colonized should not imply any prior belonging; it keeps the inside inside and the outside in its improper place where it should belong, therefore guaranteeing the community's racial hygiene and preventing miscegenation.

To recapitulate: the narrative of *La Peste* is predicated upon a system of conceptualization of opposed terms such as inside/outside, same/other, universal/specific, history/space. By suppressing one of the terms of this binary, the Camusian text reproduces the very Fascism it purports to combat. Dib represents what is located outside the margins of Camus' universalizing discourse, the historical reality of a people reduced to a nomadic life and vagrancy. The aestheticization of difference in *La Peste* neutralizes the diremption between the universal and colonial. As I shall argue, postmodern tropology of the *flâneur* and the nomad replicates this disjunction between space and history, inside and outside, universal and specific, between colonialism and post-colonialism. I appropriate the metaphor of the plague to represent the conflict between the European city and its colonial margins, as well as establish the points of tangency and intersection between the discourses of coloniality and post-coloniality. The trope circulates within a supplementary economy; it captures the subject positioning of the colonized, the foreigner, and the migrant, as super-added additions to be excluded. This supplementary economy works in two opposed ways. Firstly, it contains the representatives of the outside, of foreigners, of others, by imposing a *cordon sanitaire*. Since it relies solely upon appearance, the outer face of the subject, racism is indiscriminate in its project. The interpellation of second-generation immigrants into the subject position of migrants because of a perceived 'ethnic' and racial difference is an instance of such ideological containment. If anything, post-colonial critics pathologize this ideological containment; they represent the unassimilable face of difference. As I will argue, the whole discourse and debate surrounding the ethics of difference expressed *vis-à-vis* foreigners in metropolitan centres is perversely the symptom of such ideological interpellation and containment. Secondly, this economy also operates through a process of assimilation which neutralizes the foreignness of the foreigner. Post-colonial critics seem to be caught in the paradox of a so-called 'authenticity' which is nothing but inauthenticity. They deploy a so-called difference (so-called because it is the improper artifice of racist discourse) in order to claim a sameness which they cannot espouse without relinquishing their sense of difference. According to Sartre, this inauthenticity is 'bad faith'.

I do not want to get into a discussion of the issue of authenticity and inauthenticity: my main concerns in this chapter are twofold. Firstly, my intention is not only to divulge what is hidden behind the gloss of universal representation, but to highlight the diremption between the universal and colonial on the one hand, and between the discourses of colonialism and post-colonialism on the other. Secondly, I want to analyse the racism embodied in the metaphor of *La Peste* in order to grasp not merely the discourses which governed a system of apartheid in colonial Algeria, but the discourses which led to the ghettoization of difference in post-colonial urban centres.

Subjects on the move: cultural nomadism in the post–colonial city

Before discussing the terms of this diremption, let me first refer to Anne McClintock's, Ella Shohat's and Arif Dirlick's critiques of post-colonialism which Stuart Hall summarizes succinctly thus: (a) post-colonialism attempts to universalize and depoliticize conflict by blurring the distinction between colonizer and colonized; (b) it is a ragbag in which different discourses, irrespective of their historical and political specificities, have been put together – colonialism, neo-colonialism, third-worldism, gender and sexuality; (c) it is at one and the same time a term of periodization and a political project which attempts to overcome the regime of coloniality; (d) it is a discourse which 'consecrates Western hegemony'.[23] Hall engages with all these arguments, putting forward a different critical gloss on the concept which in his view describes the shifting relations in global economy and politics. According to him, post-colonialism marks a transition from the age of colonization to post-decolonization.[24] The concept 'refers to a general process of decolonization which, like colonization itself, has marked the colonizing societies as powerfully as it has the colonized (of course in different ways) ... one of the principal values of the term "postcolonial" has been to direct our attention to the many ways in which colonization was never simply external to the societies of imperial metropolis'.[25] Like Iain Chambers, Hall presents to us the city as a trope hypostatizing the geopolitics of colonialism and post-colonialism. As he puts it: 'The notion that only the multi-cultural cities of the First World are "diaspora-ised" is a fantasy which can only be sustained by those who have never lived in the hybridised spaces of a Third World, so-called "colonial" city'.[26] I will return to elaborate upon this point. It suffices to note that Hall invokes here the hybridized nature of the so-called colonial culture or the syncretism and diaspora-ized character of Western culture only to reject narratives of origin which were deployed as mobilizing discourses against the advances of Western colonialism. He rejects any possibility of returning to origin, because of what he calls the 'transcultural' effects of colonization which are in his view irreversible.[27] He acknowledges that there are differences between the colonizing and colonized cultures, but argues that the relationship between the two never operated in a binary fashion, and can no longer operate in this fashion in a post-colonial era.[28] He insists that the post-colonial marks the transition from a relation marked by difference to *différance*, i.e., a shift from an anticolonial relation (which assumed a binary form of representation) to one which is decentred.[29] According to him, the concept of post-colonialism urges us to reinterpret this binary form of representation as a form of transculturation, or what he calls 'cultural translation' which is troubling all the terms of conceptual opposition, i.e. the relation between inside/out, same/difference, self/other, colonizer/colonized, centre/periphery, here/there, home/abroad, etc.[30] Most significantly, Hall avers that post-colonialism reinscribes '"colonization" as a part of an essentially transnational and transcultural "global" process – and it produces a decentred, diasporic or "global" rewriting of the earlier, nation-centred imperial grand narratives'.[31] I agree with Hall that the deconstruction of essentialist binary concepts at the level of theory does not necessarily mean that these concepts are displaced politically.[32] He warns against falling into 'playful deconstructionism', 'the fantasy of a powerless *utopia* of difference',[33] only to fall into the same trap. He argues that 'Colonialism ... was only intelligible as an event of global significance – by which one signals not its universal and totalising, but its dislocated and differentiated character'.[34]

Here I want to take issue with Hall for offering us an apologia of colonialism. In order to dismiss narratives of origin and eschatology of history Hall
retrieves the concept of colonialism by giving it a positive character. He goes
on even to condone colonialism by arguing that 'Understood in its global and
transcultural context, colonization has made ethnic absolutism an increasingly
untenable cultural strategy'.[35] Hall's argument that identity is 'in process' does
not further the debate on cultural difference. It forecloses the discussion by
displacing the terms and the location of the debate. His poststructuralist
agenda positing the notion of the subject in process subsumes and obfuscates
the cultural, historical and racial specificities which come to define this notion
as difference. Such a conception of identity in post-colonialism strives towards
ideologizing difference, by equating a universal definition of the subject as
always-already decentred with the subject of difference as subjected to the
politics of colonialism. It is sad to see Hall confuse this universal definition of
the subject with the subject of colonialism. If the subject is always-already
decentred, Hall does not tell in what ways is the subject of colonialism and
post-colonialism different. The *a priori* notion of subject as always-already
decentred means that the subject was decentred in the past, that it is decentred in the present and that it will always be decentred in the future; this
notion, let us remind Hall, did not prevent the expropriation of the colonized
subject in the past, and has not put an end to the prevalent racism festering in
Western metropolitan centres which both Hall and Chambers celebrate as sites
of cultural difference.

Chambers also espouses the same definition of the subject as Hall.
Migrancy expresses for Chambers the simultaneity of differences, the collusion
of inside/outside, the disruption of the manichean discourse of centre/periph-
ery.[36] Migrancy is a state which exceeds the limits of geopolitics; it captures
poststructuralist thinking caught on frontiers which cut across different languages, religions and cultural geographies.[37] In the modern city, he maintains,
we are all migrants, nomads travelling across different worlds.[38] As he puts it:
'The boundaries of liberal consensus and its centred sense of language, being,
position and politics, are breached and scattered as all our histories come to
be rewritten in the contentious languages of what has tended to become the
privileged *topos* of the modern world: the contemporary metropolis'.[39] He
argues that to be an outsider lost in the drift of contemporary life is a condition of postmodern life; the transitoriness of existence in the metropolis is
vehicled by the metaphor of the nomad. Chambers conflates the discourses of
post-coloniality with postmodernity: to the condition of the uprooted migrants
in urban centres of Europe he 'add[s]' the increasing nomadism of modern
thought'.[40] He declares that '[n]ow that the old house of criticism, historiogra-
phy', Cartesian thinking with its 'intellectual certitudes' and 'grammar of
authenticity' are in a state of 'ruin', 'we all find ourselves on the road'.[41] He
presents the nomad in a Baudelairean guise, as a figure living in a state of
uprootedness. According to Chambers, this metropolitan figure is an inventor
of new languages re-accentuating what hitherto used to be the language of
the master.[42]

These new languages constitute the *sine qua non* of post-coloniality.[43] Its mode, in Chambers' view, is 'citation, re-inscription, re-routing the historical';[44] bringing about the dispersal and deconstruction of the discourse of modernity. Consequently, identity becomes contingent, 'diasporic', formed 'on the move', permanently in a state of 'transit', not having a 'final destination'.[45] In 'the metropolitisation of the globe', Chambers contends, 'The migrant's sense of being rootless, of living between worlds, between a lost past and non-integrated present, is perhaps the most fitting metaphor of this (post)modern condition'.[46] From his standpoint, it is no longer possible to maintain the authenticity of a given culture in a diasporic world, just as it is no longer possible to 'go back home'. It is not useful any more to oppose a Western hegemony and a subaltern other.[47] In his parlance, the city exceeds the logic of centrality. He argues for experiencing and thinking the cultural monopoly of the centre in terms of contamination and hybridity. As we have seen, these terms pertain to a problematic of racism imaged in the metaphor of the plague and to the rhetoric of pharmacy which endeavours to expel difference.

Chambers' migrant is similar to Kristeva's 'foreigner'. An examination of her views could shed some light on Chambers' theoretical problems. Upon the Freudian notion of *unheimlich*,[48] she elaborates that the uncanny is what is repressed within us, that is to say, the displacement of the uncanny strangeness from the outside has led to its location inside not what was regarded as familiar (one's own and proper), but inside what is strange (improper) and beyond its symbolic origin. The *unheimlich*, i.e. uncanny strangeness, is a return to the sites of the repressed in order to familiarize and clear up the feelings of estrangement which the Foreigner provokes.[49] The uncanny strangeness is, therefore, 'a psychic law which allows us to encounter the unknown and elaborate it in the process of *kulturarbeit*, the task of civilization'.[50] This task is the underlying principle which governs the rules of discourse and which produces politics. If the notion of 'civilization' pertains to the city (polis) and citizenship to the state of being policed and civilized, politics is the material expression of this psychic law. It is an agency which polices, regulates and disciplines difference; it is a law which contains that which impinges on the contours of identity. Racism is therefore the product of this law and cannot be dissociated from the notions of politics and discourse. Xenophobic political sentiments, Kristeva argues, 'contain, often unconsciously, that apprehension of the frightened rejoicing that has been called "*unheimlich*", that the English call uncanny, and the Greeks quite simply ... *xenos*, "Foreign"'.[51] The uncanny strangeness introduces 'the fascinated rejection of the Other that is at the heart of "our self"'; and 'when we escape from or struggle against the Foreigner, we are fighting our unconscious – that "improper" side of our impossible "own [proper]" self'.[52] Cultural difference is conceived in the unconscious, that 'foreign land of borders and alterities ceaselessly constructed and deconstructed'.[53]

To resolve the conflicts which perpetuate the politics of difference, Kristeva proposes an ethics of psychoanalysis which could 'integrate foreigners' and 'welcome them to that uncanny strangeness which is as much theirs as it is ours'.[54] Whilst she positions this Other at the centre of Western self, Kristeva intimates that the 'liquidation of the foreigner could lead to the liquidation of the psyche'.[55] Emptying the psyche leads either to the murder of that improper Other (the foreigner) or to the eradication of one's (own and proper) unconscious. This implies either the *mise-en-abîme* of psychoanalysis or the end of politics. Whilst inviting us 'not to reify the foreigner',[56] Kristeva could not help but neutralize the difference of the foreigner. The whole discussion of the uncanny foreignness is made confusing by her failure to distinguish between *who* is Othered and *what* is Othering. The foreigner seems to be caught in the double bind of the Other as the subject position of that which is excluded and as a metaphor representing the unconscious (i.e. it is at one and the same time the excluded outsider and the uncanny strangeness which excludes this Othered outsider). The foreign (Other) is thus condemned not to have its Others and not to have its 'own and proper' self. Kristeva never deals with the alterity of the foreigner *per se* but scrutinizes the 'Other scene' within us, i.e. Western Sameness: the foreigner is thus within us and is never conceived as a reality outside Western self. Kristeva fails to perceive that difference has its own and proper identity. Difference is neutralized and lost in her conception of the foreigner. Kristeva's foreigner is at one and the same time the 'agent' and 'agency' of exclusion, the foreigner *per se* and the Western psyche from which this foreigner is expelled.

Like Kristeva, Chambers fails to comprehend the foreignness of the nomad. He appropriates the metaphor of the foreigner to delineate the estrangement of the postmodernist subject and the 'nomadism' of Western thought. Like Kristeva, Chambers neutralizes the cultural difference of the nomad. For Chambers, the desert 'holds the key to the irruption of other possibilities: that continual deferring and ambiguity of sense involved in the travelling of ... people who come from elsewhere ... modernity or post-, advanced capitalism, the industries and networks of trans-national communications, the city, the end of the millennium'.[57] As he buries his head in the sand of his postmodern desert, Chambers ignores the conditions of the immizerated and overexploited immigrants, the subjects of colonial and post-colonial racism ghettoized in the inner cities. His main concern is not the conditions of an overexploited Third World from which he appropriates orientalizing metaphors, but the impossibility of maintaining a stable epistemological view.

Critics like Chambers argue that we are now witnessing the paradox of the periphery moving to occupy the centre, that migrants are reversing the old colonial trend. The question is: has the manichean nature of colonial discourse which used to characterize the relationship between centre and periphery changed? Before discussing this question, let me briefly refer to a major shift which occurred in critical and cultural theory, a movement from the discourse of modernism to that of postmodernism, from the study of centred ideology to analyses of decentred subjects. This movement coincided with a shift from colonial to post-colonial discourse, a movement away from a critical interrogation of colonial discourse to an aestheticization of difference. What the 'post' in post-colonialism and postmodernism signifies is the decentring and instability of identity as a cultural sign. Both Hall and Chambers epitomize this shift in cultural theory and argue for this decentring of the subject. We need to untangle two interweaving questions in Hall's and Chambers' critical accounts: is the ambivalence of the notion of the subject consequent upon the decentring of the centre? Or is this instability the outcome of the colonial encounter? Hall and Chambers fuse and confuse these two problems. However, what the global postmodern in their critical discourse negates is what it in fact promotes: a discourse of alterity, what is required for the subject to become an agent – identity. Hall's and Chambers' rejection of any normative, fixed, narrative of identity is premised on the old liberal humanist assumption that we are all different. However, this notion of 'all-difference' contains the paradox of perpetuating a totalizing definition of identity which it seeks in the first instance to reject, the paradox of abolishing difference. For what gives difference its signification is identity, rather than identicity, the state of being same. Inherent within their post-colonial theorizing is a teleology striving for a universal end. The assumption that identity is decentred and unstable is in effect predicated upon a universalist project implicating both centre and periphery in a discourse of decentred sameness.

Just as the periphery has realized the sense of its empowering difference, of being other amongst cultural others, the post-colonial and postmodern positionality of Hall and Chambers abolishes any privileges which a position of difference might offer. Any attempt on the part of the colonized to reclaim their past, their history, is dismissed as nostalgia. Benjamin never advocated a disjunction between past and history, for history is an active agency through which the past passes into the future. If history can be characterized in Valéry's terms as 'backing into the future', then narratives of identity and cultural belonging are always already construed from the temporal perspective of a past-present.[58] 'To articulate the past historically', as Benjamin puts it, 'does not mean to recognise it "the way it really was". It means to seize hold of a memory as it flashes up at a moment of danger.' He warns that 'every image of the past that is not recognised by the present as one of its own concerns threatens to disappear irretrievably'.[59] He perceives a continuity between past and present. The interruption of this continuum, as Albert Memmi and Fanon argue, mummifies the colonized culture. As the periphery folds back onto the centre, the periphery becomes identified with the centre in post-colonial/postmodern theorizing. This margin which is a product of racism and colonial legacy is conflated with the larger margin. What is lost in this conflation is the history of an oppressed Third World. The example chosen from French colonialism establishes that vagrancy, homelessness, and nomadism are not stylized metaphors but represent the condition of the colonized. The idea that the margin is in the centre overlooks that their oppression is the outcome of their exclusion from history. Any narrative which presents them as victims of colonialism is viewed as dated, inappropriately perpetuating the logic of dichotomy, the binary opposition of centre/periphery. Here is where Hall's and Chambers' theorizing converges with colonial discourse; it looms large as a prescriptive system, imposing itself upon others, reducing 'actual' difference to silence with respect to difference as such. Their theorizing, like the discourse of postmodernity upon which it draws and which claims to speak on behalf of humanity as a universal and undifferentiated entity, runs the risk of reproducing the same modernist ideology, perpetuating the very Fascism it denounces because it suppresses the specificities of difference. As we have seen, *La Peste* illustrates this paradox.

Global postmodernism, like the universalism it dismisses as monolithic, contains the same paradox: it produces a universal structure of undifferentiated differences, monads, vagrant nomads dispossessed of any sense of cultural belonging, an absolute difference copying the arbitrary cultural capital of the postmodern West. P. Ricoeur warns us against this sort of universalism which threatens to destroy cultural difference. He argues that globalization is in the process of undoing cultural closures which defined traditional societies, and that this phenomenon of universalization is an acute destruction of the ethico-mythic core of humanity which contains and maintains the wealth of cultural diversity.[60] Hall accepts as unproblematic the notion of a decentred subject, but, interestingly, argues: 'in order to maintain its global position, capital has to negotiate and by negotiate I mean it had to incorporate and partly reflect the differences it was trying to overcome. It had to try to get hold of, and neutralise, to some degree, the differences'.[61] Hall fails to see that the postmodern vision of a decentred world, of 'nomadic subjects in transit', to use Braidotti's expression,[62] is the mythology and ideology used to perpetuate the cultural capital of the West.[63] The crux of the problem is in which culture are we decentred and on whose terms? 'The right to difference' in postmodern and post-colonial theorizing is, according to Martine Charlot, 'a concession made to certain minorities by the dominant to the dominated on the condition that the hierarchical relations be maintained and safeguarded. The right to difference could never lead to equality. The important questions to ask in this respect are: is this right really claimed, and by whom? And who gives it its ideological content?'[64]

Critics consider 'this idea of decentring and the announced demise of the subject close to and corresponding with the conservative neo-liberal emphasis on the supremacy of the blind forces of the market'.[65] The decentring of the subject seems to be a deceitful ideology promoting, under the banner of cultural pluralism and global postmodernism, a neo-colonialist, capitalist discourse. It is a mythology which impoverishes the signification of difference, and robs it of its historical and cultural definition. Not only does this ideology hide the nature of the relationship between dominant and dominated within the urban West, but, also, mystifies the nature of the relationship between the West and its peripheries, ignoring what lies outside its fortified frontiers which protect it from the influx of invading migrants. Let me appropriate the spatial metaphor of the enclosed, walled city of Oran in *La Peste* and its universalizing discourse which suppresses the history of the marginalized colonized to describe the gap between postmodern and post-colonial theorizing on the one hand, and the political reality of those who are ghettoized in metropolitan centres and located outside Europe's demarcated and enclosed settings on the other. The fortified and policed frontiers of the West are, like *La Peste*'s *cordon sanitaire,* not to be crossed by foreigners.

La Peste carries within the fertile grounds of its universal narrative the seeds of colonial Fascism which are now germinating in postmodern, post-colonial, Western cities. Here, I disagree with critics like I. M. Young who proposes these Western metropolitan cities as a kind of decentred locus privileging cultural difference. These cities celebrate the logic of capital predicated upon a metaphysics of presence which excludes difference and perpetuates the plague of racism. Difference is faceless in these cities. What gives it a face and a voice is not legal citizenship but cultural membership. This membership identifies and differentiates: it gives a voice and a face; and insofar as migrants are rejected on the basis of their (difference) physical appearance, identity cards are devoid of their cultural and symbolical value. Immigrants are interpellated into a permanently contingent subject position[66] which perpetuates their identity of being homeless,[67] of not belonging, of facing a wall of racism which will be for ever erected against their integration.

Notes

1 J. Derrida (1981) *Dissemination,* Athlone, London, 144–51.

2 J. Donald, (1996) 'The citizen and the man about town', in S. Hall and P. du Gay *Questions of Cultural Identity,* Sage, London, 174.

3 A. Camus (1964), *Carnets* (Janvier 1942 – Mars 1951), Gallimard, Paris, 18.

4 M. Dib (1957) *Le Métier à Tisser,* Seuil, Paris, 162.

5 A. Nouschi (1962) *Le Nationalisme Algérien,* Minuit, Paris, 37.

6 Ibid., 37. Cf. A. Nouschi (1974), 'Observations sur le prolétariat rural en Algérie', in *XXIV Congrès International de Sociologie,* vol. 1, Office des Publications Universitaires, Alger, 1018–31.

7 R. Quilliot, 'Notes et variantes', in A. Camus (1965), *Théâtre, Récit, Nouvelles,* ed. R. Quilliot and L. Faucon, Bibliothèque de la Pléide, Gallimard, Paris, 1967–8. R. Grenier provides the same account about the typhus epidemics as that of Quilliot. He reads the epigraph to *La Peste* as a transposition of a historical situation which occurred in Oran – the plague of typhus – into a metaphysical issue. R. Grenier (1987), *Albert Camus Soleil et Ombre,* Gallimard, Paris, 146–7.

8 R. Quilliot (1962), 'Albert Camus' Algeria', (ed.) in G. Brée, *Camus,* Prentice Hall, Englewood Cliffs, 42.

9 Ibid., 42.

10 A. Césaire (1972), *Discourse on Colonialism,* Monthly Review Press, New York, 14.

11 Ibid.

12 See J. M. Jacobs (1996), *Edge of Empire,* Routledge, London, 20.

13 See Tahar Ben Jelloun (1984), *Hospitalité Française,* Seuil, Paris, 124.

14 David T. Goldberg (1996), *Racist Culture,* Blackwell, Oxford, 200.

15 Ibid., 200. René Girard notes that society, when threatened by a calamity, such as 'the plague, famine, foreign invasions', must have at its disposal a *scapegoat.* René Girard (1986), *Violence and the Sacred,* John Hopkins University Press, Baltimore and London: 94–5. Derrida compares the *pharmakos* to a scapegoat; the exclusion of *pharmakos* is therapeutical in nature, it keeps evil 'out of the body of city'. Derrida, *Dissemination,* 130.

16 Ibid., 128–33.

17 Girard, *Violence and the Sacred*, 74.

18 Ibid.

19 Ibid., 76.

20 A. Memmi (1994), *Le Racisme*, Gallimard, Paris, 78. Memmi contends that the racist aspires to the idea of pure/perfect state. The question is the following: is this state to be rediscovered in the past or is it a new order to be inaugurated in the future? A nostalgia or future aspiration? He argues that the future is 'perceived as a projection from the past and the past is constructed in relation to the future'. It is at one and the same time a lament for having lost the state of a golden era and a will to retrieve it. Racism is a violence directed against any perceived intruders, outsiders, who threaten the imaginary unity and integrity of the community.

21 R. Girard, *Violence and the Sacred*, 79.

22 I want to point out the contradiction in Derrida's argument that the *pharmakos*, i.e. the 'representative of the outside represent[ing] the otherness of the evil that comes to affect or infect the inside by unpredictably breaking into it' could not be constituted and kept in the inside. See Derrida, *Dissemination*, 133.

23 S. Hall (1996), 'When was the post-colonial? Thinking at the limit', in I. Chambers and L. Curti (eds), *The Post-colonial Question*, Routledge, London and New York.

24 Ibid., 246.

25 Ibid.

26 Ibid., 250.

27 Ibid., 246–7.

28 Ibid.

29 Ibid.

30 Ibid.

31 Ibid., 247.

32 Ibid., 249.

33 Ibid.

34 Ibid., 250.

35 Ibid.

36 I. Chambers (1994), *Migrancy, Culture, Identity*, Routledge, London and New York, 6.

37 Ibid.

38 Chambers echoes Kristeva's view that 'we are all Foreigners, we are divided'. J. Kristeva (1991) *Etrangers à nous-mêmes*, Gallimard, Paris, 268. This notion is ambivalent: it expresses at one and the same time notions of belonging and non-belonging, homeliness and unhomeliness. This notion is akin to Kristeva's concept of the foreigner: it refers to that which used to belong but which is excluded through the process of repression.

39 I. Chambers, *Migrancy, Culture, Identity*, 14.

40 Ibid., 18.

41 Ibid.

42 Ibid., 23.

43 Ibid.

44 Ibid.

45 Ibid., 25.

46 Ibid., 27.

47 Ibid., 79, 83.

48 Freud defines the uncanny as 'that class of the frightening which leads to what is known of old and long familiar', that which 'ought to have remained secret and hidden but has come to light'. S. Freud (1988). The 'uncanny' in his *Art and Literature*, Penguin Books, Harmondsworth, 363–4. Quoted also in Kristeva, *Etrangers à nous-mêmes*, 270.

49 Kristeva, *Etrangers à nous-mêmes*, 278–80.

50 Ibid., 280.

51 Ibid., 283.

52 Ibid.

53 Ibid.

54 Ibid., 284.

55 Ibid., 281.

56 Ibid., 284.

57 Chambers, *Migrancy, Culture, Identity*, 84.

58 P. Valéry (1962), *History and Politics*, Bollingen Foundation, New York, 127. W. Benjamin shares the same views on history as Valéry. Benjamin portrays the angel of history's face 'turned towards the past', echoing Valéry's conception of history 'backing into the future' (p. 249).

59 W. Benjamin (1992), *Illuminations*, Fontana Press, London, 247.

60 P. Ricoeur (1955), *Histoire et vérité*, Seuil, Paris, 292.

61 S. Hall (1991), 'The local and the global', in A. D. King (ed.). *Culture, Globalization and the World-System*, Macmillan London, 32–3.

62 R. Braidotti (1994), *Nomadic Subjects*, Columbia University Press, New York and Chichester.

63 Abdelmalek Sayad (1991), *L'Immigration ou les paradoxes de l'altérité*, De Boeck-Wesmael, Bruxelles. According to Sayad, the immigrant is defined essentially by his or her labor force which is provisional, temporary, and in transit. It is inconceivable to be an immigrant and unemployed. The immigrant's leave to stay is therefore temporary, determined by his/her condition of employment which could be revoked at any moment (p. 61).

64 M. Charlot (1981), *Les Jeunes Algériens en France*, Ed. CIEM, Paris. Cited in Tahar Ben Jelloun, *Hospitalité Française*, 88.

65 J. Larrain (1994), *Ideology and Cultural Identity*, Polity Press, Oxford, 154.

66 Sayad, *L'Immigration*.

67 P. Bourdieu, 'Preface' to Sayad, *L'immigration*, 9. He describes the migrant as *atopos* without place, displaced and unclassifiable. 'Neither a citizen nor an outsider, not really Same or Other, the migrant occupies the place of the Bastard', he is at the frontiers of the social being and non-being. As an absent presence, the migrant urges us to rethink a discourse which equates Nation with State, justifying the pretence to found citizenship on a notion of linguistic community, culture or race, and also to problematize the assimilationist discourse which allocates to the State the task to 'educate', 'assimilate' and 'produce' the Nation on hegemonic premises which could not conceal a chauvinism of the Universal.

Chapter 5

Denizens, citizens, tourists, and others: marginality and mobility in the writings of James Kelman and Irvine Welsh*

Willy Maley

In this essay I shall focus on the literary representation of two major Scottish cities, Edinburgh and Glasgow, through the work of their most contentious contemporary writers, Irvine Welsh and James Kelman. Like Roddy Doyle's Barrytown novels, set on the fringes of Dublin, Welsh and Kelman are producing a literature of the outskirts whose chief protagonists inhabit the ill-defined edges of the city.[1] I would like to tell a tale of two cities, Edinburgh and Glasgow, which is also, as with every city, a tale of two cultures, or two classes – in fact many cultures and many classes. In the last century Glasgow has passed through three stages, from being Second City of the Empire, after London, to being a centre of socialist agitation as the hub of Red Clydeside, to its promotion as European City of Culture in 1990.[2] The transition from imperialist complicity, through masculinist workerism, to post-industrial heritage museum has been far from smooth. Glasgow's secondary status is further complicated by the fact that it is also the second city of Scotland, after its capital, Edinburgh. Yet Glasgow remains a centre on the margins, a city which, because of its history of heavy industry, tobacco lords, and Irish immigration, is often seen as 'unrepresentative' of Scotland, a city without a country, or one whose country is outwith its nation, a city 'north of the border' whose roots lie in an expansionist state whose putative centre is located far to the south.

In Glasgow, three worlds collide: the world of Union and Empire; the world of socialist activism; and the world of cultural consumerism. But these three dimensions do not exhaust the city. There is, in addition to these three worlds, a Third World too. The Third World first and foremost refers to economic status rather than geographical space, and there are vast areas of the Third World in the so-called 'West'. Glasgow has long been synonymous with poverty and violence, and this fourth dimension is a thread running through the grand narratives of Empire, Socialism, and Culture. These diverse but complementary ways of seeing Glasgow, in the media and in fiction, from Saatchi and Saatchi to James Kelman, are as impoverished and violent as those applied to any city, each one acting as a boundary, or city limit. The image of Edinburgh, as Scotland's commercial and legal centre and tourist capital, rich in international cultural events, is equally limiting.

What are the limits of the city, its boundaries, historical and geographical? One approach, a temporal or sequential or diachronic one – let us call it *chronologocentric* – would see in the city a history, a development, a progression to a point where one could speak of the post-industrial, postmodern, post-colonial city. Another approach, a more topical or synchronic or anachronistic one, might envisage the city as a series of discrete areas, with political boundaries that have remained fairly stable over time. On one level, of course, there are no limits to the city, to what the city can represent, or how it can be represented, but one cannot really think about city limits without thinking about citizenship, about the inhabitants of a city, and the frontiers they encounter – linguistic, social, spatial, historical. Everything comes back to classes, nations, literatures.

In talking about the city we are not necessarily speaking the same language, nor would any of us agree on the changing place of the city in history or modernity. The entry under 'City' in Raymond Williams' *Keywords* distinguishes between *urbs* as city in the modern sense, and *civitas*, implying citizenship, or citizen, 'which is nearer our modern sense of a "national". *Civitas* was then the body of citizens rather than a particular settlement or type of settlement'.[3] In an essay entitled 'Urban signs and urban literature: literary form and historical process', Richard Lehan does try to tie literary forms in with the historical development of the city, but in that essay there is a tendency to see literary history as sequential, as chronological, which to me seems to narrativize too much, and to minimize tradition, the cultural archive, literary influence, the complex dialectic of text and context, historical hangovers, and so on.[4] I share Lehan's conviction that 'the better we read the city, the better we are reading the literary text' (p. 112), but the emphasis he places on realism and narrative, on a historical reading of the city in literature, seems to me to be flawed, as is his privileging of London as the first commercial city, and the notion of the commercial centre as exemplary of the modern city, which seems to me an impoverished notion compared with Williams' idea of interlocking and differentiated versions of the city that have formed over time existing in a single space, parallel and plural cities co-existing, inhabiting different times.

Of course, every city is structured along class lines. There is always segregation and apartheid. In *The Condition of the Working Class in England* (1845), Friedrich Engels explored the system of exclusions that operated in Manchester:

> I know very well that this hypocritical plan is more or less common to all great cities; I know, too, that the retail dealers are forced by the nature of their business to take possession of the great highways; I know that there are more good buildings than bad ones upon such streets everywhere, and that the value of the land is greater near them than in remoter districts; but at the same time I have never seen so systematic a shutting out of the working-class from the thoroughfares, so tender a concealment of everything which might affront the eye and the nerves of the bourgeoisie, as in Manchester.[5]

Some of the most notorious and vexed passages in *The Condition of the Working Class in England* concern the Manchester Irish. Although in their political writings Marx and Engels were broadly sympathetic to the Irish, many critics have taken Engels to task for seemingly succumbing to racial stereotyping in speaking of the living conditions of Irish immigrants. In a problematic conflation of class and colonialism Engels locates the Irish in the poorest region of the city:

> But the most horrible spot ... lies on the Manchester side, immediately south-west of Oxford road, and is known as Little Ireland. In a rather deep hole, in a curve of the Medlock and surrounded on all four sides by tall factories and high embank-ments, covered with buildings, stand two groups of about two hundred cottages, built chiefly back to back, in which live about four thousand human beings, most of them Irish. The cottages are old, dirty, and of the smallest sort, the streets uneven, fallen into ruts and in part without drains or pavement; masses of refuse, offal and sickening filth lie among standing pools in all directions; the atmosphere is poisoned by the effluvia from these, and laden and darkened by the smoke of half a dozen tall factory chimneys. A horde of ragged women and children swarm about here, as filthy as the swine that thrive upon the garbage heaps and in the puddles. In short, the whole rookery furnishes such a hateful and repulsive specta-cle as can hardly be equalled ... The race that lives in these ruinous cottages, behind broken windows, mended with oilskin, sprung doors, and rotten door-posts, or in dark, wet cellars, in measureless filth and stench, in this atmosphere penned in as if with a purpose, this race must really have reached the lowest stage of humanity. This is the impression and the line of thought which the exterior of this district forces upon the beholder.[6]

Engels appears to blame the victims here, and to naturalize the poverty of the Irish, or rather to racialize it. This tension or slippage in discourses on and of the city between the urban working class and colonial others who are per-ceived as outsiders irrespective of how long they have been in the city is a recurring feature of representations of the city in modern literature. One thinks here of Joyce and the figure of Bloom.

In a Scottish context, there has long been a tendency to see Glasgow as the city of immigrants, of industrialism, and of political ferment, and Edinburgh as the cultural and commercial centre. But in his ground-breaking book *Scottish Journey* published in 1935, Edwin Muir devoted separate chapters to Edinburgh and Glasgow, in which he described each city in terms of their socio-spatial division, and their exemplary and archetypal nature.[7] *Scottish Journey*, is a piece of travel writing by a native Scot that reads at times like an instance of colonial discourse, which is not at all surprising since there is a long tradition of viewing lower-class citizens and colonial subjects through the same linguistic lens, using the ethnographic present and engaging in a com-mon process of othering. Indeed, on one level the process of colonialism was a displacement of class, and the descriptions of the London poor in the six-teenth century were applied first to the Irish, and then to Native Americans, so that social status and ethnicity are intimately bound up together.

Scottish Journey is an example of internal colonialism, a piece of travel writing relating a journey into the interior of a modern European nation that reveals a culture of primitivism and poverty. Muir had been brought up on a farm in Orkney, and had moved to Glasgow in his teens at the turn of the century. A translator of Kafka, he was a widely travelled European figure. The ways in which Muir depicts Edinburgh and Glasgow would surprise any modern reader who expected to discover Glasgow in its industrial heyday or Edinburgh as the 'Athens of the North', as it is often styled in promotional literature. Instead, Muir paints two very distinctive portraits that leave no stone unturned in their efforts to convey the extremities in which natives of these cities dwell. In a letter penned in 1940 Muir lamented: 'It seems to me, looking at Scottish life, that discouragement is everywhere in it'.[8] Muir's vision of Scotland as a whole is profoundly pessimistic, but he reserves his greatest pessimism for Scotland's urban centres, Edinburgh and Glasgow. Muir appears ahead of his time in his accounts of these two Scottish cities. He anticipates many of the debates of the 1980s and 1990s. I want to quote at some length two passages, one on each city. First Edinburgh, then Glasgow:

That first half-blind glimpse of Edinburgh happened by chance to catch one thing about it which anyone accustomed to cities would probably not have seen: that it is a city of extraordinary and sordid contrasts. The tourist's eye is a very specialised mechanism, and it is quite capable of such apparently impossible feats as taking in the ancient monuments and houses of Edinburgh without noticing that they are filthy and insanitary. Yet the historical part of Edinburgh, the part most frequented by visitors, is a slum intersected by ancient houses that have been segregated and turned into museums and training-colleges. Most of the Canongate is a mouldering and obnoxious ruin. The stone of the houses looks diseased, as if it were decaying not with old age, but with some sort of dirty scurvy produced by poverty, filth and long-continued sorrow. The street itself, on fine days, is thronged with groups of young men. But the crowds are stationary, for every inducement that might make them leave the corner where they stand has long ceased to exist. In Princes Street people walk, for they have the money, and along with it hope and a host of effectual desires; but in the Canongate they stand, or when they are tired of that sit on the pavement in groups with their feet out on the causeway. Some of them must have stood and sat like this for a decade … In spite of its proud display, then, Edinburgh cannot hide away its unemployed or its poor. Yet as it is a city which must keep up appearances, there are certain rules which it does not like to see broken. It accepts the unemployed groups in the Canongate without visible annoyance; but when about a year ago a procession of the unemployed stopped in the town on their way to London, and slept for the night in Princes Street gardens, there was general indignation, in the tea-rooms, the tram-cars, and the columns of the local newspapers; for people's sense of propriety was outraged.

A town is like a very big and inefficiently yet strictly run house. The work in this house is done in the most haphazard way; good servants are ill-treated and badly paid, and dishonest servants praised and coddled; and the refuse which every big house continuously produces is not decently disposed of and hidden away as it is in most big houses, but barefacedly dumped some distance away in full view of the public yet where the master is not likely to stumble into it. All this is done with the most cynical inefficiency; but on the other hand the servants have to submit to the strictest regulation, both in their working hours and their leisure. They have their quarters, for instance, to which they must keep. They must on no account sprawl about in the drawing-room, even in their spare time. If their windows should happen to look out on the general refuse-heap, it is merely a geographical accident. Their windows look one way, their master's another. By this parable I merely wish to point out a fact which has often astonished me: that is, the complete success with which, in a large town, everything is kept in its place. There are streets in Edinburgh which correspond exactly to the drawing-room and the servants' hall. The people one meets in the first are quite different from the people one meets in the second. The crowds that walk along Princes Street, for instance, are a different race, different in their manners, their ideas, their feelings, their language, from the one in the Canongate. The distance between the two streets is trifling; the difference between the crowds enormous. And it is a constant and permanent difference. You never by chance find the Princes Street crowd in the Canongate, or the Canongate crowd in Princes Street; and without a revolution such a universal American Post is inconceivable. The entire existence of Edinburgh as a respectable bourgeois city depends on that fact. Nothing more than a convention is involved, but the conventions on which a society rests easily become sacred; and so a wholesale invasion of Princes Street by the poor would be felt not only as an offence against good taste, but as a blasphemy. That is why the temporary presence of the unemployed there was so deeply resented and feared. (pp. 9–11)

What Muir does is to make Edinburgh strange and unfamiliar by challenging the impressions of the tourist or first-time visitor who may see only the castle and the main thoroughfares. Muir's Edinburgh is a city of contrast, and of starkly visible poverty. Oddly, Muir, having undone the stereotype of Edinburgh as a civilized cultural counterweight to Glasgow's heavy industrialism and notorious squalor, then presents that city as a place apart, unScottish in its make-up:

Glasgow is in every way the most important city in modern Scotland, since it is an epitome of the virtues and the vices of the industrial regions, which comprise the majority of the population. A description of Scotland which did not put Glasgow in the centre of the picture would not be a description of Scotland at all.

Yet at the same time Glasgow is not at all a typically Scottish town; the worst of the many evils with which it festers were not born of the soil on which it stands or of the people who live in it – a mixed population of Lowlanders, Highlanders, and Irish – but of Industrialism; and in writing about it I shall be writing about a particular area of modern civilisation which is independent of national boundaries, and can be found in England, Germany, Poland, Czechoslovakia and most other countries as well as on the banks of the Clyde. This No Man's Land of civilisation comprises in Scotland an area which, though not very large in extent, is very densely populated. In one way it may be said that this area *is* modern Scotland, since it is the most active and vital part of Scotland as well as the most populous: the proof of its vitality being that it influences rural Scotland in all sorts of ways, while rural Scotland has no effective influence on it. But from another point of view one may say that it is not Scotland at all, or not Scotland in particular, since it is merely one of the expressions of Industrialism, and Industrialism operates by laws which do not recognise nationality. To say that is to say that Scotland is in the same position as most other European countries, except those which are still mainly agricultural. One part of its life is traditional and closely bound to the soil; another part is modern and has no immediate bond with the soil. Glasgow is consequently far more like Manchester than like Edinburgh. It is more like a manufacturing town anywhere in Europe than like a Scottish town which has preserved the old traditions of life. If the picture is sordid, then, the reader should remember this. (pp. 102–3)

Glasgow, untypical as it is with regard to the rest of Scotland, Edinburgh included, is more typical of modern Scotland than its rural hinterland. Muir reads Glasgow as an example of industrialism rather a product of Scottish culture or history:

If one were set down in the middle of Glasgow without knowing what town one was in and without seeing a single factory or slum, one would know by looking at the people passing that one was in an industrial city. Certainly one would see many remarkable and eloquent faces, from which one could read more perhaps than from the polite masks of Edinburgh; but a great number of them would show that characteristic distortion which one finds in industrial regions, and there alone. (p. 156)

As well as comparing Glasgow to Manchester, Muir, like Engels, alludes to the Irish, and he does so in a very peculiar and tangential way, as though he does not want to engage with the historical realities of Glasgow's Irish heritage:

I have kept largely to the economic side of Glasgow life, because it is the essential one: the other aspects of modern towns have become journalism. I actually intended at one point to say something about the churches, and in particular about the furious clashes between Orangemen and Catholics which fill the cells of the lock-ups after every St. Patrick's Day. But these things do not matter; they have only a fictitious importance; and to try to understand an Orangeman's state of mind in any case would not only be extraordinarily difficult, but also quite profitless, for the Orange superstition is surely one of the most insensate of existing superstitions, and also one of the most uninteresting. Unfortunately the Orange demonstrators and marchers belong mainly to the working class, just as the Catholic ones do. This feud causes a great deal of trouble, and has not even the excuse of being justified by interest. It is sheer inane loss: a form of hooliganism under the cover of something too silly even to be called an idea. But in the final count it comes to almost nothing. The fundamental realities of Glasgow are economic. How is this collapsing city to be put on its feet again? (pp. 161–2)

It is ironic that Muir, a poet and literary critic, wants to insist that the 'fundamental realities of Glasgow are economic', and is reluctant to see in resistance to Irish immigration anything other than a sectarian 'feud', or 'a form of hooliganism'. There is further irony in the fact that 'hooligan' is an Irish term, derived from the name of a notorious Irish family living in the south of London in the late nineteenth century.

Muir's text is crucial because it is an early example of an account of Edinburgh and Glasgow that focuses on deprivation and on the simultaneous visibility and marginalization of sections of their populations. It is only with the advent of Irvine Welsh that the two cities are once again being compared, as Welsh's fiction is seen to be a response both to the opening that Kelman created in terms of language and representation, and to the conventional and conservative view of Edinburgh as a tourist attraction that has none of Glasgow's markers of poverty or social exclusion. Edinburgh and Glasgow can be instructively surveyed through the fiction of Irvine Welsh and James Kelman. Kelman and Welsh map out the bleak existences of those who live in the outlying schemes of their respective cities, who are seldom cited, rarely seen or heard, and they do so in the local language(s) of the working classes. Language matters in speaking of the city, as does place in the social sense. City universities are interesting spaces in this regard. Often the catering, maintenance and security staff live in the city, and speak a vernacular, dialect, or non-standard variety of the national language – not necessarily the same one – while academics live in the suburbs or the West End and often speak the standard – again, not always the same.

Kelman does not deal directly with class as such. His art sets out rather to tackle the issue of class from the perspective of individual human beings, a strategy that could be said to entail an adoption of a bourgeois standpoint, the individual itself being a construct of middle-class culture. Kelman asks:

> What actually is the proletariat? Or for that matter the bourgeoisie? How do you recognise a class of folk? Or a race of people? You recognise them by general characteristics. When we perceive a member of a class we are not perceiving an individual human being, we are perceiving an idea, an abstract entity, a generality; it is a way of looking that by and large is the very opposite of art.[9]

This artistic individualism could be one of the tensions in Kelman's work. It is certainly one difficulty that I have with him. Catherine Belsey has described the risks implicit in an assertion of selfhood that does not take adequate account of the fact that the self is a social and historical invention, a cultural construct, and one that serves the interests of a conservative political order:

> The quest for the truth of the self, our own and others', endlessly fascinating, is precisely endless, since the subject of liberal humanism is a chimera, an effect of language, not its origin. Meanwhile, the social and political are placed as secondary concerns – naturally, since our democratic institutions are so clearly expressive of what we essentially are. In the subject's hopeless pursuit of self-presence politics can safely be left to take care of itself. And we can be sure that the institutions in question will in consequence stay much as they are.[10]

This process of self-fashioning also entails acts of othering which defeat its allegedly enlightened aims. As a writer, Kelman wants to maintain close links with his roots, his origins, his culture, his working-class background, yet the characters he creates in his fiction find themselves out on a limb, isolated from the communities from which they arise. In the eyes of Hines, for example, his colleagues and the Union and the working class in general are 'a bunch of bastarn imbeciles' (p. 180). Welsh shares this contempt for the conventional, conforming, and collaborationist working class, who appear as 'draftpaks', 'schemies', or 'straight pegs' in his work. Thus we have writers who wish to remain within their native cultures but who invent individuals who are very much alienated from that culture. Hand in hand with this disdainful attitude to the working class in general goes a suspicion of tourists and travellers, those preoccupied with sightseeing and with cultural consumption. In *Trainspotting* there are attacks, physical and verbal, on tourists, notably Begbie's outburst at the Canadian visitors to Edinburgh whom he and Renton encounter on the train to London. Begbie is enraged at those who see only the castle in the air and not the predicament of the population living in the moat:

> These burds ur gaun oantay us aboot how fuckin beautiful Edinburgh is, and how lovely the fuckin castle is oan the hill ower the gairdins n aw that shite. That's aw they tourist cunts ken though, the castle n Princes Street, n the High Street.[11]

Although Begbie serves a structural function in the book as the irredeemable villain of the piece, there is a context to his violence and resentment which has to do with class and colonialism. He is the product of damage.

How Late It Was, How Late (1994), Kelman's Booker Prize-winning novel, opens with Sammy, the ex-convict, waking in an alley after a drinking binge and squinting into the sun, thinking he is seeing some alien creatures whom he concludes are tourists:

> Fuck it. He shook his head and glanced up the way: people – there was people there; eyes looking. These eyes looking. Terrible brightness and he had to shield his own cause of it, like they were godly figures and the light coming from them was godly or something but it must have been the sun high behind them shining down ower their shoulders. Maybe they were tourists, they might have been tourists; strangers to the city for some big fucking business event. And here they were courtesy of the town council promotions office, being guided round by some beautiful female publicity officer with the smart tailored suit and scarlet lips with this wee quiet smile, seeing him here, but obliged no to hide things; to take them everywhere in the line of duty, these gentlemen foreigners, so they could see it all, the lot, it was probably part of the deal otherwise they werenay gony invest their hardwon fortunes, that bottom line man sometimes it's necessary, if ye're a businessman, know what I'm talking about.[12]

Sammy decides to play his part in this scenario he has imagined, seeing himself 'part of another type of whole', and opting for 'municipal solidarity'. He catches sight of the tourists again, only this time they're not tourists, but 'sodjers', or policemen (p. 3). There is a link between the two, because tourism is business and business wants to police the working class, to either keep them out of sight, or keep them in their place. Kelman was active in campaigning against Glasgow's facelift for its year as European City of Culture in 1990. He was part of a group of artists and thinkers who set up an organization called Workers City, determined to retain the notion of Glasgow as a workplace rather than a tourist attraction.

It is erroneous to try to construct a single entity or personality out of something as complex, heterogeneous, variegated and differentiated as a city, to give it an identity that denies its multiple realities, its competing histories. Tom Leonard has written:

> An address to a city-as-a-person ... anthropomorphises beyond the personal con-
> flict on which urban trade is actually based, to a single 'personal' unity which can
> be seen as 'natural' or aesthetically invigorating. The anthropomorphic functionalist
> image serves to create a bogus unified 'personality': it does so by leaving the streets
> clear of those whose opinions, if actually listened to, might spoil the image of a
> healthy and unified 'body' politic.[13]

Mindful of Leonard's caution, a useful comparison can nonetheless be drawn,
in terms of the representation of the disenfranchised and disaffected, between
Kelman's Glasgow and the Edinburgh of Irvine Welsh. Where Kelman concentrates largely on individuals who display integrity, moral and political, Welsh's
characters come from a lower class – lower working class – and tend to
eschew the socialist–anarchist rhetoric of Kelman's protagonists. Where James
Kelman writes of Glasgow, Welsh's native territory is Edinburgh, specifically
the outlying 'schemes', or housing projects, of Leith, Muirhouse, Pilton, and
Wester Hailes. Welsh's language is in some ways even more uncompromising
than Kelman's, his characters' consciousnesses even more divided and fragmented. The swearing is again a strong feature, and the use of Scots makes
fewer concessions to the reader brought up on a strict diet of standard
English. Where in Kelman there is always the minimal, never the maximal violence in terms of grammar and lexis – a bit of but at the end of a sentence, a
few phonetic transcriptions – Welsh goes to town on the tongue.

The kind of writing that Kelman and Welsh represent has a radical and progressive dimension, but both writers, in attacking cultural impositions and
positions of dominance, have a tendency to succumb to stereotyping and to
repeat the very chauvinism that they are trying to counter. Take the most
notorious passage in *Trainspotting*, when Renton denounces his friends and
fellow Scots:

> Ah hate cunts like that. Cunts like Begbie. Cunts that are intae baseball-batting
> every fucker that's different; pakis, poofs, n what huv ye. Fuckin failures in a coun-
> try ay failures. It's nae good blamin it oan the English fir colonising us. Ah don't
> hate the English. They're just wankers. We are colonised by wankers. We can't even
> pick a decent, vibrant, healthy culture to be colonised by. No. We're ruled by effete
> arseholes. What does that make us? The lowest of the fuckin low, the scum of the
> earth. The most wretched, servile, miserable, pathetic trash that was ever shat into
> creation. I don't hate the English. They jist get oan wi the shite thuv goat. Ah hate
> the Scots. (p. 78)

Having railed against Begbie for directing his anger at those who are different
– 'pakis, poofs, n what huv ye' – Renton goes on to describe the English as
'effete arseholes'. His preference is for a 'decent, vibrant, healthy culture'.
There's nothing wrong with self-loathing – it can have salutary political effects
– but the representation of others as 'effete' plays into the hands of precisely
the homophobic and exclusivist perspective that Begbie articulates.

We see parts of each city through the fiction of Welsh and Kelman that are
more likely to crop up in court reports than tourist brochures. Dorothy
McMillan has pointed to the system of social apartheid that obtains in
Glasgow:

The means of separating rich and poor, powerful and powerless, in the city have changed radically in this century. The invention of suburbia supported by trains and motor-cars worked to protect the managerial and professional classes from too disturbing proximity to urban degradation. But the *coup de grâce* to the notion of the readable city comes with the development of outer city schemes: it is no longer necessary to bypass the others, their existence need scarcely be recognised at all. For why would anyone go to Drumchapel, Easterhouse, Castlemilk and so on unless they lived there ...? It is in this gap between the inner-city and this undiscovered territory that James Kelman imagines his Glasgow.[14]

Kelman and Welsh map out the bleak existences of those who live in the outlying schemes of their respective cities, who are seldom cited, rarely seen or heard, and they do so in the local languages of the working classes. In Welsh's *Marabou Stork Nightmares*, Roy Strang, a working-class youth from Muirhouse, the housing scheme in which Welsh was raised, enjoys a short stay in South Africa before his family is forced to return due to the behaviour of his father. Far from being the 'City of Gold' his mother promised him, Roy finds life in Johannesburg as drab and bleak as life in Edinburgh: 'Close up, downtown Johannesburg just looked like a large Muirhouse-in-the-sun to me' (p. 61). Back home in Scotland, Roy reflects:

> South Africa was a sort of paradise to me. Funnily enough, I felt at home there; it was as if it was the place I was really meant to be, rather than shitey Scotland. When I thought back to Edinburgh I recollected it as a dirty, cold, wet, run-down slum; a city of dull, black tenements and crass, concrete housing schemes which were populated by scruffs, but the town still somehow being run by snobs for snobs ... Edinburgh to me represented serfdom. I realised that it was exactly the same situation as Johannesburg; the only difference was that the Kaffirs were white and called schemies or draftpaks. Back in Edinburgh, we would be Kaffirs; condemned to live out our lives in townships like Muirhouse or So-Wester-Hailes-To or Niddrie, self-contained camps with fuck all in them, miles fae the toon. Brought in tae dae the crap jobs that nae other cunt wanted tae dae, then hassled by the polis if we hung around at night in groups. Edinburgh had the same politics as Johannesburg; it had the same politics as any city.[15]

In this passage, social differentiation within Edinburgh is collapsed into racial differentiation in Johannesburg. We are urged to recognize 'the same politics'.

Marabou Stork Nightmares, Welsh's second novel, is a Manichean allegory on three levels, in which a comatose, hospitalized young man recollects his Edinburgh childhood while at the same time fantasizing about a safari park hunt for a predatory bird that preys on flamingos. Welsh explicitly draws an analogy between the plight of Scotland's urban poor and the victims of apartheid in South Africa, a comparison whose appropriateness has been challenged by some critics, who see in such comparisons a strong element of appropriation.[16] But as ideas about Otherness become more sophisticated, questions of class, race, gender and national identity are being increasingly applied across cultures, and social and political segregation is seen to have various analogues. One problem is that the post-colonial field is largely conceived of in non-European terms, with the notable exception of Ireland, but Scotland too now has a claim to post-colonial status.[17]

In *The Busconductor Hines*, the story of Rab Hines, the busconductor, a member of a dying profession, imminently outmoded, and his desperate efforts to become a bus driver, Kelman writes about 'the District of D', or Drumchapel, one of Glasgow's largest peripheral housing estates.[18] Like *Trainspotting*, *The Busconductor Hines* is about a city in which the only people with freedom of movement are the tourists, and they choose to see only those parts the civic authorities want them to see. The new peripheral housing estates are off the map and out of the question.[19] These urban borderlands are not bound up with particular local economies or specific industries, but are merely dumping grounds for surplus labour. In *Trainspotting*, Renton and Begbie find themselves in Leith Central, a disused railway station which is set to become a supermarket (pp. 308–9). An old man sees the two friends and decides to make light of their presence there: 'What yis up tae lads? Trainspottin, eh? He sais, laughing uncontrollably at his ain fuckin wit' (p. 309). The joke is that there are no trains. Their easy access to Edinburgh cut off, the residents of Leith are out on a limb.

Kelman's characters, whether poor or blind or mad, have similar problems getting around. For the denizens of Drumchapel, in *The Busconductor Hines*, living on the edge of the city means being stranded in an economic wilderness. An unreliable public transport network and a lack of private cars effectively shuts them off from the heart of the city. A move inward, towards the town centre, is desirable, and preferably to Knightswood, an older and more prestigious housing estate. Rab Hines, the busconductor whose days are numbered, knows he lacks momentum. Hines' quest for promotion to driver, his relationship with his wife and son, and the imminent demolition of his house tend to get lost in the 'strange kettle of coconuts' that is his head. Hines dreams of escaping to Australia and starting a new life there, but like Roy Strang's visions of South Africa these dreams of leaving are mere exoticist and orientalist delusions. A key moment in the novel comes when Hines takes his son out, and Paul complains of sore ears with the cold. Hines, in response, assures Paul that it is by no means as cold in Glasgow as it is in the Arctic, and proceeds to tell him about the Eskimos:

> The Eskimos son, they wear a lot of fur and that to keep the cold out. Wrap it about their ears and toes. See if they didnt, that'd be them, finished. They have to go about the whole year wearing them as well because of that fucking weather they've been landed with. They eat whales and stuff. Use up every bit of their bodies – oil out the skins; and this oil they make into various items, fuel and that. Short people with stumpy legs though maybe it's the furs makes them look so fucking stumpy I dont know. If they unwrapped all their clothes and that they might be skinny underneath. Skinny by christ. Aye but that's it about the ears son I'm no kidding you, they have to wear stacks and stacks of clothes. Never catch a cold either. I mean imagine somebody from this bloody dump going up to where they live son, they'd be dead in a matter of moments – pneumonia or some fucking thing. Unless they started doing the same as the locals. And vice versa down here I mean they'd

probably wind up catching a disease. Really desperate. Poor auld fucking Eskimos son it makes you sick so it does. (p. 69)

It is really desperate, and it does make you sick, but then ethnocentrism and existentialism go hand-in-hand, with Western crises of consciousness mapping onto subjection of the Other. Self-fashioning and subjectivity are purchased through slavery and exploitation. Later, Hines returns to this theme of someone worse off than his son or himself. Thinking of his son's future, and 'a million things', he reflects:

> There you are son that'll sort you out, put you on the vertical path, climb aboard the social transport. Ding ding, fares please, and fuck the eskimos. The eskimos. The poor auld fucking eskimos by christ it makes you sick, the way things happen, 1 minute you've got the ball at your feet and the next you're fucked. (p. 101)

On the surface, Kelman's novel deals with a specific historical moment when there was a crucial change in the public transport system with the introduction of driver-only buses, dispensing with the services of a separate conductor. Conductors like Rab Hines must move on or upgrade to driver in order to survive. The whole novel revolves around the question of social mobility, 'social transport' as Hines puts it. Ding. Everybody wants to get on, but can you get on and at the same time stay close to your culture? In the end, one might ask whether 'fares please, and fuck the eskimos' has to be the cry of every working-class intellectual. To get on, one must (or must one?) leave the other behind. It is a situation that gives rise to 'the same politics' as those which obtain in Welsh's *Marabou Stork Nightmares*, where, in a moment of identification that is also an act of appropriation, social apartheid eclipses racial apartheid. Within the orbit of their urban arenas of ambition, the Eskimos and Kaffirs of Glasgow and Edinburgh not only see themselves as others – they see others as themselves.[20]

Notes

* I would like to thank Azzedine Haddour for encouraging me to develop my thoughts on the city visions of Kelman and Welsh along these lines.

1 See Roddy Doyle (1992), *The Barrytown Trilogy: The Commitments, The Snapper, The Van*, Secker & Warburg, London. For a careful critique of this new urban pastoral see Shaun Richards (1992), 'Northside Realism and the Twilight's Last Gleaming', *Irish Studies Review*, **2**, 18–20.

2 On Glasgow's civic renewal and cultural repackaging see Neil McInroy and Mark Boyle (1996) 'The refashioning of civic identity: constructing and consuming the "New" Glasgow', in *Scotland*, **3** (1), 70–87. For a general overview of literary representations of Glasgow see Moira Burgess (1998), *Imagine a City: Glasgow in Fiction*, Argyll Publishing, Glendaruel.

3 Raymond Williams (1976), *Keywords: A Vocabulary of Culture and Society*, Fontana/Croom Helm, London, 14.

4 Richard Lehan (1986), 'Urban signs and urban literature: literary form and historical process', *New Literary History*, **18** (1), 99–113.

5 Friedrich Engels (1845), *The Condition of the Working Class in England*, cited in Terrell Carver (1981), *Engels*, Oxford University Press, Oxford, 16.

6 Cited in K. Marx and F. Engels (1978), *Ireland and the Irish Question*, Lawrence and Wishart, London, 48.

7 See Edwin Muir (1985), *Scottish Journey* [1935] Flamingo, London. Future references to this work will be by page number in the text.

8 Cited in Douglas Dunn (ed.) (1992) *The Faber Book of Twentieth-Century Scottish Poetry*, Faber & Faber, London and Boston, xxxiii–iv.

9 James Kelman (1992), *Some Recent Attacks: Essays Cultural and Political*, AK Press, Dunfermline, 11.

10 Catherine Belsey (1985), *The Subject of Tragedy: Identity and Difference in Renaissance Drama*, Methuen, London, 54.

11 Irvine Welsh (1993), *Trainspotting*, Secker & Warburg, London, (Norton, New York, 1996) 115. Future references to this work will be by page number in the text. For an incisive reading of this novel in terms of language and identity see Alan Freeman, (1996) 'Ghosts in sunny Leith: Irvine Welsh's *Trainspotting*', in *Scottish Studies: Publications of the Scottish Studies Centre of the Johannes Gutenberg Universität Mainz in Germsheim*, vol. 19: *Studies in Scottish Fiction: 1945 to the Present*, Peter Lang, Frankfurt, 251–63. For a nuanced interpretation of Welsh's construction of the city see Gill Jamieson (1997) 'Fixing the city: arterial and other spaces in Irvine Welsh's fiction', in Glenda Norquay and Gerry Smyth (eds) *Space and Place: The Geographies of Literature*, Liverpool John Moores University Press, Liverpool, 217–26. According to William Drummond, Ben Jonson saw the Scottish capital as a lookout post for Empire: 'In a poem he calleth Edinburgh "The heart of Scotland, Britain's other eye".' See George Parfitt (ed.) (1974) *Ben Jonson: The Complete Poems*, Penguin, Harmondsworth, 472. Welsh plays upon Edinburgh's imperial past and cosmopolitan pretences throughout his work.

12 James Kelman (1994) *How Late It Was, How Late*, Secker & Warburg, London, 2–3. Future references to this work will be by page number in the text. I have attempted a reading of this novel around questions of obscenity and othering in W. Maley (1996) 'Swearing blind: Kelman and the curse of the working classes', *Edinburgh Review*, **95**, 105–12.

13 Tom Leonard (1987) 'On reclaiming the local and the theory of the magic thing', *Edinburgh Review*, **77**, 42–3.

14 Dorothy Porter McMillan (1991) 'Imagining a city', *Chapman*, **63**, 47.

15 Irvine Welsh (1996) *Marabou Stork Nightmares*, Vintage, London, 75, 80.

16 For a subtle and searching engagement with this issue see Alan Freeman (1996) 'Ourselves as Others: *Marabou Stork Nightmares*', *Edinburgh Review*, **95**, 135–41.

17 For a recent attempt to map out the space of Scotland in post-colonial terms see Michael Gardiner (1996) 'Democracy and Scottish postcoloniality', *Scotlands*, **3** (2), 24–41.

18 James Kelman (1984) *The Busconductor Hines*, Polygon, Edinburgh (Orion, London, 1995). Future references are by page number in the text.

19 On Kelman's demonstration of interiority and spatial constraints through language see Graeme Macdonald (1996) 'Writing claustrophobia: Zola and Kelman', *Bulletin of the Emile Zola Society*, **13**, 9–20.

20 For a trenchant encounter with ideas of otherness in a Scottish context see Berthold Schoene-Harwood (1998) '"Emerging as the Others of Ourselves" – Scottish multiculturalism and the challenge of the body in the postcolonial condition', *Scottish Literary Journal*, **25** (1), 54–72.

Chapter 6

Suburban tales: television, masculinity and textual geographies

David Oswell

The Buddha of Suburbia was written by Hanif Kureishi and originally published as a novel in 1990.[1] It was produced in the UK as a television serial in November 1993 on BBC2. The story focuses on the life and experiences of the main protagonist, Karim Amir. Karim's story is an escape and a quest. It is a story of identity. But his story does not work to (re)affirm his *true* identity. Rather Karim's journey is a quest only inasmuch as it is a search for an escape from, or a way out of, identity. In the opening paragraph of the novel, Karim declares:

> My name is Karim Amir, and I am an Englishman born and bred, almost. I am often considered to be a funny kind of Englishman, a new breed as it were, having emerged from two old histories. But I don't care – Englishman I am (though not proud of it), from the South London suburbs and going somewhere. Perhaps it is the odd mixture of continents and blood, of here and there, of belonging and not, that makes me restless and easily bored. Or perhaps it was being brought up in the suburbs that did it. (p. 3)

Karim is restless about his belonging. This restlessness concerns his 'ethnicity', 'race' and national identity. However, although not apparent from this quote, but evident elsewhere in the novel and television serial, there is also a restlessness concerning his sexual identity. From the above quote it is clear that hybridity and suburbia are identified as reasons for Karim's restlessness. Roger Silverstone (1997) has argued that the dynamics of suburbia are essentially hybrid: the hybridization of time, of space and of culture. He argues that in suburbia distinctions 'have lost their past and are defined only in the present, embodied perhaps in the ghostly images of family albums', that suburban identities have 'both reconcile[d] and den[ied] the essential differences between urban and rural life' and that 'differences of position in the system of production have gradually been overlaid and replaced by the differences of position grounded in the system of consumption' (pp. 8–9).

In this chapter I want to look at a type of manhood which has emerged within a series of overlapping discourses of suburbia and, primarily sexual, hybridity. I want to look at this discursive space in relation to a particular story: the story of suburban man-in-the-unmaking. I, first of all, look at the construction of the suburban in terms of its *textual geography*, but also in terms of, what might be called, an ethic of suburban disclosure. I discuss the suburban in its widest extension to include television. I then consider how *The Buddha of Suburbia* displays a form of masculinity in the unmaking, which is caught up in the *geography of the text*, a site of hybrid restlessness. I finally conclude by showing how such a cultural geography of suburban man is contested, as we might imagine, from more conservative sites of discursive production. Underlining my chapter is an argument that the suburban is an important site of cultural analysis and contestation both for and within media and cultural studies.

The suburban

Kureishi was brought up in the South London suburb of Bromley. Bromley is, Silverstone notes, an 'exemplary' suburban space. It was home to H. G. Wells and to a railway line *en route* to London. But it was also home to David Bowie and to a strand of the 1970s London punk scene known as the Bromley Contingent (including Siouxsie Sioux and Billy Idol). Bromley is, Simon Frith has declared, 'the most significant suburb in British pop history' (1997: 271). This was a place that young people turned their backs on, only to return once older (Silverstone, 1997). Bromley is, James Saynor has suggested in a review of *The Buddha of Suburbia* (*The Observer*, 31 October 1993), the 'global capital of shoe shops and *Daily Telegraph* readers'. Moreover, Karim, he argues, is 'a prisoner of a world located at the precise midpoint between heaven and hell'.[2]

Such criticism of suburbia is common. As a site of pleasure or hostility, the suburban is either ridiculed or loved for the *petit bourgeois* mentality of its inhabitants, the ordinariness of its architectural quirks (the stone-cladded walls and neo-classical pillars), the embeddedness of a kitsch cultural taste or sensibility (the garden gnomes and the importance of the car in the drive-way) and the underlying desire of its population for both privacy and community surveillance (the net curtain, for example, both curtails curious prying eyes looking in and yet allows those inside to freely observe what's going on outside the home). On a cultural level, the hostility to these harm-less quotidian pleasures, in part, derives from an avant-garde perspective which mocks the ordinariness of a taste lacking the finesse of high culture and the authenticity of a folk culture (cf. Bourdieu, 1984). On a political level, this is mirrored in terms of a vanguardist left political culture which derides those housed in the suburban as no more and no less than *petit bourgeois*: a class without hope of revolution.[3]

A particular mentality, then, is engrained within this space. In 1909 C. F. G. Masterman, in his portrait of English social structure, made a distinction between 'The Conquerors', 'The Suburbans' and 'The Multitude'. He defined 'The Suburbans' in the following manner:

> They form a homogeneous civilisation – detached, self-centred, unostentatious – covering the hills along the northern and southern boundaries of the city, and spreading their conquests over the quiet fields beyond. They are the peculiar prod-uct of England and America: of the nations which have pre-eminently added

commerce, business and finance to the work of manufacture and agriculture. It is a life of Security; a life of Sedentary occupation; a life of Respectability; and these three qualities give the key to its special characteristics. Its male population is engaged in all its working hours in small crowded offices, under artificial light, doing immense sums, adding up other men's accounts, writing other men's letters. It is sucked into the City at daybreak and scattered again as darkness falls. It finds itself towards the evening in its own territory in the miles and miles of little red houses in little, silent streets, in number defying imagination. Each boasts its pleasant drawing-room, its bow-window, its little front garden, its high-sounding title – 'Acacia Villa' or 'Camperdown Lodge' – attesting unconquered human aspiration. (Masterman, 1960 [1909]: 57–8)

Masterman's discourse is ambivalent in its account. He both looks down on the suburbans' diminutive dwellings and yet recognizes the worth of their aspirant desires. In representing this class, which consists of accountants, managers, clerks, solicitors and other pen-pushers of a growing metropolitan administrative centre, Masterman offers an image of the suburban (as character and geographical terrain) which is still resonant in contemporary imagination. On hearing that I had moved to Ealing, a few years ago, a friend of mine, who lived in a Georgian terraced house overlooking Clapham Common, stated quite bluntly, 'But that's so suburban!' And it is, with its rows of predominantly Edwardian architecture. Even in 1881, when the geography of Ealing looked quite different, W. S. Clark referred to it as one of the 'Suburban Homes of London' (quoted in Burnett, 1986: 193).

The emergence of the suburban can, in part, be traced historically to a gendered division of labour between 'home' and 'work' and to the development of better transport systems in the nineteenth century. The train allowed men to travel to work in the city, while women and children stayed in the domesticated surroundings of suburbia. The city was a place of masculine work and spectacle. But the city also carried the connotations of poverty and overcrowding. Suburbia was built as a way to get away from the lower orders, dirt, noise, disease and crime of the inner city. Mrs J. E. Panton, in her book *From Kitchen to Garret*, wrote in 1888 that the suburbs were a place for neither rich nor poor and that 'smuts and blacks are conspicuous by their absence' (quoted in Burnett, 1986: 192). Similarly, suburbia needed to be protected from any working-class incursions. The Royal Commission on the Housing of the Working Classes stated in 1885: 'The selection of the workmen's escape routes to the suburbs should be strictly limited … Other districts which are not spoilt should not be thrown open to the working classes, otherwise these districts will become spoilt too' (quoted in Burnett, 1986:192).

It is understandable, then, that the cultural politics of the suburban have not been seen as desirable. For example, in an interview, black gay independent film-maker Isaac Julien states: 'Suburbia, to put it crudely, has been the place where white people moved to get away from blacks … It's something I can't really take seriously – especially in relation to 'gay culture' – what would be my investment in appealing to suburbia or countryside places?' (Julien, 1993: 127).

But the distinctions and boundaries that constitute the suburban are more complex. The suburban is not a unitary phenomenon. For example, geographers and sociologists have long discussed the escape to the suburb as a form of leapfrog (Wilmott and Young, 1960; Rex and Moore, 1967; Rex, 1968). Rex and Moore, in their important study *Race, Community and Conflict* (1967), argued that as the urban business district expanded, the professionals and 'captains of industry' moved to the suburban outskirts, followed by the lower-middle classes who moved further out, followed lastly by the working classes. Others subsequently argued that urban/suburban distinctions are not simply determined by class (Couper and Brindley, 1975; King, 1997). What is at stake here is not so much the actual geography of suburbia, but the way the suburban becomes an important marker of cultural distinction.[4]

The cultures of suburbia are not simply limited to the physical geography of the suburban. Importantly, suburban sensibilities extended outwards to both city and country, but also inwards to domestic design and architecture. As US media historian Lynn Spigel has argued 'domestic architecture of the period was itself a discourse on the complex relationship between public and private space' (1992: 101). In the US the post-war suburban housing boom was linked to a reinvention of domestic space. In media and cultural studies significant work has investigated this relationship in the US context (e.g. Haralovich, 1988; Spigel, 1992); there has been less research concerned with the British context.

In Britain the post-war reconstruction of housing began in earnest with the publication of the Dudley Report in 1944. There were seen to be two main faults in pre-war council housing: lack of variety and lack of sufficient space for modern living. There was a shift away from an earlier design for popular housing: scullery, for washing dishes and clothes; living-room/kitchen, for cooking, eating, socializing and generally living; and the parlour/front room, in which the best furniture was kept and which was used only on special occasions (Boys *et al.*, 1984). Although the Dudley Report was 'working against the background of the "thirties"' (Ministry of Housing and Local Government, 1961: 1), architects and designers in the late 1940s were beginning to reconceptualize contemporary domestic living. Howard Robertson, in *Reconstruction and the Home*, declared that:

> The conception of living is in process of being closely scrutinized, and at times drastically revised. Introduction of improved mechanical services, obligatory stress on labour-saving, and an all-round reduction in the size of dwellings have focused attention on open planning and 'space-making' within a limited compass. More and more ingenuities go to compensate for less and less super-footage. (1947: 10)

The modern home was opened up to the outside world. It had larger windows. It was lighter. Bright colours were used in the decor. Furniture was made to look lighter in weight, thinner and lower to the ground. Different rooms were no longer distinct and separate from each other, but were made to merge into one another. The concept of 'open planning' was deployed in order to create, as *Ideal Home* referred to it, 'space through unity' (*Ideal Home*, April 1960).

The conceptualization of the modern home broke away from the cellular design which we can, to use Foucault, trace back to the eighteenth century:

> The house remains until the eighteenth century an undifferentiated space. There are rooms – one sleeps, eats and receives visitors in them, it doesn't matter which. Then gradually space becomes specified and functional ... The working class family is to be fixed; by assigning it a living space with a room that serves as a kitchen and dining room, a room for parents which is the place of procreation, and a room for the children, one prescribes a morality for the family ... the little tactics of the habitat. (Foucault, 1980: 48–9)

The increasing visibilization of domestic space, although it shifted away from the cellular ordering of rooms, increased the intensity of the mechanism of the cell. It focused, now, on the individual and on freedom within the home. For example, the Council of Industrial Design journal *Design* stated in 1959:

> Focus on space, a key word, space that gives freedom. Destroy the distinction between rooms. The home is subservient to life in the home. Banish the cold formality of front parlours that attempt to impress callers – then stand unused, to collect dust ... Push back the wall, bring the kitchen in, dissolve divisions that separate life into compartments ... Allow freedom to change and space to move. (cited in Attfield, 1989: 219)

But talk of freedom belied a greater intensification of the disciplinary tactics or techniques that Foucault identifies, such that power worked through the freedom of individuals rather than in opposition to them (cf. Rose, 1989a). Most notably, domestic Taylorism (the breaking down of domestic activities into their temporal and spatial elements) in Britain and the US provided an intense disciplinary arrangement and administration of the suburban home, as well as offering freedom from domestic drudgery.

The knot between domestic modernity and the suburban was tied securely through consumption. The disruption of the boundaries between public and private – a consequence of the increased visibilization of the suburban home – allowed the home to become a site of consumption and a means of display to other aspirant consumers in the suburban neighbourhood (Silverstone, 1994; Spigel, 1992). Suburbanization and the modernization of the home both fuelled consumer demand for new utilities, such as refrigerators and televisions, and provided an object of address. The suburban home offered a set of overlapping spaces and times for the machinations of advertising and one of the central means of addressing this constituency was through television.[5] Moreover, it was the 'housewife' and 'mother' who personified this address (Haralovich, 1988; Spigel, 1992). The housewife was tied into the home through the demands of cleanliness, aesthetics, childcare and family health.

For Silverstone, television plays a central role in the culture of suburbia. It articulates domestic modernity and the suburban, but it also extends its address beyond the physical locality of suburbia. In Britain, television's address to suburban domesticity was tempered by a public service ethos which construed the audience as neither highbrow nor lowbrow, but middlebrow. As a consequence of the Reithian struggles over 'culture' in the 1920s and 1930s in relation to radio broadcasting, television has consistently steered a course through 'high culture' and 'mass entertainment'. Or, to put it more accurately, high culture and mass entertainment have been configured, until recently, within an address to a middlebrow sensibility. BBC radio addressed 'the ordinary English family' who lived in 'Acacia Avenue or Laburnum Grove, the tree-lined suburbs of Greater London and the Home Counties' (Scannell and Cardiff, 1982: 168). It addressed, not a mass audience, but a public comprised of individuals in families whose 'home was an enclave, a retreat burrowed deeply away from the pressures of work and urban living, with radio as part of that cosy, domestic warmth' (ibid.). Paddy Scannell has shown how an ethos of public service broadcasting has emerged through attempts to speak as if speaking to someone in the privacy of their own home. Broadcasters attempt 'to affiliate to the situation of their audience, and align their communicative behaviour with those circumstances' (Scannell, 1991: 3). For Scannell, such an endeavour 'produces the world, and endlessly reproduces it, as ordinary: as familiar, knowable, recognizable, accessible to all' (Scannell, 1988a: 3). In addition, it has been argued by Alison Light, in relation to literary culture in the 1920s and 1930s, that such a middlebrow culture can be defined in terms of 'conservative modernism', which is wholeheartedly suburban and 'whose apparent artlessness and insistence on its own ordinariness has made it peculiarly resistant to analysis' (Light, 1991: 11).

Radio and television broadcasting in Britain has, thus, consistently pitched its middlebrow taste and sensibility to an imagined suburban audience (Silverstone, 1994). The appeal to Acacia Avenue and Laburnum Grove has continued from the Reithian traditions of radio in the 1920s and 1930s to the quotidian humour of the television sitcom (Medhurst, 1997).[6] For example, the BBC sitcom *Happy Ever After* (written by John Chapman and Eric Merriman), which began in 1974 and spawned a sequel series entitled *Terry and June*, exemplifies the suburban as familial and ordinary. In the first show, Terry, played by Terry Scott, worries about his daughter, Susie, who has recently moved from Purley to Chelsea, the centre of swinging London. In the exchange between Terry and his wife, June, played by June Whitfield, the distance between the metropolitan and the suburban is clearly marked through a process of making the city Other. London is constituted, primarily by Terry, as dangerous, cosmopolitan and sexual. In the opening sequence to the episode Terry says, 'An eighteen year old girl in London is highly vulnerable, to say the least.' London is a site of news stories: 'You've only got to read any newspaper to see something else has happened to an eighteen year old girl.' It is the target of possible nuclear attack: 'It's a known fact that an atomic bomb could wipe out the entire London area in twenty seconds.' It is a place of sex: 'That girl of yours is going to turn into a right raver.' And it is filled with fears of and desires for the 'foreign': 'I wouldn't like to say what goes on in some of those Chinese restaurants on the King's Road.' All this talk takes place in the sitting room. The metropolitan Other is both invoked and sucked from the suburban sitting room, as dirt hoovered from the carpet (cf. Douglas, 1970). Hegemonies of race, gender and sexuality are neutralized in their embedding within the domestic infrastructure of the suburban: the design, clothes and mode of speech. The sitting-room encodes a suburban taste: smart, but not ostentatious; modern, but not avant-garde; tasteful, but not snooty. The clothes are casual, but not messy, and Terry's sporty V-neck jumper exemplifies this visibly. Similarly, the mode of speech through which, for example, sex is articulated resonates with this particular style. Sex surfaces, not as a full-on discussion, but as a series of hints and allusions: a series of tit-bits, enough to understand, but not enough to be vulgar or bohemian. There is a kind of play between the hidden and the revealed, the private and the public. For example, reflecting on the time available now that the children have moved out, Terry, in apparent innocence, says, 'This is just the time for us to have a good look at each other to see what we've got.' June responds, 'We've still got what we've always had.' Terry then adds, 'Yeh, but I haven't had a look at it lately.' Just as introspection turns to bodily investigation and pleasure, so an innocent language is revealed as bawdy exchange. Everything, and nothing, is explicit.

These codes of dress, design and speech form a kind of ethical practice of the self (Foucault, 1984a) predicated on the manner of disclosure (i.e. the process of translation between private and public). This suburban discourse and disclosure, however, constitutes a fixity of social relations between gender, race and public/private, rather than a hybridity. Moreover, the textual and conversational features of *Happy Ever After* align an ethics of the self with a suburban imagined community. There is an intimacy between viewer, other similar imagined viewers and viewed. Television, if you like, reveals the private life of the suburban – the goings-on behind the net curtain – only inasmuch as it establishes a particular community across space within the regular repetition of the schedule (Scannell, 1988b). Just as the train allowed the suburbanite to venture into the city to work and to return safely home unharmed by the working classes and other 'races', so television allows the suburbanized viewer to travel in a world of sitcoms in which the working classes and foreigners are in their place.

Central to this ethics in television sitcom is its normalization of sexuality (for the purpose of this chapter, specifically masculine sexualities). Andy Medhurst has recently argued that this suburban televisual space is shaped by a heterosexual familialism: 'Of all the hegemonies of suburbia, it is the hegemony of heterosexuality that cuts deepest, bites hardest, and the reason is evident: "The family is crucial both to the decision to move to the suburbs and to the whole suburban way of life"' (Medhurst, 1997: 266). Furthermore, John Hartley (1987) has argued that this familialism (although he does not mention its heterosexual normativity) provides the framework for television institutions to imagine their suburban audience. Hartley argues that television is formed as a 'paedocratic regime'. Television imagines its audience as a collection of families (including children) and as childlike because the figure of the child offers a point of identification for the lowest common denominator, for the mass audience. In Hartley's later writings he has discussed the modern public sphere, as it is formed through television, in terms of its feminization:

'Contemporary suburbia is the physical location of popular reality, a newly privatized, feminized, suburban, consumerized public sphere' (Hartley, 1996: 156).

It is the television audience (as imagined audience), then, which since the 1950s has provided the ground, in the West at least, within which a suburban culture can be cultivated. To ignore such a culture, however complex, or to misrecognize its extensiveness, might appear arrogant. In the closing sequence of the television serialization of *The Buddha of Suburbia* the main characters are seated in an Indian restaurant on the day of the 1979 General Election. A television screen in the corner presents an image of Margaret Thatcher. Karim states, 'Nobody is going to vote for that cow – she's too suburban!' To which Eva, Karim's father's lover, retorts, 'Don't we live in a suburban country?'[7]

Masculinity and travels in hybridity

Karim's hostility and arrogance toward the suburban resonates with earlier literary constructions of suburban masculinity. Mr Polly, for example, 'hated Fishbourne, he hated Fishbourne High Street, he hated his shop and his wife and his neighbours – every blessed neighbour and with indescribable bitterness he hated himself' (Wells, 1963 [1910]: 18). His masculinity is asserted through his desire to escape from the feminine, the domestic and the suburban.

Medhurst shows how this relation between gender and suburbia is common in British popular culture. Suburbia is represented as a space of 'men who dream of escaping routine and rut, women who unquestioningly accept or at worst, embody them' (Medhurst, 1997: 241). Medhurst draws attention to the cases of, not only Mr Polly, but also Billy Liar and the television character Reggie Perrin. These characters are seeking to escape from suburban domesticity and to pursue their escape through fantasy. These characters – who forever dream of leaving for exotic lands and adventures or even just a place which lacks consistency and monotony – are firmly cemented in the suburb. It is, in many ways, not surprising that Kureishi cites his influences as Kingsley Amis' *Lucky Jim* and the early novels of Evelyn Waugh. *The Buddha of Suburbia* is, as James Saynor in the *Observer* (31 October 1993) has stated, a mix of the 'mundane and the mindbending'. Nevertheless, Karim, although in many ways similar to these earlier constructions of suburban masculinity, presents a different story of identity and space.

Kureishi's representation of suburbia is not the fixed uniformity of Edwardian Dunroamins, but a feast of cultures: a mix of Asian and white English, late hippy, glam rock and punk, S/M and wife swapping. Karim's flight is from a suburb and family, which is hybrid. Karim is an actor and his performances draw on these particular suburban and familial resources.[8] For example, while working for a radical theatre company, Karim is asked to explore his ethnicity in the development of a character for a play. Matthew Pyke, the theatre director, says to Karim: 'You need somebody from your own background, somebody black.' To which Karim replies, 'I'm not sure I know anyone who's black.' His answer, a side-stepping of the self, refuses Pyke's positioning. But when pushed, Karim uses his Uncle Anwar to study. Uncle Anwar is an elderly, superstitious corner-shop keeper. When he presents his characterization of Anwar to the theatre group, Karim is accused of misrepresenting 'black people': 'Your picture is what white people think of us.'[9] He is asked to rethink his character. He then takes his new friend Changez as a subject to study. Changez is unwilling to be represented: 'You entered my wife. Now promise you will not enter me by the back door.' Nevertheless, Karim betrays Changez's trust and in the play the audience are amused at Karim's performance. Although the reproduction of these stereotypes is problematic (inasmuch as they could be read as a reinforcement of such stereotypes), the fact of their reiteration displays the extent to which these stereotypes are masquerades (cf. Doane, 1982; Butler, 1993). It is through the performance of identity that the veils of such identity can be exposed. Bhabha has discussed this in relation to the figure of the 'mimic man' (Bhabha, 1985).[10] What is interesting about *The Buddha of Suburbia* is that, in accordance with Silverstone's analysis of the suburban, Karim's world is hybrid. Moreover, Karim's escape does not leave this world behind, but foregrounds the suburban as a space of performance and hybridity. In the opening episode of the television serial, Karim goes with his father, Haroon, to Eva's suburban home. Haroon has been asked to guide Eva's guests to spiritual enlightenment. Haroon is clearly faking. Haroon plays the mystic to the sitting-room gathering in order to escape the dull routinization of the office in which he has no control. He plays upon their prejudices in a bid for freedom. Eva is not foiled by this, rather she is complicit in the construction of Haroon. She plays the otherness of an exotic and romantic masculinity in her bid for metropolitan status. While the guests are left chanting, cross-legged in the sitting-room, Haroon is fucking Eva and Karim is giving Charlie, Eva's son, a hand-job in his attic bedroom.

Kureishi thus avoids deconstructing suburban masculinity from a position of authenticity. Creamy (Jamila's nickname for Karim, which carries the connotations of both sex and race) is not searching for psychological explanations. He is not searching for 'the inner room'. His ostensible antipsychologism is deployed as a narrative technique for critiquing a 1970s and early 1980s form of identity politics.[11] This form of identity politics, aided by an earlier existentialism, constructed 'political commitment' through the managing of one's self in terms of one's 'correct credentials' and one's place within the essentialized categories of class, race, gender and sexuality (cf. Mercer, 1990). It drew upon the language and the techniques of a psychotherapeutic discourse. Talking about oneself, or even better expressing oneself to others through role-play, was a way of 'getting in touch' with one's *real* being. The *truth* of one's identity was discussed and acted upon in terms of one's ability to express and confess. In the field of sexual politics Foucault has rightly identified the technique of the confessional as a modern form of power/knowledge through which our conduct is governed (Foucault, 1979).

Instead of seeing this as a reactionary form of privatization or an unhelpful obsession with the self, to the extent that the 'political' is ignored, the language of psychotherapy, which gained credibility across the political spectrum, provided a way of governing the relationship between the personal and the political and setting in place new regimes of expertise (Rose, 1989b). The disclosure of self is construed as the discovery and fixing of identity.

The critique of this earlier identity politics is brilliantly played out in *The Buddha of Suburbia* within the context of the radical theatre group, mentioned earlier, which uses the language of therapy as politics. In one scene, in the television serial, an actor, by the name of Richard, talks about his coming out and confesses his desire for black men and his disgust of white men. Pyke, also referred to in the novel as 'the Judge', asks the actors to reflect upon the way in which their position in society had been fixed. He asks Karim what he thinks, to which Karim replies that he doesn't think he thinks deeply enough about these things. Karim throws the question back to Richard. Pyke intervenes and asks Karim, 'Who do you fuck, Karim? Blacks or whites? Do you feel comfortable sleeping with white women?' Karim replies, 'I don't know. I'll fuck anybody!' The cast burst into sympathetic laughter. Creamy, through articulating his diverse sexual desire, implicitly sidesteps the technique of the confessional and disarticulates the question of sexual desire from the question of identity.

For Karim, sex is not tied to romance nor is it desired as an object in itself, rather it is a means of becoming-other-than-oneself. In this sense, Karim is very different from the earlier versions of suburban masculinity discussed by Medhurst (1997). The episodic narrative of Karim's story is marked by his multiple desires for, or encounters with, both men and women: Charlie, Helen, Jamila, Terry, Eleanor, Matthew and Marlene. The construction of Karim draws upon the figuring of bisexual masculinity in popular 'glam rock' discourses in the 1970s (e.g. Mick Jagger, David Bowie, Elton John and Rod Stewart). The use of Bowie's music within the television serial is a clear indication of this. However, *The Buddha of Suburbia* draws upon this reservoir of past images and reconstructs that masculinity and refigures it in relation to a notion of multi-ethnicity. The *NME*, in a review of the serial, quotes Bowie: 'It is our innate sense of mix that could be our saving grace, if only some elements would stop refusing to recognise what a fantastic cultural invention we're becoming.'[12] Although the review downplays the question of sexual ambivalence and emphasizes the problem of 'Englishness', it is clear that *The Buddha of Suburbia* does not simply repeat the figure of masculine bisexuality as 'withdrawn aesthete'.

However, it is difficult not to see Karim both as modern-day dandy, flamboyant and transforming himself in the process of acting, and as *flâneur*, traversing the suburban in the manner of Baudelaire's 'passionate spectator'.[13] Both camp and queer, Karim seeks pleasure and reinvention. Foucault, in his later work on the aesthetics of existence, reads Baudelaire's description of the dandy as a way of understanding and rethinking the relationship between self and community in terms of self-creation: 'To be modern is not to accept oneself as one is in the flux of the passing movements: it is to take oneself as an object of a complex and difficult elaboration: what Baudelaire, in the vocabulary of his day, calls *dandyisme*' (Foucault, 1984b: 41). Foucault talks about the practices of the self through which we can rethink ourselves and move beyond the historical limits of our existence. The critical practice of genealogy, he argues, makes possible a way of thinking outside of the normalizing relations of power/knowledge. Critical thinking becomes a way, not of affirming identity, but of problematizing our sense of who we are. Although Karim is in no way presented as being conscious of his past and conscious of his self-invention outside the limits of 'normal' sexual identity, the character of Karim, in being placed in the 1970s and in suburbia, begins to rework earlier formations of masculine identity. Through Karim we reinvent a relationship between ethics, identity and politics. The dandyism of Karim allows us to reinvent a politics of selfhood. The dandy is, as Baudelaire reminds us, only transitory. This figure of masculinity merely signals the *possibility* of a better future, of different forms of loving and romance. The process of self-creation and of recognizing the full weight of the modern does not mean that we should adopt the attitude or the ethos of the dandy. On the contrary the dandy is only a pose to be studied. It is a figure to be worked through in the face of modernity. It is a mask, which partially and fleetingly defines a masculinity which refuses to be named.[14]

Karim is neither 'straight' nor 'gay' nor simply 'bisexual'. His love for Charlie, Jamila, Terry and Eleanor is not prescribed within fixed categories of identity and community. His sexual identity is not explained in terms of his personality, but in terms of his playing with identity. The masculine self-formation of Karim has a close affinity with masculine self-formation in 'queer' discourses. The title song, sung by Bowie, even declares: 'Screaming along in South London. Vicious but ready to learn. Sometimes I feel that the whole world is queer. Sometimes … but always in fear.' However, 'queer' has been construed as a predominantly metropolitan formation in opposition to the suburban (or, for that matter, rural). This metropolitanism is particularly problematic as queer theorist Alan Sinfield notes:

> And it depends on where you live. In rural Shropshire say … just to keep going at all may require more courage than getting arrested on a demo in central London. And if you are struggling to be Gay, the last thing you want is someone living in more fortunate circumstances telling you that you don't measure up because you can't think of yourself as Queer. (Sinfield, 1992: 26)

How might we deconstruct this relation between selfhood and space? How might we analyse the performativity of the self, the geography of the text and textual geography? Jo Bristow offers some useful insights in his analysis of textual pleasure. He has argued that 'the chance of imagining what it is like to be other identities' allows the possibility of 'creating alliances with them' (Bristow, 1989: 79). It provides a way of thinking about and acting upon the relationship between pleasure and politics, such that identity is no longer fixed and essentialized. Bristow argues: 'The opportunity exists for everyone to be camp and to practise sado-masochism' (ibid.) This 'opportunity' exists not simply in *actual* encounters, but also in the *imaginary* space (or the imagined communities) of the text. For Bristow, the meeting of others through the space of the text can be seen, in some sense, in terms of the process of self-invention. For example, in *The Buddha of Suburbia* we are able to identify with Karim, the narrator, and through this identification we are able to cruise other characters and form ourselves in the process. Bristow argues that the text is a 'cruising-zone'. It is 'a site where desire charts its spaces for others to share the pleasures of contact: contact between those who are able to identify each other but who have not – as with most cruisers – ever met before' (ibid.: 78–9). In the process of this self-invention we identify with others and we become a little bit different: 'we re-present the desire to find out who we "are" in an unpredictable location where unforeseen jouissance may be created' (ibid.: 79). We become constituted within other communities and constituencies. However, in *The Buddha of Suburbia* the locations which are visited are precisely predictable. They are predominantly suburban and domestic. The difference lies in the way these homely spaces are reconfigured within a series of hybrid desires and narratives. Bristow's analysis of the text as 'analogous to an urban map of sexual desire' is only partially useful in describing masculine self-formation in *The Buddha of Suburbia*. Firstly, the characters cruised by Karim are not *of* a fixed community. None of the characters cruised is simply 'straight' or 'gay'. For example, Karim sleeps with Jamila who sleeps with Joanna. The focus is much more on sexual practices than on sexual identities. Secondly, Karim is uncomfortable in the constituencies within which he finds himself. He is clearly troubled by Terry's revolutionary politics, with Matthew and Marlene's sexual libertarian therapies and Charlie's S/M scene. He is made uneasy by Eleanor's aristocratic community of stories and interests: 'it was her stories that had primacy, her stories that connected to an entire established world' (Kureishi, 1990: 178). This sense of unease is located for Karim within power relations of class and race. Karim's 'past wasn't important enough'. It was not 'substantial' (ibid.) It was the story of the suburbs, his Dad and his Mum. Although Bristow's analysis of gay identity and textual pleasure shifts away from a notion of the formation of alliances as a conscious political project and towards a more sustained link between pleasure and politics, the metaphor of 'cruising' is deployed within the context of the urban and also leaves unquestioned the gendered nature of this metaphor. The practice of cruising, for example, is a particular form of metropolitan masculine gay sexual practice. As with queer politics the sexual geography of the suburban is left unexplored.

The Buddha of Suburbia, especially in its televisual version, addresses precisely this non-metropolitan constituency. The serial deals with the everyday of the suburban. For example, Karim's romantic encounters are not, on the whole, played out within avant-garde, fetishized, S/M scenarios. And yet its ordinary appeal and attempt to work within a consensual politics does not impurify its radicalness. *The Buddha of Suburbia* works within the predictable space of the suburban and embeds a different set of masculine desires.

Rather, in locating this address within the suburban, the text begins to reconstitute the cultural geography of this previously uncontested domain and to problematize contemporary straight suburban masculinity.

Contested spaces

Chris Savage-King, writing in the *New Statesman and Society*, argues that '[t]he conviction of the suburbs that it houses no one special means that everyone works overtime at self-creation'.[15] However, the construction of a suburban hybridity was, in relation to *The Buddha of Suburbia*, contested in other sites of textual production. The disclosure of a forbidden ethics of the self within a televisual discourse agonized journalists from the stronghold of traditional suburbanism.[16] One scene in particular upset the conservative critics of the tabloid press. In Episode 3, Karim's love for Eleanor is manipulated within an orgiastic world of sexual experimentation and power relations. Matthew, the theatre director, and Marlene, his wife, use sex to 'discover themselves'.[17] Matthew says at one point to Karim: 'I'm on the look-out for a scientist – an astronomer or nuclear physicist. I feel too arts-based intellectually' (Kureishi, 1990: 190). In this scene Matthew, the Judge, *gives* his wife to Karim, as if she were a gift, and *takes* Eleanor. Karim is clearly troubled by this situation. Then Eleanor, white English and upper-class, *offers* Karim to Matthew. The final scene of this episode shows Matthew smiling with pleasure as Karim sucks his cock.[18] The positioning of Karim within this scene vividly portrays the complex network of power relations.

The tabloid press interpreted this scene in the television serial through the lens of a prurient, populist and moralist gaze. In this discourse, sexual invention and pluralization are displayed as both an object of desire and derogation.[19] The *Sun* displayed a nude shot from the serial on its front page and carried the headline, 'Shame of BBC's Porn Play' (17 November 1993). The *Daily Mail* splashed 'A Full Frontal Assault' on top of a story which included 'BBC bosses refused to make eleventh-hour cuts to last night's episode of the controversial sex drama *The Buddha of Suburbia* despite public concern over explicit scenes' (18 November 1993). This tabloid blast of disgust ricocheted within the wedding-caked walls of suburban England.

In the press, gendered sexual identities and modes of conduct were normalized within the terms of 'taste and decency'. The *Daily Mail* (18 November 1993) gave space to a number of its writers to respond to the serial. Graham Turner, 'married with three children', said: 'This is a seedy, sordid, violent and brutalising film.' Concerned about the 'degrading' sex scenes and the 'moral tone' of the programme, Turner stated that 'if it attracts large audiences, I fear for the mental health of those able to sit through it'. Likewise, Mary Kenny, 'married with two children', stated: 'It depressed me profoundly, when I began to think of what the BBC once stood for: culture, decency, dignity. And now here it is portraying the sex act explicitly.' *The Buddha of Suburbia* is seen to transgress the codes of a suburbanized public sphere, as Hartley calls it. Its form of disclosure is not cultured, decent and dignified, but explicit. Moreover, its explicitness reveals that there is no true self to be revealed.

The Broadcasting Standards Council (BSC), one of whose tasks is to monitor 'taste and decency', received a number of complaints about this scene and others.[20] The Council did not seem to take into consideration the way in which the *Sun* and the *Daily Mail* had hyped-up the concern. Nevertheless, it agreed that, in Episode 1, Karim's masturbation of 'a male friend' did not exceed 'limits reasonable for the portrayal of the incident within a work of this kind broadcast later in the evening and preceded by a warning' and it did not uphold complaints about 'nudity, homosexuality, nor the lovemaking between Karim and Eleanor'. However, it did uphold 'complaints directed at the act of group sex, believing it to have been treated at greater length than was necessary' (BSC February 1994: 26).[21] The actors in this space come together over the single, but contested, problematic of the manner of disclosure and it is suburbia which provides a key popular representational site for such a translation between private and public. Suburbia becomes, in this sense, a key representational site for the troubles of liberalism.

References

Attfield, Judy (1989) Inside pram town: a case study of Harlow house interiors. In Attfield, Judy and Kirkham, Pat (eds) *A View from the Interior: Feminism, Women and Design*, The Women's Press, London.

Baudelaire, Charles (1964 [1863]) The painter of modern life. In *The Painter of Modern Life and Other Essays*, Phaidon Press, London.

Bhabha, Homi (1985) Signs taken for wonders: questions of ambivalence and authority under a tree outside Delhi, May 1817. In Barker, Francis (ed.) *Europe and its Others*, vol. 1, University of Essex, Colchester.

Bourdieu, Pierre (1984) *Distinction: A Social Critique of the Judgement of Taste*, Harvard University Press, Cambridge, Mass.

Boys, Jos *et al.* (1984) House design and women's roles. In Matrix, *Making Space: Women and the Man-Made Environment*, Pluto Press, London.

Bristow, Joseph (1989) Being gay: politics, identity, pleasure. *New Formations*, (9), 61–81.

Burnett, John (1986) *A Social History of Housing*, Methuen, London.

Butler, Judith (1993) *Bodies That Matter*, Routledge, New York.

Couper, M. and Brindley, T. (1975) Housing classes and housing values. *Sociological Review*, **23**, 563–76.

Doane, Mary Ann (1982) Film and the masquerade: theorising the female spectator. *Screen* **23** (3–4), 213–34.

Douglas, Mary (1970) *Purity and Danger*, Penguin, Harmondsworth.

Foucault, Michel (1979) *The History of Sexuality*, vol. 1, Allen Lane, London.

Foucault, Michel (1980) The eye of power. In Gordon, Colin (ed.) *Michel Foucault: Power/Knowledge*, Harvester Press, Brighton.

Foucault, Michel (1984a) What is Enlightenment?. In Rabinow, Paul (ed.) *The Foucault Reader*, Pantheon, New York.

Foucault, Michel (1984b) On the genealogy of ethics: an overview of work in progress. In Rabinow, Paul (ed.), *The Foucault Reader*, Pantheon, New York.

Frith, Simon (1997) The suburban sensibility in British Rock and Pop. In Silverstone, Roger (ed.) *Visions of Suburbia*, Routledge, London.

Haralovich, Mary Beth (1988) Suburban family sitcoms and consumer product design: addressing the social subjectivity of homemakers in the 1950s. In Drummond, Philip and Paterson, Richard (eds) *Television and Its Audience*, British Film Institute, London.

Hartley, John (1987) Television audiences, paedocracy and pleasure. *Textual Practice*, **1** (2), 107–28.

Hartley, John (1996) *Popular Reality: Journalism, Modernity, Popular Culture*, Edward Arnold, London.

Hebdige, Dick (1979) *Subculture: the Meaning of Style*, Methuen, London.

Julien, Isaac, (1993) Performing sexualities: an interview. In Harwood, Victoria, Oswell, David, Parkinson, Kay and Ward, Anna (eds) *Pleasure Principles: Politics, Sexuality and Ethics*, Lawrence & Wishart, London.

King, Anthony D. (1997) Excavating the multicultural suburb: hidden histories of the bungalow. In Silverstone, Roger (ed.), *Visions of Suburbia*, Routledge, London.

Kureishi, Hanif (1990) *The Buddha of Suburbia*, Faber & Faber, London.

Lebeau, Vicky (1997) The worst of all possible worlds. In Silverstone, Roger (ed.) *Visions of Suburbia*, Routledge, London.

Light, Alison (1991) *Forever England: Femininity, Literature and Conservatism between the Wars*, Routledge, London.

Massey, Doreen (1991) Flexible sexism. *Environment and Planning D: Society and Space*, **9**, 31–57.

Masterman, C. F. G. (1960 [1909]) *The Condition of England*, Methuen, London.

Medhurst, Andy (1997) Negotiating the gnome zone: versions of suburban British popular culture. In Silverstone, Roger (ed.), *Visions of Suburbia*, Routledge, London.

Mercer, Kobena (1990) Welcome to the jungle: identity and diversity in postmodern politics. In Rutherford, Jonathan (ed.) *Identity: Community, Culture, Difference*, Lawrence & Wishart, London.

[Parker Morris Report] Ministry of Housing and Local Government (1961) *Homes for Today and Tomorrow*, HMSO, London.

Oswell, David (1998) True love in queer times: romance, suburbia and masculinity. In Pearce, Lynne and Wisker, Gina (eds) *Fatal Attractions and Cultural Subversions: Scripting Romance in Contemporary Literature and Film*, Pluto Press, London.

Rex, J. (1968) The sociology of a zone of transition. In Pahl, R. (ed.) *Readings in Urban Sociology*, Pergamon, London.

Rex, J. and Moore, R. (1967) *Race, Community and Conflict*, Oxford University Press, Oxford.

Robertson, Howard (1947) *Reconstruction and the Home*, The Studio, London.

Rose, Nikolas (1989a) Governing the enterprising self. Paper delivered at Conference on the Values of the Enterprise Culture, University of Lancaster, September.

Rose, Nikolas (1989b) *Governing the Soul*, Routledge, London.

Scannell, Paddy (1988a) The communicative ethos of broadcasting. Paper presented to the International Television Studies Conference, London.

Scannell, Paddy (1988b) Radio times: the temporal arrangements of broadcasting in the modern world. In Drummond, Philip and Paterson, Richard (eds), *Television and Its Audience*, British Film Institute, London.

Scannell, Paddy (1991) The merely sociable. Paper presented to the Language Workshop.

Scannell, Paddy and Cardiff, David (1982) Serving the nation: public service broadcasting before the war. In Waites, B., Bennett, T. and Martin, G. (eds), *Popular Culture: Past and Present*, Open University Press, London.

Silverstone, Roger (1994) *Television and Everyday Life*, Routledge, London.

Silverstone, Roger (1997) Introduction. In Silverstone, Roger (ed.), *Visions of Suburbia*, Routledge, London.

Sinfield, Alan (1992) What's in a name? *Gay Times* (May).

Spigel, Lynn (1992) *Make Room for TV*, University of Chicago Press, Chicago.

Wells, H. G. (1963 [1910]) *The History of Mr Polly*, Pan Books, London.

Wilmott, P. and Young, P. (1960) *Family and Class in a London Suburb*, Routledge & Kegan Paul, London.

Wolff, J. (1985) The invisible flâneuse: women and the literature of modernity. *Theory, Culture and Society*, **2** (3), 37–46.

Notes

1 This article was based on a paper for the City Limits conference, Staffordshire University, September 1996, and draws on another work (Oswell, 1998).

2 A review of *The Buddha of Suburbia* in the *Daily Telegraph* refers to the 'spiritual desert of suburban Kent' and its domestic 'banalities' (4 November 1993).

3 At the time of my rewriting this piece, Andrew Moncur wrote an article for *The Guardian* entitled 'Net curtain rises on suburban revolt' (4 November 1998). The article was a spoof on the revolutionary politics of suburbia.

4 Of course, questions of culture have been discussed extensively within geography, but cultural geography has been less discussed within media studies.

5 Vicky Lebeau, in a fascinating re-reading of Wilmott and Young's classic study of class and suburbia in the 1950s, argues that the move to suburbia and its enforced domestication provided the condition for television's popular appeal (Lebeau, 1997).

6 Silverstone argues that the quintessential suburban television form is soap opera, but I would concur with Medhurst who suggests that it is the sitcom that has this privileged position. Beth Haralovich discusses the relationship between suburbia and the sitcom in the US in the 1950s (Haralovich, 1988).

7 At stake is not living in the actual, physical locality of suburbia, but living within the lifeworld of television's imagined suburban community.

8 The flight and yet also the drawing from the suburban is described by the director of the television serial, Roger Michell. He argues that they tried to juggle with 'the wish to be loyal to the suburbs in the face of a sneering metropolis, and the need to depict the frustration of a place you moved from' (cited by Chris Savage-King, 'The Buddha of Suburbia', *New Statesman and Society*, 5 November 1993).

9 The adoption of Anwar's character is also caught up in the generational conflicts so typical of the suburban (cf. Lebeau, 1997).

10 Homi Bhabha defines hybridity as follows: 'Hybridity is the sign of the productivity of colonial power, its shifting forces and fixities; it is the name for the strategic reversal of the process of domination through disavowal (i.e. the production of discriminatory identities that secure the "pure" and original identity of authority). Hybridity is the revaluation of the assumption of colonial identity through the repetition of identity effects' (Bhabha, 1985: 97).

11 This earlier discourse is ostensibly metropolitan. But it can also be read as the

articulation of a 'new *petit bourgeois*' discourse against a traditional *petit-bourgeois* discourse. Pierre Bourdieu, for example, argues that this new class fraction is made up of therapists, television producers and other professions. He shows that 'whereas the old morality of duty, based on the opposition between pleasure and good, induces a generalised suspicion of the "charming and attractive", a fear of pleasure and a relation to the body made up of "reserve", "modesty" and "restraint", and associates every satisfaction of the forbidden impulses with guilt, the new ethical avant-garde urges a morality of pleasure as duty ... The fear of not getting enough pleasure, the logical outcome of the effort to overcome the fear of pleasure is combined with search for self-expression and "bodily expression" and for ... immersion in others ... and the old personal ethic is thus rejected for a cult of personal health and psychological therapy' (1984: 367).

12 *NME*, 6 November 1993. In this sense the use of Bowie's music also becomes a way of refiguring the star-image of Bowie and of reinventing his less-than-radical past. For example, Hebdige quotes Bowie, reported in *Temporary Hoarding*, a Rock Against Racism periodical: 'Hitler was the first superstar. He really did it right' (Hebdige, 1979: 61).

13 Baudelaire argues in his essay on 'The painter of modern life': 'For the perfect *flâneur*, for the passionate spectator, it is an immense joy to set up house in the heart of the multitude, amid the ebb and flow of movement, in the midst of the fugitive and the infinite' (Baudelaire, 1964 [1863]: 9).

It has been argued that the *flâneur* is a specifically masculine position from which women have been traditionally excluded (due to women's exclusion from public space and to their being the object, not subject, of a voyeuristic gaze) (cf. Wolff, 1985; Massey, 1991).

14 Foucault misreads Baudelaire when he focuses on the dandy as the heroic figure of self-creation (Foucault, 1984b: 41). For Baudelaire, C. G. (Constantin Guys, the artist who remains anonymous) is not a dandy because this mask is too blasé, insensitive and unaware of the 'world of morals and politics' (Baudelaire, 1964 [1863]: 7, 9). Likewise, C. G. has 'an aim loftier than that of a mere *flâneur*' (ibid.: 12).

15 *New Statesman and Society*, 5 November 1993. Savage-King argues that 'Karim ... views Bromley as a phantasmagoria. The suburbs, as a cliché, are reinvented, and shown more as a seat of possibility, less a death sentence' (ibid.)

16 My argument then is that suburban television is not necessarily, as Silverstone argues, hybrid, but that its hybridity is contingent and contested.

17 In the novel Karim describes Marlene and Matthew as 'more like intrepid journalists than swimmers in the sensual'. He perceives them as separated from the world and that 'their obsession with how the world worked just seemed another form of self-obsession' (Kureishi, 1990: 191).

18 The novel presents this scene slightly differently. Matthew inserts his cock in Karim's mouth without being invited. Karim gives his dick a 'South London swipe' and Matthew withdraws. It is only later that Eleanor puts Matthew's fingers, recently withdrawn from Eleanor's cunt, into Karim's mouth and suggests that the two men touch each other. Karim has little he can say with Matthew's fingers in his mouth (ibid.: 202–4).

19 This is similar to the manner of disclosure found in *Happy Ever After*.

20 Other complaints were made to the Council concerning Karim's masturbation of Charlie in Episode 1 and the blasphemy of phrases such as 'Jesus fucking Christ'.

21 The BBC, in its statement to the BSC, argued that the programme was about 'the
    search of a young man … for his sexual and cultural identity'. It continued by say-
    ing that 'The theme was explored frankly within the context of a satirical treatment
    of attitudes prevalent at the time, seeking to convey the current moral confusions
    and uncertainties' (Broadcasting Standards Council, February 1994).

Chapter 7

Ethical transgressions beyond the city wall

*Christopher Stanley**

... An intervention in a continuing struggle to 'ground' a cultural politics of difference upon an ethics of alterity. Figuring the relation between identity and respect in the city. It is a struggle which may never be resolved because the ground is always the impossible possibility as the aporetic. But it is a struggle driven by a political and ethical exigency determined through a desire to reconfigure being in the city as a being in difference together in a just community within the city. The city in this chapter is the ground on which a political and ethical engagement occurs the terms of which are proscribed by the relation between proximity and communication in the realization of aporia.

The aporia is the impassable threshold. In this chapter the aporia determines the form of the negotiation of engagement in both thought and practice. As an inclination toward the impossible satisfaction of a desire (lack) and of the completion (closure) of this exigency to articulate the relation between the ethical and the political in the expectation of an always becoming just community in the city, the aporia is the risk of writing and thinking. An inclination toward an epistemological and ontological other space for being–thinking– writing is the expression of an attempt to elude the represented parameters of the epistemological and the ontological. The inclination as a falling toward an impossible ground implies the aporia. For what might be articulated as the ethical and political topos of a non-violent community of justice-in-difference can never be described-inscribed in terms of a real configuration or grounding. The description would always be inauthentic and an appearance of a reality – confined within the violent orders of simulation. There is only the provision of a version of a possibility always and already inauthentic because inculcated as part of the consequences of the strategy which is the violent tyranny operative in the sovereignty of the processes of representation. These processes demand another violence in the process of judgement as the compulsion to inscribe this possibility of configuration in terms of values is always ordered between parameters always excluded.

The confrontation or compulsion by an exigency, the fulfilment or response to or of which cannot be represented, but which is grounded in a city, might be thought of as utopic but which is heterotopic and therefore elusive in terms of inscription and the installation of value. The tyranny of sovereign representation is not only of a violence of judgement and of inauthenticity and simulation but also one of working. A working toward a just community is a process in which community and justice are located in a static field which is already colonized and subjected through representation. There cannot be a 'working' through which community and justice are to be inscribed or grounded through a discourse of ethics and politics.

This does not mean that the ethical and political elements within such a configuration of community and justice are utopic but rather that the possibility, the arrival and the actuality of this configuration and grounding of community and justice, which repels the violence of judgement and the erasure of difference in the immanence of representation, is always elsewhere and imminent, suspended and momentary. The legislator and the interpreter cannot occupy the space of this community of justice-in-difference because to do so would already mean that this space has ceased to be heterotopic and has become completed and colonized through the tyranny of representation and devalued through the violence of valuation. The open place of possibility becomes the closed space of totality. Where representation necessarily misrepresents we find the political moment when the strategy nature of closure is revealed. Such closures are moments in a politics of articulation, an echo of something. But for every closure there is a remaindered excess excluded in the process of judging the moment of closure. That which is refused is the refuse of refusal marking the constant incision into the parameters of closure. The refuse is the location of an intimation that the event and act of community and justice is a suspended possibility or a trace of an inclination toward a spirit left in mourning whose trace is always exorcized by the ravages of the closure of ethics and devalued through the rhetoric of politics but whose presence as a trace in the city cannot be erased but it leads elsewhere.

Part 1

The city is the space and place through and against which this thinking and writing occurs.

The city is polis, metropolis, megalopolis and heteropolis.

The city is the polis, metropolis, megalopolis and heteropolis of stelai, acropolis, agora, domus, taberna.

The city is polis, metropolis, megalopolis and heteropolis of the veneration of the dead, the celebration of the future, the activity of governmentality, the place for dwelling and the space for the sustenance of trade.

City visions determined by the number five. From Thebes to Auschwitz to L.A. (New) Jerusalem? Too unreal cities.

The millennial city is the conclusion of a lineage stretching back to antiquity and the birth of the city, through the imposition of sovereignty to the obligation of the contractual bond, through modernity and the construction of the subject enabling concentration camp-gulag and the partial project in social justice as utilitarian policy. The five principles of classical form persist through inversion, appropriation and perversion. At this still point in time these cities become fractured sites encompassing aspects of both a neglected modernity which now appears anachronistic yet persists, and an incomplete postmodernity reliant upon nostalgia, pastiche, irony and a completed post-industriality produced through technology and globalization. The tensions which were always latent are now apparent:

homogeneity and heterogeneity, completeness and incompleteness, separateness and apartness, the smooth and the striated, risk and safety, the restricted and the general, bridge and door, presence and absence, sharing out and sharing in, friend and foe, civilian and idiotes, law and transgression, deviance and dissent, with-in law and with-out law:

enabling a multiplicity of subject-positions to be exposed through hybridity against a tendency toward the eradication of differences and the flattening of meaning through the processes of banal violence of representation. These cities are the sites and sights of the play of culture and of habitus, of being-in-the-world and of that being as a being-with and as a dwelling and belonging in the city as a concrete, plastic and virtual form enabling multiple occupancy and multiple value to be attached and manipulated in struggles utilizing tactics of inversion, appropriation and perversion expressed as refusal, indifference and disappearance. It is the cityscape interpreted through a psychogeography which provides an alternative cognitive mapping enabling the unrepresentable, the sublime, the inarticulate and the silenced. It is no more saying that received notions of space filled with universals by individuals must be replaced with a more nuanced version that acknowledges dimensions of identity-desire, community-justice reflecting inscribed structures of power:

space, law, society, regulation, government, sovereignty, contract:

the injustice of judgement which desires sameness but can only process through singularity in enacting through violence the force of law.

Without a fixed reference point – no matter how remote or tangential – writing and thinking through community and justice would be ou-topic investment whereas the reference points searched for are within heterotopic interstices within the city – the unity of the real and the imagined within the inside but as an outside which cannot be assimilated and sometimes beyond sight because too near (pan-oramic). The city as the site of tension in terms of the production of a counter-production as transgression, the ludic and the indifferent occurring within its fractures ('beyond the city wall' in the absence of a city wall – the paradoxical inner city as outside city) provide a source for witnessing and makes the maintenance of another fevered archive important: other voices, other stories, other tears.

To be 'aligned' through a commitment to a just community where the cultural politics of difference is 'grounded' on the ethics of alterity and in realization of the bind or paradox which this commitment ensures, the role of the subject confined to writing and thinking through community and justice is that of the witness who can only glimpse that which demands testimony without judgement. The task of the by-stander (at the point of judgement) is to bear witness to the unrepresentable and the sublime from the margin and to enable other voices to tell other stories through others' tears. These timeless stories from the city are of violent trauma (terror, fear, loss of hope, redundant expectation) and violent desire (passion, yearning, dreams, love) and the inclination for something other – the transformation of the jurisdiction of dissent into a non-violent community of difference 'togethered' in-through a justice of otherness:

civitas purgatorio to civitas peregrina.

(One way to record the dissonance-resonance of these stories is to re-configure the official line of communication or ordering of discourse so that deviance becomes dissent and crime becomes transgression. To replace nouns with verbs – to be in process (a movement toward). These stories would then assume previously absent aspects (such as gender and ethnicity) which had formerly been erased or aligned in a particular repressive variation of representation. This process of re-configuration of discourse and re-alignment of genealogy undermines the authority of the legislator and the interpreter affording an alternative manifestation of genealogy and discourse to that which had been subject to closure, enabling another voice to be heard out-with the law, in the crypt of another archive yet to be inscribed, but for a moment heard as a sharing of voices in an ecstatic community on a common ground before the law – the affirmation of the absent presence of proper names.)

Part 2

Community is the antagonistic supplement of modernity: in the metropolitan space it is the territory of the minority, threatening the claims of civility; in the transnational world it becomes the border problem of the diasporic, the migrant, the refugee. Binary divisions of social space neglect the profound temporal disjunction – the translational time and space – through which minority communities negotiate their collective identifications. For what is at issue in the discourse of minorities is the creation of agency through incommensurable (not simply multiple) positions. Is there a poetics of interestical community? How does it name itself, author its agency? (Bhabha, 1994: 228, 231)

It is this incommensurability that arises from and constitutes the spatialities of sub-jectification that work through and on 'the street'. It is out of the incommensurability of different spatialities that both spaces of resistance emerge as social practice and the moments of utopianism can insert themselves as political projects of discursive closure. (Keith, 1995: 308)

Part 3

The heteroglossia of heterotopia, the poetic politics of interstitial – the intangible glimpsed from the corner of the eye, an authenticity the apparent appearance of which disappears as soon as the glimpse hardens to the gaze, the visible becomes the invisible (a city vision). The negation of judgement through a traumatic liminality. The trace of this traumatic liminality can be figured in the cultural representation, the art(e)fact, a gathering of voices into a plasticity of form translated into cultural commodity but serving to infect through its representation of the incommensurable the order of being in the city through a disruptive interruption. And then consigned to another archive.[1]

....

An art installation called *House* created in the London Borough of Tower Hamlets by Rachel Whiteread. It is from this piece of 'work' that notes can be jotted regarding a non-violent configuration of community and justice. *House* provides a mechanism of entry into a process of tracing an unwritten history and an uncertain future. *House* can be read (and has already been read – interpreted to death?) as a possible site for the spirit of Antigone, and it is the relation between Antigone and *House* that is pursued here in terms of a (new) relation between community and justice. The troubled spirit of Antigone within the walls of *House* is a powerful image through which to construct a non-violent mechanism of entry into excavating the ground of an alternative genealogy and discourse of community and justice. In this sense *House* is read and a further metaphor is produced in the process. This metaphor is 'grounded' in both a political-ethical and material-mental reality (dwelling in the city). Metaphor is connected to both materiality and meaning and to deny these connections is doomed to failure and absurd – the use of metaphor must not deny its own construction but must itself be investigated for its con- nection to the material and the mental. In figuring Antigone 'in' *House*, in 'constructing' this lineage, what is provided is both a formerly absent continu- ity between antiquity and the contemporary (that there is a trace which has formerly been violently closed but which persists and cannot be entirely cov- ered over) and that the issues of place and dwelling and of an incomplete or suspended mourning are potential tools through which to realize an account of community and justice which provides a partial incision into these processes of closure and silence. This form of analysis does not provide a theoretically coherent account (save perhaps in terms of the writing and thinking through the work of community and justice in a process of radical contextualization), rather it serves as a statement of allusion or an interstitial poetics which might be utilized in further incursions into the processes of closure and forgetting in an alternative thinking and writing balanced between the tension of 'working' and 'unworking' the occupation of which is all that can be offered in terms of the function of the witness listening to other voices in the living archive – how the work to be realized requires no action and is an in-operativity.

In discussing *House* as a possible marker in a configuration of a non- violent grounding of community and justice through the relationship between *House* and Antigone, the argument will take the form of an analysis in the affirmation of loss in terms of the act, event and place of mourning for com- munity and-as justice Before the Law. Since this configuration of community and justice is imminent (as opposed to the immanence of society and law) and always elsewhere (as heterotopic not utopic, as present absence as opposed to absent presence) the trace of *House* and the eternal return of Antigone point to the significance of memory and of the retrieval or excavation of that which has been forgotten or the spirit of that which has been repressed but which always returns. *House* is no longer and Antigone is myth (the scapegoat), but similarly to the problem of appearance and approach, the residue of them, or their essence, are keys to a remembrance of a possibly before the trauma of diremption (the bridge becomes the door) and exclusion and the arrival of figures of possibility occurring elsewhere in an immemorial (common) place: they are tropes. This argument is developed through a discussion of the relationship between community and justice in terms of mourning and dwelling in an ethical and political dimension of interruption and suspension (*paralogy* and *phronesis*). *House* and Antigone are not figures which can directly point to ways of thinking and writing community and justice in terms of a supplementary and unnecessary account, rather their place on the borders of a constant occurrence and possibility enable them to intimate that there are events and moments beyond and before the law of injustice which is the law of the city, which inevitably becomes a violent point of origin within the city wall.

Part 4

> Let us return to the boundary wall of the city of Athens – here, just outside the boundary we find mourning women: Antigone, burying the brother of her fratricidal brother in defiance of Creon's decree, witnessed by her reluctant sister, Ismene, who urges her to desist in the name of conformity to the law of the city; and the wife of Phocion, gathering the ashes of her disgraced husband, with her trusted women companion, who, as look-out, bears the political risk in her own contorted posture (Rose, 1996: 35).

What might it mean to discover Antigone in the London Borough of Tower Hamlets tomorrow? The injustice suffered by Antigone because of Creon's decree is repeated every day with varying degrees of intensity by the inhabitants of the inner cities of megalopolis. Antigone's ethical transgression of Creon's rule in preference to 'another justice' is a conflict replicated in a myriad minor instances in a struggle for survival which incurs the performance of a rupture. We no longer breach a rule regarding the burial rites of a dead brother but the homeless, the single parent, the care-in-the-community client, the asylum seeker, the long-term unemployed, the geriatric, the disabled and all those others who are not complete participants in the routine and regime of metropolis and have become statistics within the margins of the officially visible, perform their own breaches of the laws of the city every day in the effort of survival, despair, exhaustion and indifference. Through processes of representation, nomination and location these others are the subjects of the laws of a city which grants rights, duties, status and opportunities differentially in terms of materiality and sociality. There are civilians and others, subjects of the city and those subjected to the city.

A precarious balance has been struck between the external notification of the presence and absence of these others. Through coercive regulatory policies and tactics these others assume presence only as a marginal problem, at worst as statistics in the battles of banal rhetoric between redundant ideologies. The scale of this problem is managed through the assumption of the absence of many of these others, they disappear and are forgotten, trapped in piss-drenched doorways and on neglected council estates (the detritus, the refuse, the excess: where politics and ethics is radically reconfigured around friend and foe?). Social order – alignment within determined representations and values – is maintained through processes of classification which invest in social disorder at the margins. The violence of social disorder manifest as criminality and deviance and broadcast as the fear of crime is measured by the violence of social ordering through the allocation of subject-position in a process of disciplinary procedure and invasive surveillance. Deregulatory initiatives, as the foundation of the neo-liberal political-economy schema dominant in post-industrial societies, are the contemporary manifestation of the collusion between dominant economic and cultural interest groups in insidious conspiracies of a repressive governmentality (the turn of liberal democracy to unheard totalitarianism) ensures that moral and economic indeterminacy save in terms of the hidden order of representation through corporate media manipulation occurs not only within the spheres of economic production but also within the sphere of cultural production.

The distinction between the inside of the city and the outside of the city becomes increasingly distinct, even if the metaphor remains problematic and the extent of the city is beyond panorama and the enlightenment fear of 'them' and 'us' artificial because of the displacement and fluidity of subject-positioning. Nevertheless, the 'inside' of the city maintains its veneer as a *civitas*, a space of moral and economic negotiation, mutual interest and similarity of cultural capital, alignment between indeterminable places constituting the habitus of the possessed. It becomes indistinguishable from other global cities subject to the processes of postmodernization: technological supremacy, the internationalization of capital, the sovereignty of the service economy and the emergence of risk and anxiety as the dominant cultural and economic dynamic. The 'outside' of the city as the margin between urbanity in the city and ironic urbanity in the non-urban has become *civitas peregrina*, a space of moral and economic neglect, mutual interest through conflict and diversity between dreadful ethnic enclaves and forgotten, de-industrialized empty spaces – the space of the smallest and the most singular, the *idiotes*. The inner city, within the shadow of the inside city, is the outside city which the processes of postmodernization have passed by or colonized differentially through entrepreneurial re-development and cultural gentrification. This is the territory of the dispossessed urban poor that at once haunt the millennial city and at the same moment are absent in the account. It is in this other space, this 'outside' city which is the inner city, moulded through the physical violence of deprivation as opposed to the allegorical violence of affluence, where can be located an intervention on the relationship between justice and community practised through the performance of rupturing ethical transgressions.

Part 5 _____

The law of the city and the city of the law have changed very little since the establishment of the first cities of antiquity. The sites of the law – the agora, the parliament palace, the prison, the court, the police station, the benefit office, the council chamber, the archive – are familiar. From antiquity to modernity these buildings might have changed architecturally but their functions as the residences of authority and legitimation are the same. The scope of the legitimatory function of these buildings of the law of the city might have diminished because of the turn of postmodernity and globalization. There has been a simultaneous intensification of governmentality as surveillance of the everyday activities of the citizens has increased together with an apparent removal of governmentality as the source of legitimation, authority and coercion has been delegated to the quasi-public organizations and specialist practitioners in coalitions of commercial interests in hegemonic power blocks. The law of the city simultaneously becomes more violent in its authority in the sense of responding to the threat posed by the dispossessed (discretionary powers in terms of both social welfare and policing) and at the same time more subtle in a shift toward the governance of representation, that cultural capital and cultural goods are now more significant than commodities and possessions. The control of the circulation of representations as sources of value and therefore of legitimation – dreams, desires, aspirations – is a strategic regulatory goal achieved through advertising and media manipulation in the organization of leisure spectacles as cultural reference points. The law of the city is a law which purports to deliver social justice but which in fact delivers injustice in the maintenance of an unjust political economy reliant upon material differentiation achieved through an apparent material equality. The mechanisms of this law are at once most subtle and at the same time obscene as law cedes to regulation and there is the turn to a law of movement, visuality and narrative despite the absence of the narrative certainty which a variation of postmodernity suggests. It is a law which strives to erase difference whilst being reliant upon difference for the scope of its jurisdiction because operative through calculation, a jurisdiction the border of which defines a territory without-law but which is a territory constructed by and through law. Just beyond the boundary is where Antigone tended the body of her brother and where _House_ was created. An act and a monument no longer present but becoming part of a fabric of experience, the immemorality of which is the grounding of another community and another justice on a common ground, the boundary of which marks an opening into an affirmative heterotopia, without-of the law as opposed to out-with the law but existent only because Before the Law.

> Antigone is a figure of tombs and shrouds, a watcher of bodies, a guardian not of Law, but of flesh ... The domain of this sun king is the _agora_ and public life, the sphere of public discourse and action. Creon is the lord of war and peace. But he transgresses the law of the _oikos_, of the household, of hearth and bed, of table and bath. Antigone's kingdom (_basileia_) is not of Creon's world (_polis_). (Caputo, 1993: 167)

Part 6

The desire for monuments and memorials within the city uncannily persists. It is this persistence to leave a mark and define the scope of a territory for the future which I will discuss in terms of an inversion of the 'monumental tradition' when confronted with the absent monument of an 'immemorial continuity' – *House*, a memorial which sought to interrupt the legislated account of memory and in so doing marked the return of a repressed memory. Monuments – memorials – are erected in an act of remembrance. It is one of the contradictions of the postmodern condition that as memories are erased through the processes of social fragmentation and the collapse of systems of legitimation consequent upon the rapidity of technological and scientific change, postmodern culture is reliant upon nostalgia, parody and pastiche which reinvent memories through their reliance upon memory. In terms of an affirmative postmodernism this enables a continuity of remembrance especially in the sense that there are events and names which should never be forgotten, in other ways these edifices are uncanny and their inscriptions illegible and unintelligible, a precarious tension between the ordering of memory and the enabling of other memories and an other's memory.

The advent of modernism provided a moment of variation to the memory map of the city. The modernist planners and ideologues had desires to forget as well as to remember. The old city, its old monuments and its traditional significance were too solidly implicated with the economic, social and political problems of the old world to justify retention. An act of forgetting might take the form of an erasure of the city itself (literal and figural) in favour of a *tabula rasa* that reinstalled nature as a foundation for a dispersed urbanism and made its monuments out of the functions of modern life (the advent of the skyscraper). This process was another version of urbanism, symmetrical but counter to that of earlier forms. The twist within this turn within modernism, and that which provided a point of rupture, was the recognition of the presence of an absence. The figure-grounded city of modernism was established on the erasure of the two fullnesses of being: the city and its monuments. The disturbance of the presence of absence is that which provokes the anxiety of the modern city and which the irony and nostalgia of the postmodern attempt to redress. The nostalgia of the postmodern is premised upon an attempt to retrieve the fullness of being by retrospective memory, nostalgia for a moment that points forward to an event that never happened. There is a third absence and the introduction of a paradoxical anxiety founded on the psychoanalytic of the uncanny. This inadvertent breach in the interpretation of memory as displayed in the schizophrenic gestures of a monumentalism forced to confront the ruptures of the material forms of the city enables the anxiety of the uncanny to be interpreted affirmatively because it disturbs ordered notions of space and the proper spaces of proper stories letting in, through strategies of appropriation and inversion, echoes of memories which had been interrupted.

These are the traces which *House* (no less violently) exposes, the traces of the *oikos* which the monuments of the *polis* intended to deny and displace in favour of a universal truth without difference. This passive and indeterminate exposure reveals Antigone in the walls of *House* beyond the 'city wall' recalling and pointing to a justice and community grounded before the law of the *agora*, still in silence and before language, still in mourning which is the act and event of being-together in sharing this lack (trying to remember that which has been interrupted) through an un-working – an ethics of justice operative because of the constancy of the injustice of the city which displaces dwelling forcing us to appear together as a community of strangers.

> We wander surprised but not shocked by the continuous repetition of the same, the continuous movement across vanished thresholds that leave only traces of their former status as places. Amidst the ruins of monuments no longer significant because deprived of their status systematic status (and often their incorporeality) we walk on the dust of inscriptions no longer decipherable because lacking so many words, whether carved in stone or shaped in neon, we cross nothing going nowhere. (Vidler, 1993: 156)

Part 7

At this juncture, always on the edge of the disaster (of the implosion of the city, of the end of history, of the triumph of a representation of homogeneity, of writing and thinking), it is possible to attempt to make an interim statement of account, to make a statement of empathy to be lodged in another archive. It is a statement which might be interpreted as a potential supplement[2] to the configuration of a non-violent grounding of the establishment of a community of difference enabled through an alterity of justice. The statement will proceed in four parts. First, the location of *House* within an heterotopic topography. Second, the nature of *House* in terms of artistic technique. Third, *House* as an anti-monument (and therefore its relation to the psychoanalytic of the uncanny) which simultaneously displaces fixed notions of memory and evokes a tomb as a resting place. Finally, and by way of an interim conclusion, the relation between *House* as a crypt, Antigone and the process of mourning in terms of the displacement of the oppositions house and home, space and place, *polis* and *oikos*. This analysis enables a consideration of the processes in the disturbance of memory and the suspension of mourning as possible elements in a configuration of a non-violent grounding of community and justice the basis of which is a passive dwelling in proximity to others in a common place before the law.

(a)

House was completed on 25 October, 1993. Its completion was based on the alteration of the structure of a building called 193 Grove Road in the London Borough of Tower Hamlets, in the neighbourhood of Bow, East London, the last house of what had once been a Victorian terrace. The area was in the process of re-development: Tower Hamlets is one of the poorest boroughs in

London with high unemployment and severe housing problems. Across the road from the site of 193 Grove Road are three different churches (Baptist, Jehovah's Witness and Church of England). A little way down Grove Road there are some wooden sculptures adjacent to a stretch of empty park land. Carved from trees blown down after a storm in 1987, they stand in forlorn arrangement. Towering over Grove Road, with its churches and refurbished flats, loom three concrete high-rise blocks. Even more imposing is the shadow of the Canary Wharf Tower, the tallest building in England and the ironic symbol of the transformations which were wrought upon the East End of London through the 1980s. Another building can be seen on the skyline, another emblem of global capitalism, the NatWest Tower in the centre of the City of London. Bow is guarded on two sides by these watch towers of entrepreneurial enterprise. This process of surveillance ensures that Bow, which is beyond the 'city wall', is still within sight, an area deregulated to the extent that it has been developed differentially in terms of economic and cultural initiatives and neglected in terms of social policy. As a community, Bow is heterogeneous and fractured, as differential processes have meant rapid change in parts and continuity in others – a contested space, a space capable of juxtaposing in a single real place several spaces, several sites that are in themselves incompatible and incommensurable – heterotopic.

(b)

Rachel Whiteread's artistic technique used to create *House* and previous works such as *Ghost* and *Closet* is a process of casting. *Closet* was the cast of a wardrobe and *Ghost* was the cast of a room. *House* was the cast of a house. To be precise, Rachel Whiteread casts the negative of a space. In *House* there was a series of independent concrete boxes which formed the negative of each room, which, when exposed through the removal of the original skin, had adequate rigidity to act as a completely stable unit once exposed. The exposed unit provided an excellent representation of each room forming *House – 193 Grove Road. House* was finished on 25 October 1993 and demolished on 11 January 1994.

(c)

House (was) an anti-monument. In physical form *House* was monumental. The simple process of exposing the interior spaces of an ordinary dwelling through the use of grey concrete gave it a monumental aspect. Its location on the edge of a new park space exposed it in the same manner that monuments such as war memorials are set apart from everyday activities and assume a resonance beyond their actual physical dimensions serving as markers of collective memory or as sites of legitimation. But *House*, despite its monumental resonance, was an anti-monumental structure. First, its destruction only months after its creation (despite attempts to 'save' it) immediately undermined its monumental status. Second, it was created as a public work of art in the tradition of Joseph Beuys and others, not as a civic memorial. Third, it provoked reactions that are not usually attributed to monuments, it was recognized as the cast of a house but it was unfamiliar – this grey monolith de-contextualized from the orthodox locations for public art – was 'uncanny'.

The use of monuments in the city is to 'last in time' and to signify the importance of whatever memory they wish to pass on to the future and to ensure that they will endure in the future. In addition, the monumental tradition is public – and this includes tombs and signs of covenant. The established meanings of particular monuments might be changed in the course of time because of social, cultural or political forces but the original intention of the monument is static and constant. *House* simultaneously displaced and exposed the role of the monument in the processes of the ordering of social memory. In this sense it was iconoclastic, disturbing the tradition of the monument as an aesthetic expression of the archive and the crypt of a universal truth of legitimacy. In the process it enabled the expression of other memories which had not been expected, provoking an anxiety of response which was uncanny because it forced a confrontation with that which had been yearned for but repressed. Its blank inscription and temporal instability destabilized received interpretations of monumentality. In utilizing a domestic structure (a private house) and exposing to the outside the inside of this private domain, *House* confused established notions of the public and the private spheres. The ideological appropriation of *House* by local politicians was premised on the interpretation that this piece of public art was a comment on social housing and on the destruction of communities – not messages usually relayed by monuments. Such was the multiplicity of meanings available through *House* that both Left and Right could appropriate it in such a way in their manipulation of the folk memory of the East End.

The iconoclasm of *House* resided in its power to provoke the psychoanalytic of the uncanny. In opposition to the certainty of interpretation and familiarity, *House* disturbed the unconscious and expressed repressed fears and neurosis. Affirmatively, in the textbook account of psychoanalytic practice, enabling repressed memories to be confronted through naming them serves to alleviate their force as the locus of neurosis and this is how *House* can be utilized. *House* forced the observer to confront anxieties and the inevitable sense of loss and fear associated with childhood, domesticity, the home and the first recognition of belonging in a community through dwelling and therefore the first recollection of law, injustice and justice.[3] In Freudian terms the 'experiencing' of *House* was uncanny because space was exposed which is usually closed and access to an order of memory blocked through the installation of neglected memories – it was an instrument of trauma and neurosis, the admission of which enabled another's memories to emerge, signified through the presence of the figure of Antigone forever suspended in the act of mourning and the slippage of the law of the *polis* to the justice of the *oikos*. *House* re-inscribed the collective significance of mourning – necessarily suspended – in evoking the injustice of the intervention of the space of the law into the place of justice in community through the identification of that which has become *unhomely* which should have remained *homely*.[4] The 'unhomely' is the conscious state generated by the presence of the homeless, the stranger and the alien, the figure without dwelling or whose place of dwelling has been erased through the space of law. This figure without community and as the sufferer of injustice in the name of law demands justice (hospitality) and a place of belonging in *civitas peregrina*.

Part 8

House was a temporary piece of public art. Its destruction ensured its significance as a postmodern cultural icon was guaranteed in that postmodern culture, if it can be sensibly defined at all, is eclectic and disposable, nostalgic and ephemeral, an aspect of the waste and refuse accumulated and recycled within the post-industrial economy. The intention that *House* should be a temporary installation served to heighten the poignancy of its potential meaning in terms of the politics of public housing provision, the ideology of community initiatives and the deregulatory strategies through which the inner city has been neglected and differentially deprived. In this way the destruction of *House* created a 'new' memory at the very moment that its physicality and location provoked other memories through the disruption of oppositions between fixed conceptions of public and private, space and place and of house and home. Not only did the form, but the form and its destruction displaced memory. As a form, *House* served to provoke questioning between the relation between houses and homes, a relation which is contingent upon the acceptance of oppositions between space and place, public and private. Houses are socio-economic units within a public sphere whilst homes are cultural-psychoanalytic units within a private sphere. The house is therefore a space whilst the home is a place. In exposing the negative space of a home to the outside world, *House* exposed the naiveté of our investment in the security vested in the home. If the home is the first site-sight of our experience and therefore of memory then to expose this sacred closed place to the outside and onto the street is to interrupt the process of remembrance in a violent action which intimates the violence inherent in the domestic sphere – that the home is not only the site of the maternal and our first belonging but also of the intervention of the law of the father as the public law and the feeling of injustice which this intervention of the law brings (the colonization of the *oikos* by the law of the *polis* – from home to *agora*). It is the violently contested nature of the house which we forget in our mourning for the childhood of the home, a violence which is repressed through the forgetting of the proper names of justice and community. But the repressed always returns in the margins of the law. *House* was a trace of this return.

The figure who signifies this violent diremption and eternal return (of the political and ethical life of community) is Antigone. It is Antigone discovered at the site of *House*, the site of the meeting between justice and law, the maternal and the paternal, the home as the house of death. Antigone is forced onto the streets of Tower Hamlets beyond the city wall, no longer able to return and therefore without a dwelling and without 'proper being' because not belonging, a stranger, a foreigner, demanding justice and community, forever wandering in the interstices of the city which have now become the spaces 'beyond the wall' of the city in megalopolis. It is Antigone called to appear Before the Law but whose thought and actions in the name of justice cannot be heard or recognized by the Law and therefore she is constituted through Law as an absence, denied a place and allocated a non-space in an heterotopia which is the site of a suspended or delayed moment in the grounding of another's justice and community, of the place of dwelling in the domain of the *oikos*:

> The *oikos* is the kingdom of nurturing and feeding, of shelter and rest, of growth and recuperation, of eating and sleeping, of birth and death. The *oikos* provides a place to be born and nurtured, a place to grow old, a place to die. It is the sphere of new-born flesh, living flesh, dying flesh. (Caputo, 1993: 167)

In a postmodern account a return to the domain of the *oikos* is not possible. This is what *House* signifies. We may desire a return but when confronted by its consequences we realize that this place has become a space which is empty and subject to law and a different archive: an empty tomb. But we know there cannot be a return of the loved object, no matter how hard we try and that our attempts only encounter the uncanny, and this is the importance of the process of mourning.[5] To find ourselves out on the street means that we have to encounter justice and community in *civitas peregrina*, the trace of the place of the *oikos* in the real world of the public sphere. To be 'out on the street' and homeless is to suffer as an out-law, denied a dwelling and therefore a proper being, denied a voice and therefore a space before the law in which to be heard. It might appear as a traumatic revelation and it would be tempting to return as being-in-law as being-in-common but this would mean only the completion of the totalization of the jurisdiction of law which cannot calculate otherness. A possible configuration of a non-violent grounding of a community of difference secured through a justice of alterity would assume an interruption or suspension of a collective mourning between strangers called before the law but unable to bring themselves within the sight-site of the law and therefore appearing in a place which is the heterotopia shaped through the poetics of the interstitial. At this moment of singularity and difference in being-together as being-in-common as being-in-difference in the absence of a shared language but in the presence of a shared loss (justice, community, humanity) is the prospect of a sharing of voices cracking through tears. This sharing in mourning as our appearance-together would be the basis of communication which would establish proximity and from that proximity the grounding of a just community through the responsibility to recall the proper names of justice and community in the name of a forever absent other who cannot be named. This passive union secured through the recalling of a story of a lost presence – how the other who is un-named was sacrificed to the law – which through the telling become present, is enabled only through the violent interruptions to the process of mourning as a remembrance imposed through the presence of the law forcing us onto a common ground without the law, out of the *House* and into the street with Antigone. 'Our' inclination should be to be out-with Antigone, discovering ourselves always in the act of communication through mourning in the absence of thought and writing and therefore un-working in the shadow of the disaster.

References

Bhabha, H. (1994) *The Location of Culture*, Routledge, London.
Caputo, J. (1993) *Against Ethics*, Indiana University Press, Bloomington and Indianapolis.
Derrida, J. (1995) *Mal d'Archive: une impression freudienne*, Editions Galilée, Paris.
Freud, S. (1985) *Art and Literature*, Penguin Books, Harmondsworth.
Heidegger, M. (1971) *Poetry, Language, Thought*, Harper Colophon, New York.
Keith, M. (1995) Shouts of the street: identity and the spaces of authenticity. *Social Identities*, **1–2**, 297–315.
Rose, G. (1996) *Mourning Becomes the Law: Philosophy and Representation*, Cambridge University Press, Cambridge.
Vidler, A. (1993) *The Architectural Uncanny*, MIT Press, Cambridge, Mass.

Supplementary sources

Blanchot, M. (1986) *The Writing of the Disaster*, University of Nebraska Press, Lincoln.
Blanchot, M. (1988) *The Unavowable Community*, Station Hill Press, New York.
Gould, P. and Olsson, G. (eds) *A Search for Common Ground*, Pion Press, London.
Jabes, E. (1993) *A Foreigner Carrying in the Crook of His Arm a Tiny Book*, Wesleyan Press, Hanover.
Lingwood, J. (ed.) *Rachel Whiteread's House*, Phaidon Press, London.
Nancy, J.–L. (1991) *The Inoperative Community*, University of Minnesota Press, Minneapolis.

Notes

* This essay is for Sharon.

1 The library as another source of law: from Plato to Freud and beyond, the structure of the archive as a function of memory has been directly and indirectly bound to shifting techniques of reproduction. The advent of electronic technology not only resists ancient puzzles about the relation between recording and remembering but also poses new questions about the ways in which prosthetic devices both supplement and constitute subjectivity. The word 'archive' means by the commencement of memory and the command to remember. The archive is the location of the policing of the memory of the law through the statement of the correct hermeneutic process of retrieval, the archive of justice is always yet to be instituted and inscribed (see Derrida, 1995).

2 A supplement in the Derridean sense, that there is always within a text a supplement that works to contaminate any claim to a fundamental truth based on an essential identity or purity of a concept; neither presence or absence, the supplement is the play of difference. Supplement can here be related to radical

contextualization in that the project is to subvert the binaries because neither abstraction nor empiricism (in knowledge production), neither theory nor practice (in political action), neither universalism nor relativism (in aesthetic judgement) can resolve any of Kant's three fundamental questions of philosophy: What do I know? What should I know? What do I want? The 'answer' to all such questions depends on abolishing the *a priori* and spelling out the 'grounds' on which they are made (speaking in the 'proper name'), accepting that the traces of historicity and spatiality are always constitutive features of the processes of subject and object formation.

3 Belonging inhabits a space between being and becoming: of being-together before being singular. Belonging is here both the recognition of the singularity of being and at the same moment of this recognition the becoming of proximity through the acceptance of separation between self and other. Belonging is located in a system of tension between the experiences of dwelling and the uncanny, between strangeness and familiarity. Belonging is the moment of inclusion which must also be the moment of exclusion. The distance of apartness in a Heideggerian reading is the space through which existence in the world is confirmed. Dwelling is an interval in the process of being and becomes a characteristic of the interlocutor of subjectivity and objectivity (Heidegger, 1971).

4 Freud writes that 'what is homely becomes unhomely' and traces this through the etymology of the word *heimlich* – a word, the meaning of which develops in the direction of ambivalence until it coincides with its opposite – *unheimlich*. The uncanny is the experience of recognition when confronted by a space which is filled but which is expected to be empty: 'There is nothing new or alien, but something which is familiar and old-fashioned in the mind and which has become alienated from it only through the processes of repression ... something which ought to have remained hidden but which has come to light' (Freud, 1985: 363–4).

5 'By insisting on the rights and rites of mourning, Antigone and the wife of Phocion carry out the intense work of the soul, that gradual rearrangement of its boundaries, which must occur when a loved one is lost – so as to let go, to allow the other fully to depart, and hence fully to be regained beyond sorrow. To acknowledge and to re-experience the justice and injustice of the partner's life and death is to accept the law, not to transgress it – mourning become the law. Mourning draws on a transcendent but representable justice, which makes the suffering of immediate experience visible and speakable. When completed, mourning returns the soul to the city, renewed and reinvigorated for participation, ready to take on the difficulties and injustices of the existing city. The mourner returns to negotiate and challenge inner and outer boundaries of the soul and the city: she returns to their perennial anxiety' (Rose, 1996: 36).

Chapter 8

Dancing bodies in city settings: construction of spaces and subjects

Valerie Briginshaw

One of the attractions of the postmodern city is that because 'it is mainly about the production of signs and images' it provides an exciting arena for representations. It has been likened to 'a theatre, a series of stages upon which individuals could work their own distinctive magic while performing a multiplicity of roles' (Harvey, 1989: 5). The city's plasticity allows and invites interactions of many kinds which can open up possibilities for a shifting kaleidoscope of identities. It is perhaps not surprising that this 'soft city of illusion, myth, aspiration, nightmare' (Raban, 1974 quoted in Harvey, 1989: 5), which fires the imagination, has become a favoured setting for recent dance texts.

In this chapter the use of cities as settings in three contemporary dance videos is examined.[1] The chapter is based on the premise that concepts such as space and bodies are constructed. As Lefebvre states: 'each living body is space and has space: it produces itself in space and it also produces that space' (1991: 170). The city is a particular kind of constructed space and its construction is inextricably bound up with constructions of subjectivity. Consequently, in the analysis of the dance texts particular attention is paid to the ways in which the dancers' interactions with the urban environments contribute to these construction processes.

The three dance video texts analysed here are: *Muurwerk* (1987; 27 mins), a female solo set in a Brussels alleyway, choreographed and performed by Roxanne Huilmand and directed by Wolfgang Kolb; *Step in Time Girls* (1988; 24 mins), choreographed by Yolande Snaith and directed by Terry Braun, where three women from different periods – 'contemporary', 'wartime' and 'Victorian', are seen inhabiting and coming and going from the same London flat; and *Duets with Automobiles* (1993; 13 mins), choreographed by Shobhana Jeyasingh and directed by Terry Braun, where three female classical Indian Baratha Natyam dancers perform in three empty London office buildings.

The dance videos

'Muurwerk', translated, means brickwork or, literally, wall work, and there is certainly a sense in which Roxanne Huilmand, the solo dancer, is 'working' with the walls of the alley in the old centre of Brussels in which *Muurwerk* is set. She opens the piece by walking down the alley away from camera, drag-

ging the palm of her hand along the walls, the only sound is that of her footsteps echoing in the empty space. What initially appear to be everyday
pedestrian movements such as walking and leaning on a wall transform into
more obvious contemporary dance as Huilmand begins to rhythmically roll
and spin her body along the walls as well as running and kicking against
them. These movements are all incessantly repeated. This style of dancing,
consisting of recurrent patterns of often violent gestures and actions, frequently hurled in the face of the audience in a quite confrontational manner,
usually by young athletic dancers, often clad in boots, has come to be known
as 'Euro-crash'. Dutch and Belgian choreographers such as Wim Vandekeybus
and Anne Teresa De Keersmaeker established this style of dancing and choreography in their work of the mid 1980s. Roxanne Huilmand is a former
member of Anne Teresa de Keersmaeker's company Rosas and Wolfgang
Kolb, who directed *Muurwerk*, has also directed several films and videos for
Anne Teresa de Keersmaeker and Rosas.

In *Step in Time Girls* the three female dancers portraying women from different periods – 'contemporary', 'wartime' and 'Victorian' – are filmed for the
most part on their own, inside and outside the same London flats. Very occasionally they are seen with a male partner, a small child or a baby, but never
with each other. Whilst outside the flat their movements are pedestrian; walking, running and going up and down stairs or steps, for example. When
inside, however, they repeat and enlarge movements, such as rolling along
walls and windowsills, rocking in chairs and opening and closing cupboards,
usually in a rhythmic manner. The 'contemporary' woman becomes quite athletic at times, running and jumping up on a table, propelling her upper body
through a serving hatch and hanging down from it, swinging along the wall
below. The editing of the film continually shifts scene from inside to outside,
from one woman to another, and the women are seen performing similar but
not identical movements in the same spaces. Yolande Snaith, the choreographer, who also performs as the 'contemporary' woman, is known for her
gymnastic, weighty, 'release' based style of dancing, characteristic of British
'new dance' developed in the 1970s and 1980s, particularly at Dartington
College of Arts where Snaith studied under the tutorship of Americans Mary
Fulkerson and Steve Paxton. Paxton is noted for his promotion of the athletic
duet form of 'contact improvisation'.

Duets with Automobiles is filmed in three late 1980s postmodern London
office buildings; the Ark, Hammersmith, the neo-classical Canary Wharf complex, with its long corridors and marble floors, and the glass-walled Alban
Gate, London Wall. The dance is filmed from inside and outside the buildings,
allowing the juxtaposition and interaction of the dancers with the geographically situated architecture to facilitate Jeyasingh's intention of creating 'an icon
of Indian womanhood … appropriate to urban women in the 1990s'
(Rubidge, 1995: 34).[2] The choreography combines traditional Baratha Natyam
postures, gestures and phrases with pedestrian and contemporary dance
sequences. Jeyasingh's company was formed in Britain in 1988 following her
seven-year solo career in Baratha Natyam. One of her resource packs states:

'The Company's work questions inherited notions of what Indian dance is and what its possibilities are, stressing its belief that Indian dance is open to personal, contemporary and innovative use' (Rubidge, 1993). So, although Jeyasingh's work is rooted in Baratha Natyam vocabulary and style, she develops it in various ways, evident in *Duets with Automobiles*, for example, by extending the solo form to group choreography and by using predominantly contemporary music.

Constructed spaces

Spaces, through their construction, are invested with power. The postmodern city, like the postmodern subject, is fractured and fragmented, it is falling apart and full of contradictions. It is utopian and dystopian, attracting and alienating. It is constructed as labyrinthine, free-flowing and uncontrollable, but also as containing and trapping. The effects of this for some of the different sorts of women that exist in modern, multicultural cities can be negative; they can be isolated, excluded, constrained and defamiliarized in these mainly public spaces historically constructed by men for men.

The three dance video texts discussed here provide examples of different kinds of constructed cityspaces invested with power. In *Muurwerk* the deserted alley in the centre of old Brussels can be seen as alienating and trapping; the sort of inner-city space that women are warned about and expected to avoid. This is the sort of city space that has associations of male freedom derived from the nineteenth-century *flâneur*. In *Step in Time Girls* the rooms of the London flat and its environs can be seen as unsympathetic urban spaces which isolate, contain and constrain the women that inhabit them. In *Duets with Automobiles* the office buildings have connotations of power, money and the city as a financial centre. They are created by men who are at the heart of controlling flows of finance. They are valuable and powerful properties because of where they are and what goes on in them: the control of capital; Western capital. They are masculine public spaces; women have historically been excluded from their creation and from power over their operations. In the dance video texts the choreographers and film-makers use discourses which can either reveal or hide ways in which the bodies and spaces are constructed and invested with power. I intend to explore how these three texts reveal in a problematic way the possibilities for reclaiming a way of inhabiting the city, which is turning anonymous spaces into situated places, by negotiating their negative and alienating characteristics for women and 'others'. In the process these choreographers and film-makers demonstrate the possibilities for affirmation and empowerment through defining new relationships with cityspaces.

Mutual definition of bodies and cities

In each of these dance video texts the dancers' bodies interact with the city in different ways. These interactions, which involve the physical contact of the dancers' bodies and the fabric of the city, are explored focusing on the mutual construction and definition of bodies and cities. Elizabeth Grosz's notion of the body as a surface of inscription, which she derives from Nietzsche, Foucault and Deleuze, can inform an analysis of these texts in the sense that the power invested in the cityspaces inscribes movement patterns on the women's bodies in different ways. Grosz is 'concerned with the processes by which the subject is marked ... or constructed by the various regimes of institutional, discursive and non-discursive power' (1995: 33). These values and regimes of power, determined by the ways in which the world is seen and conceived, include strands of this conception and construction such as gender. In this inscriptive process 'the body's boundaries and zones are constituted ... through linkages with other surfaces and planes' (Grosz, 1995: 34). In the dance videos the boundaries and zones of the performers' bodies are partly constituted through their linkages with the surfaces and planes of the city. In this sense bodies and cities are 'mutually defining' (Grosz, 1995: 108).

In this chapter I am arguing that through this process of mutual definition the dancing bodies become subjects and the spaces of the city settings become places. The mutual definition which occurs can affirm identities for the women contrary to what is expected of them in the cityspaces revealing the 'complex web of power, space and difference' within which the 'public spaces of the city and the private lives of its inhabitants must be understood' (Bell and Valentine, 1995: 113). The particular ways in which the dancers' bodies and the urban environments in which they are performing 'mutually define' each other are now examined, focusing on the interactions of dancers with the surfaces of the city in each of the three dance video texts.

Muurwerk

The choreography of *Muurwerk* consists largely of repetitive phrases of rolling, spinning along and jumping up and kicking the walls of the deserted Brussels alley, which appears drab and lonely in this black and white film. There are very few moments when one or other part of Roxanne Huilmand's body is *not* in contact with the alley's surfaces. The rhythmic phrases of rolling and spinning along the walls of the alley are interspersed with Huilmand rolling across the alley floor into the gutter, defiantly revealing her substantial white knickers, and with her circling round on the ground from sitting to rolling onto hands and feet. This literal inscription of body and city is emphasized in three ways: through camera close-ups; choreographically, through persistent repetition and dynamic contrasts of speed and body tension; and aurally, in the soundtrack where the brushing and scraping of body on walls and ground is heard.

As Huilmand repeatedly slides, drags and brushes her body up against the stone sides and pavement of the alley, she seems at times almost to be making love to the buildings as she rubs her torso and pelvis rhythmically up and down the walls. The camera by closing in on her movements emphasizes the intimacy of the space as it caresses the alley walls and floor. The cracks, nooks and crannies become known as they are repeatedly traversed by Huilmand's body. The filming and Huilmand's performance construct and play with 'femininity' in various ways. Huilmand veers between being seductively sensual – when the camera lingers on a close-up of her face as she brushes her hair behind her ear – and aggressively violent as she repeatedly runs, jumps up and kicks against the alley walls. She teases and titillates through the dynamics and pace of the choreography by very slowly repeating and building up patterns of rolling or spinning along the walls only to abruptly interrupt and change the action before a climax is reached. In this context I would suggest that, as Grosz (1994) states, 'the practices of femininity can readily function ... as modes of guerrilla subversion of patriarchal codes' (p. 144). It is the brazen confidence of Huilmand's apparently fearless performance of a limited but thoroughly known and worked-through vocabulary, etched with a subtle playfulness, that enables her to transform the normally alien space of an inner-city alley into, for her, an intimate place of play and fantasy.

The choreography and filming of Huilmand's performance make it seem rebellious, and new meanings are created for this body in this cityspace as they mutually define each other. The possibly expected reading of the bare facts of the situation – a young woman, dressed in a sleeveless, low backed, short-skirted frock, alone in a deserted alley, often literally pinned to the walls or floor – is averted by the choreography. Rather than appear vulnerable, as might be expected, Huilmand, because of her defiant performance, seems to be making a different sort of statement about her ability to confidently occupy and inhabit the space of the alley. Her bodily brushes with the fabric of the Brussels alley are unremittingly flung in the audience's face. Her performance, combined with the choreography and filming, has overcome the expectations associated with her gender in such a cityspace, and transformed it in the process into a non-threatening place of surfaces and textures for rolling and spinning along.

The insistent repetition of what appear to be sometimes angry, sometimes mesmeric, caresses and collisions with concrete in *Muurwerk* flies in the face of any suggestion of vulnerability, which rather looks as if it is being worked out of the dancer's system cathartically. When Huilmand rolls into the gutter and her skirt flies up revealing her knickers, she seems to be defiantly saying, 'So what!' The choreography and filming open up possibilities for change. Huilmand seems at least partly able to create her own labyrinth and, through this, discover new movement and identity possibilities and the potential of new relationships with the city and its spaces. She appears to be taking possession of the space, reclaiming it for herself. In this sense Huilmand and Kolb, the film-maker, are working with discourses of the body and film which reveal ways in which Huilmand's body and the space of the alley are constructed and invested with power. They reveal problematically the possibilities for reclaiming a way of inhabiting the city, which are turning this anonymous alley into a familiar, situated place. This is possible through the choreography and filming of Huilmand's performance which have negotiated the negative and alienating characteristics for women and 'others' of this cityspace. The boundaries and zones of Huilmand's body are partly constituted through their linkages with the surfaces and planes of the alley, which in turn construct it as knowable and negotiable. It is in this sense that bodies and cities are 'mutually defining' (Grosz, 1995: 108).

Step in Time Girls

First impressions of *Step in Time Girls* suggest that the three women, albeit from different periods, are all contained and trapped by the cityspaces they inhabit. They appear subjected by them and made powerless by being forced to repeat movement vocabularies that suggest frustration with their environment and situations. This is apparent when the women repeatedly rock back and forth in chairs, roll along walls and windowsills and stride back and forth along table and windowsill. Their containment, alienation and isolation are emphasized because they are often seen in relation to the windows, their link with the outside world. They look out of them, pace up and down in front of them and dance on the windowsill. They are also viewed from the outside; as the camera recedes, minute pictures of private lives are glimpsed, framed by the numerous identical cell-like windows of a block of flats. The space of the flat is coded 'domestic' by aspects of the choreography; the dogged repetition of tasks that the women perform, such as opening cupboards and drawers, returning a baby's spoon to its highchair tray and running up and down stairs. The choreography abstracts and emphasizes the movements through development, exaggeration and, particularly, through repetition. Different parts of the body, such as shoulders, backs and feet contact the furniture and walls and repetitive rhythms are often built up, giving impressions possibly of obsession and frustration. The movement patterns these women perform are those which the city has inscribed on their bodies through the power invested in the cityspaces that they inhabit. The choreography and filming of *Step in Time Girls*, by revealing ways in which urban spaces contain and trap, gender them traditionally feminine and render the inhabitants, also gendered feminine, powerless to a certain extent.[3]

This is not the whole story, however. Through the repetition of these movement patterns the women appear to go beyond the limitations of the cityspaces and make them their own. For example, just over halfway through the piece the 'contemporary' woman rolls along the floor of the flat and up to sit on a chair and lean back briefly. She then jumps to sit on a table where, spinning round on her bottom, she swings her feet up to stand on the table against the wall, from where she looks down at the space she has just traversed. The whole phrase is then reversed and repeated twice. With the reversal and repetition of the choreography a rhythm and phrase are established which appear to be performed and enjoyed for their own sake. The sheer energy and momentum of the dancing takes over. Jumping onto the table, spinning round on it and jumping off again are actions that, because of their audacity in the context of the domestic space in which they are performed, can be seen to be empowering and affirming, in the sense that the dancer is reclaiming and taking over this urban space for herself. It has become her place.

There are several other examples of choreography where the women appear to exceed the constraints of the space. All three women literally throw their jackets across the space of the flat at one point; the 'contemporary' woman lobs hers up against a window, picks it up and repeats the gesture, three times the 'wartime' woman hurls her jacket at a wall and then 'catches' it on her back as it slides down the wall, and the 'Victorian' woman flings her jacket across the room onto a chair. Straight after this the 'wartime' woman spins along a windowsill, jumps up to sit on it and then comes down and, whilst still holding the upright window support, jumps her hips up high in the air using her other hand to push up from the windowsill; this is then repeated. Immediately afterwards the 'Victorian' woman is seen silhouetted against an open window rocking back slowly in her chair, balancing on the back legs only, until she tips a little further back and the chair leans against the wall. She then returns forward to repeat the movement – in all it is repeated fourteen times. The last eleven times she adds a movement of her arms rising up as her body tips back, adding to the momentum. The camera focuses on her feet pushing off the floor and then dangling in mid air as she holds the balance, and on her hands as they are held suspended above her. The movement is slow; it becomes increasingly mesmeric with repetition.

At these moments it seems as if the women have appropriated the space of the flat for their own personal gymnastic or exercise routines. They are enjoying rocking, stretching, balancing and jumping up using the walls, floor and furniture as apparatus to assist them. This seems particularly the case when much earlier in the video the 'contemporary' woman literally bursts through a serving hatch, head first, stretching her arms to the side horizontally, looking, in her black top, like a thirties gymnast. She balances on her stomach briefly and then drops her upper body down to repeatedly swing it along the wall below. In the next shot she is balancing on her thighs bracing herself with her feet on the upper lintel of the hatch still swinging her torso from side to side along the wall below. She then rocks her upper body away from and towards the wall several times, before lowering herself through the hatch down onto the floor in a somersault to bring her fully into the room. These acrobatic and athletic feats go way beyond what is expected of the women in this domestic interior. Consequently this cityspace is transformed from a containing unsympathetic environment to an imaginative playground where the women revel in the experience of their bodies contacting and rebounding off walls and furniture. The fact that the three women from different periods perform similar movements in the same block of flats also suggests linkages and connections through time that add to the dancers' construction as subjects. The urban spaces of the London flat and its environs connect the three women as subjects in the city. In these ways *Step in Time Girls* reveals, through its choreography and filming, possibilities for reclaiming cityspaces as places for the women. The bodies and the spaces mutually define each other to create new meanings.

Duets with Automobiles

Duets with Automobiles opens with a shot of St Paul's Cathedral seen through a window frame with a woman, her back to camera, silhouetted in the foreground, looking out. St Paul's marks the city skyline out as London, historical London, part of the British cultural heritage. After a few seconds the woman slowly half turns to camera and then looks back at the view. She looks Indian. Placing the woman in the frame might suggest connections with British imperial history. The identity of London, and importantly the city of London, and of the Indian woman, are constructed by their juxtaposition and the role that race and history play in investing them with power. The grandeur of St Paul's architecture could be said to dominate the scene, investing London and British culture with power *over* the Indian woman. However, the next sequence of shots might suggest other readings. After the woman turns to camera, running her hand along the window ledge, there is a full-length view of St Paul's, through a glass wall this time, in a golden, possibly evening, light. The Indian dancer, back to camera, silhouetted against the cathedral, runs the palm of her hand down over its surface through the glass. St Paul's, seen in the sunset, might suggest the remnants of an imperial past and that the dancer is contemplating the future of a new hybrid existence in a city which is hers. In this reading, the power relations are different. Jeyasingh's choreography challenges the entrenched cultural values that assume a binary divide of colonial authority and anticolonialist opposition. This opening sequence sets the tone for *Duets with Automobiles* by juxtaposing the historic formality of the recognizably British public building with the intimacy of the looks and gestures of the Indian dancer, without placing these two elements in opposition. Instead, the dancer softly caresses the building, subtly stating her relationship with the space and opening up possibilities for affirming her urban identity in the process.

The downward movement of the palm of the hand on the glass wall, first glimpsed in this opening scene, is a repeated motif throughout *Duets with Automobiles*. It is usually seen on glass walls, but also on the stone walls, pillars and wooden and marble floors of the London office buildings. It acts as a kind of leitmotif that, together with the rest of the choreography and filming, marks the dancers' presence in an almost ritualistic fashion in these cityspaces endowed with power associated with capital. The three dancers build an intimate relationship with the office buildings as, alongside traditional Baratha Natyam phrases of dance, they also roll, kneel and place their heads on the floor as if asleep, run and hug rounded stone pillars, lean on them and slide the sole of a foot down them, as well as placing their heads on them as if sleeping. They also perform more naturalistic movements such as leaning on window ledges. These points of contact of bodies and surfaces of the city buildings are emphasized by close-up shots, sounds, such as those of bare feet slapping out Baratha Natyam rhythms on wood and marble floors, choreographic repetition and other visual devices such as silhouettes. By their actions the dancers transform the office buildings, and through their close contact with the floors, pillars, windows and walls of these cityspaces turn them into intimate, homely places.

There is a sense in which the presence, filming and choreography of the Indian classical dancers in the London office buildings – in these centres of capital and power – is a bold invasion of cityspace which is in turn transformed by what occurs in it. One of the ways in which the space is transformed occurs through a particular bodily relationship with the architecture when the dancers repeatedly caress and hug pillars. On the one hand, this literal embrace of the modern city building appears to be suggesting a metaphorical embrace of contemporary urban life; there is a sense in which intimate gestures such as these show the dancers making themselves at home in these offices, inhabiting them, humanizing them and making imaginary homelands out of them.[4] However, Jeyasingh has indicated that she also had in mind the image of a *yakshi*, a female tree spirit often seen in classical Indian architecture touching a pillar or the building with a part of her body, usually her feet. This 'young fecund sort of female often with very big breasts' would be standing, carved into pillars of buildings symbolizing a source of energy since 'she's the creative principle making the tree or pillar come to life' (Jeyasingh, 1997). Throughout *Duets* references are made to this tree spirit through dancers sliding their feet down or hugging pillars. Bound up in these gestures are ideas about female energy or strength from the *yakshi* that relate to notions of female subjectivity. The gestures also contain references to classical Indian architecture and they are transformed and translated by their reinscription on the pillars of a contemporary London building. By hugging and caressing the pillars in this way the dancers could be said to be enacting a ritual which transforms the building by breathing energy into it through reference to *yakshi* and, in the process, raising and generating power. The dialogues and interactions between these different cultural ideas and images illustrate the complexities of subjectivity being explored and the scope for new meanings. In this sense the choreography reveals possibilities for reclaiming the city office buildings as places, and constructing the British Asian dancers' identities as contemporary, empowered subjects of the city.

Another example of the ways in which Jeyasingh controls and redefines these cityspaces and the power associated with them occurs towards the end of *Duets with Automobiles*. A dancer kneels down on the classically (in the Greek sense) geometric designs inlaid in the marble floor of a wide office corridor and traces with her palm some of the diagonal and circular lines of the pattern. As she does so, she appears, in an act of ritualistic reverence, to be making connections with the precise curved and linear pathways she has just traced in her performance of classical Indian dance. The 'mapping' refers to a parallel historical journey whose geometrical traces remain in the cultural products of dance, architecture and design. However, Jeyasingh is also making a statement about the contemporary fluidity and movement of cultures and borders that currently allow such comparisons to be made. Talking of her work she has said she is concerned with 'the changing borders raging all around', the 'dynamism of journeys', and 'a pattern of belonging that is multi-dimensional' (Jeyasingh, 1995: 191–3). The dancer's hand on the marble floor traces connections between race, space and power. This therefore has a similar function to the opening scene described earlier, where the dancer traced her hand across the image of St Paul's Cathedral. The mapped patterns she makes have resonances with the positioning, controlling and colonizing characteristics of real maps, inherited from history and demonstrated in dance, architecture and design. But, by pointing to the fluidity of contemporary urban existence, they also challenge notions of identities and heritages fixed by borders. It is in this sense that the choreography and filming reveal possibilities for reclaiming the space and imbuing it with new meanings. As Sanjoy Roy says of Jeyasingh's work: 'By getting "under the skin" of cultural boundaries, by loosening the links between race, place and culture, her work can speak to the experience of diaspora' (Roy, 1997: 83).

Perhaps one of the most memorable moments in *Duets with Automobiles* in terms of the ways in which the choreography reinvests the city spaces with a new kind of power is when the dancer, who has just traced the pattern on the marble floor, rises to begin travelling forwards towards the camera framed by the walls and ceiling of a long corridor, emphasizing a sense of perspective as it recedes into the far distance. As she advances down the corridor slapping out Baratha Natyam rhythms with her feet on the marble floor, she is joined by first one and then the other of the three dancers. They move forward in unison, the sounds of their feet mingling with Orlando Gough's collage of rhythmic voices in the accompanying music. Their advance is emphasized by forceful arm gestures thrust directly towards the camera, sometimes led with a clenched fist. Jeyasingh has indicated that the clenched fist was an important motif throughout *Duets with Automobiles*, signifying strength and determination (Jeyasingh, 1997). The three dancers complete their powerful surge forward, exiting to camera. The impact of the phrase is further enhanced when it is repeated a little later in the piece, with two slight changes. The first is that the sequence is preceded by a long shot of the corridor, further emphasizing the grand proportions of the neo-classical concourse against which the dancer initially is hardly visible. The second change is that this time the dancers perform unaccompanied, allowing the sounds of their feet on the floor to be heard uninterrupted, reinforcing the impetus of their advancing steps. The forward approach of the dancers seems relentless; it leaves an impression of female power and strength that transforms this previously male-dominated centre of capital. The choreography and filming invest the cityspaces and the bodies of the dancers with power and the possibilities of new meanings and new identities. The spaces become particular places and the bodies, empowered subjects, where the conflicts and contradictions of gender and race are evident through the dancers' range of interactions with the architecture.

Conclusion

By placing dances in city settings, choreographers and film-makers are opening up possibilities for exploring the ways in which cities and bodies can
mutually define and construct each other. It would be possible to use the city
with its multiplicity of signs and images as an exciting arena for filmed dance
performances without taking account of the ways in which city settings and
spaces are inevitably invested with power; simply using the city as a backdrop
in an uninformed manner that takes it for granted. Some choreographers and
film-makers have done this (see Briginshaw, 1997). However, the choreographers and film-makers of the three dance video texts examined in this
chapter, in different ways, have gone beyond simply using the city as a backdrop. Through their choice of settings, the ways in which they have been
filmed and the choreography that has been set in them, they have revealed
some of the ways in which bodies produce themselves in space and in turn
produce space (Lefebvre, 1991).

In *Muurwerk* the deserted drab Brussels alley is gradually transformed
through the solo female dancer's almost hypnotic repetitive phrases of rolling,
spinning, sliding and kicking. This cityspace, with its associations of male
freedom derived from the nineteenth-century *flâneur*, is reinvested with a distinctly feminine kind of potency as, through her dancing, Huilmand weaves
her own thread of power and takes on the mantle of a contemporary
flâneuse. The anonymous space becomes an imaginative situated place of
play; of musings and teasings, suggesting new meanings.

In *Step in Time Girls* the rooms of the London flat and its environs, which
contain and confine its three female inhabitants from different eras, through
the choreography and filming reveal the potential for change. The women's
performance, by first reiterating the bounds of the spaces they inhabit, sets up
possibilities for transcending these limits. It soon becomes apparent through
the choreography and filming that these women have the potential to go
beyond the constraints of their environment. Through the repetition and
development of choreographic phrases, which increasingly visibly become
enjoyed for their own sake, the performers, by revelling in the sheer physicality of their movement experience, each in their own way, reclaim the spaces
for themselves. In so doing they each also affirm their own identities in relation to the places they inhabit.

It might be countered, however, that all the women in *Muurwerk* and *Step
in Time Girls* are doing is reiterating and repeating movement patterns
inscribed on them by the cityspaces that are invested with power; that this
repetition is reinforcing an already negative situation; that all they are doing is
making do and surviving in a situation in which they ultimately remain powerless. Although I have argued that through the repetition of their
choreography they transform the cityspaces, it might not yet be clear enough
how they can achieve this. Judith Butler, in a discussion of the constitution of
subjectivity – of the 'I' in language, and the roles that 'sex', 'race' and 'gender'
play in such a constitution – states that, just because of the associations of

such terms with sexism, racism and heterosexism, with oppressive regimes of power, this does not mean that 'we ought never to make use of such terms, as if such terms could only and always reconsolidate the oppressive regimes of power by which they are spawned' (1993: 123). 'On the contrary,' she states, 'precisely because such terms have been produced and constrained within such regimes, they ought to be repeated in directions that reverse and displace their originating aims' (1993: 123). Butler is perhaps here invoking Foucault's notion of 'reverse discourse' that he suggests was made possible, for example, in the nineteenth century when a whole series of discourses on homosexuality appeared that allowed 'homosexuality to speak in its own behalf, to demand that its legitimacy…be acknowledged' (1978: 101). In other words the repetition of discourses can sometimes, as Butler states, 'reverse and displace their originating aims' (1993: 123). She continues: 'the force of repetition … may be the paradoxical condition by which a certain agency… is derived' (1993: 124). So the women in *Muurwerk* and *Step in Time Girls* might, through their repetition of patterns of movement inscribed on them by the oppressive cityspaces, be enacting a 'rearticulation of kinship' which Butler cites in relation to the film *Paris is Burning* which, she continues, 'might be understood as repetitions of hegemonic forms of power which fail to repeat loyally and, in that failure, open possibilities for resignifying the terms of violation against their violating aims' (1993: 124). So by 'failing to repeat loyally', by playing with the movement patterns that have been inscribed on them, by enjoying the physicality of them for their own sake, the women in these dance texts are perhaps resignifying them and the oppressive cityspaces that have inscribed them in the first place. This is how they can transform those spaces.

In *Duets with Automobiles* there is perhaps the most potential for transformation because the juxtaposition of the bodies and cityspaces concerned is so radical. Placing three Baratha Natyam dancers in offices which have derived power from controlling capital and are hence at the heart of the city seems at first sight possibly an extreme imposition. However, the innovative choreography and filming, which are sympathetic and sensitive to the aesthetic concerns of the architecture, and the dance vocabulary reveal the potential for redefining the cityspaces. The performers, through their dancing and the ways in which they are filmed, seem to enact a ritual in these cityspaces invested with power. They subtly, yet firmly, assert their presence and, in so doing, they are empowered, and their identities as contemporary urban British Asian women are affirmed.

The examination of these three dance video texts has shown that because both bodies and cities can be seen as constructed, there is always potential for change, and that when bodies and cities co-exist and interact, there is also potential for their mutual construction and definition. In this context it is also important to realize that 'there is nothing intrinsic about the city that makes it alienating or unnatural' (Grosz, 1995: 109). The ways in which these dance video texts have facilitated the construction of bodies and cityspaces as subjects and places have indicated the potential for new meanings. As Grosz

(1995) states: 'the city is both a mode for the regulation and administration of subjects and also an urban space in turn reinscribed by the particularities of its occupation and use' (p. 109).

References

Bell, D. and Valentine, G. (eds) (1995) *Mapping Desire*, Routledge, London.

Best, S. (1995) Sexualising space. In Grosz, E. and Probyn, E. (eds) *Sexy Bodies*, Routledge, London.

Briginshaw, V. A. (1997) 'Keep your great city Paris!' – the lament of the Empress and other women. In Thomas, H. (ed.) *Dance in the City*, Macmillan, London.

Butler, J. (1993) *Bodies that Matter*, Routledge, New York.

Foucault, M. (1978) *The History of Sexuality*, Penguin Books, London.

Grosz, E. (1994) *Volatile Bodies*, Indiana University Press, Bloomington, Ind.

Grosz, E. (1995) *Space, Time and Perversion*, Routledge, London.

Harvey, D. (1989) *The Condition of Postmodernity*, Blackwell, Oxford.

Jeyasingh, S. (1995) Imaginary homelands: creating a new dance language. In *Border Tensions: Dance and Discourse Conference Report*, University of Surrey, Guildford.

Jeyasingh, S. (1997) Shobana Jeyasingh in interview with the author.

Lefebvre, H. (1991) *The Production of Space*, Blackwell, Oxford.

Raban, J. (1974) *Soft City*, Jonathan Cape, London.

Roy, S. (1997) Dirt, noise, traffic: contemporary Indian dance in the Western city; modernity, ethnicity and hybridity. In Thomas, H. (ed.) *Dance in the City*, Macmillan, London.

Rubidge, S. (1993) in Shobana Jeyasingh Dance Company *Making of Maps Resource Pack*, Shobana Jeyasingh Dance Company, London.

Rubidge, S. (1995) *Romance ... with Footnotes*, monograph accompanying the Shobana Jeyasingh Dance Company *Romance ... with Footnotes* video, Shobana Jeyasingh Dance Company, London.

Rushdie, S. (1991) *Imaginary Homelands: Essays and Criticism 1981–1991*, Granta Books, London.

Notes

1 This chapter draws on material and ideas from an earlier essay '"*Keep your great city Paris!*" – the lament of the Empress and other women' published as a chapter in Helen Thomas (ed.) (1997) *Dance in the City*, Macmillan, London. I should like to gratefully acknowledge help given to me by Ramsay Burt who read drafts and gave advice; responsibility for the end result however remains entirely my own. Other examples of recent dance texts that use cities as settings or inspirations include *Freefall* (1988), chor. Gabi Agis; *The Lament of the Empress* (1989), chor. Pina Bausch; *Palermo, Palermo* (1990), chor. Pina Bausch; *Circumnavigation* (1992), chor. Norbert and Nicole Corsino; *Topic II* and *49 bis* (1992), chor. Sarah Denizot; and *Dark Hours and Finer Moments* (1994), chor. Gabi Agis.

2 When discussing *Duets with Automobiles*, Jeyasingh stated she would no longer use the word 'icon', that it was too closely connected with a particular traditional image of an Indian dancer. She has 'moved on' from that idea. She also said that if she were making *Duets* today she would change the costume and make-up, removing the *bhindis* from the dancers' foreheads. The costume she would make

less silky and in a less bright colour. She stated that it would be enough for her that the dancers looked Indian through their skin colour (Jeyasingh, 1997).

3   Space tends to be feminized deriving from Plato's notions of the female *chora*. In Plato's view space is conceptualized as 'a bounded entity', 'a sort of container' associated with the female body, particularly with that of the mother (Best, 1995: 182). He states: 'it [the receptacle/space] ... is a kind of ... plastic material on which changing impressions are stamped by the things which enter it making it appear different at different times ... We may use the metaphor of birth and compare the receptacle to the mother' (quoted in Best, 1995: 184).

4   Jeyasingh is well aware of Salman Rushdie's (1991) text *Imaginary Homelands*. It is quoted in a monograph that her company has published about her work and she incorporated Rushdie's title in a title of one of her own conference papers (Jeyasingh, 1995).

Chapter 9

Moving through the city

Tim Edensor

Just as there are no places without the bodies that sustain and vivify them, so there are no lived bodies without the places they inhabit and traverse. (Casey, 1996: 24)

Like all big cities it consisted of irregularity, change, sliding forwards, not keeping in step, collisions of things and affairs, and fathomless points of silence in between, of paved ways and wilderness, of one great rhythmic throb and the perpetual discord and dislocation of all opposing rhythms, and as a whole resembled a seething, bubbling fluid in a vessel consisting of the solid materials of buildings, laws, regulations, and historical traditions. (Musil, 1954: 4)

As a centre for the intensification of a host of flows, the city is also a site for the ceaseless motion of people across and beyond its boundaries. In this chapter, I want to consider the ways in which bodies move through the city. I explore the ways in which movement can be considered as a form of performance, and then go on to look at how cities and bodies can be regulated. I follow this with an exploration of the tactics through which pedestrians can evade urban disciplinary processes, and finally, I examine the experience of pedestrians moving through an Indian bazaar.

In cultural studies, the city has characteristically been discussed in terms of its representation, its marketing, and the cultural forms and practices which it contains. However, the relationship between bodies and the city has been somewhat neglected. Bodies are not only discursive concepts and subjects of scientific analysis but are also the means through which we experience and feel the world. Yet this dominance of discourse and representation extends to discussions about the body in the city where conceptual explanations and metaphors tend to eclipse the lived bodily experience of city life. However, the senses act to inform presence and engagement to constitute a 'being-in-the-world' (Csordas, 1994: 10). And whilst there are multiple texts, representations, illusions, myths and dreams through which the city may be interpreted, bodies act upon the city, inscribing their presence through movement in a process of continual remaking. Accordingly, social relations are not only inscribed upon the body, but are produced by it. As Feld notes, 'as place is sensed, senses are placed; as places make sense, senses make place' (1996: 91). In order to explore the parameters within which bodies express themselves in the city, I want to consider movement as a species of performance.

Moving/performing in the city

It has been argued that in the postmodern city, the body has become 'primarily a performing self of appearance, display and impression management' (Csordas, 1994: 2). Certain writers (Geertz, 1993; Goffman, 1959) state that social life is thoroughly dramaturgical, and that we are constantly playing roles according to the social contexts in which we find ourselves. Thus, social action is inevitably mannered and stylized, and aims for coherence and consistency (Chaney, 1993). By extension, body posture and physical movement through space can be considered as a choreographed form of performance whereby bodies communicate meaning through stylized movements and stances. Sennett (1994) shows how the ancient Greek city-dwellers read styles of walking to identify character. An erect, steady, purposive yet slow gait signified the high status of the 'well-bred', whereas slouching connoted inferiority and sexual passivity. Thus, social identities are 'signalled, formed and negotiated through bodily movement' (Desmond, 1994: 34), through rhythms and gestures which act as markers for gender, racial, ethnic, class, and subcultural allegiances. These culturally coded patterns of behaviour, grounded in habitus and the dispositions that evolve around class, gender and ethnicity, produce distinctive gaits, ways of speaking, dress and demeanour which articulate shared forms of understanding. Performances have been described as a 'discrete concretisation of cultural assumptions' (Carlson, 1996: 16) which are situated in the relationship between performer and place, suggesting a geographically and historically located practical knowledge.

Although it is often implied, following Goffman (1959), that performance concerns the strategic, purposive management of impressions, conscious intentions are not always evident as part of our actively 'being-in-the-world'. For it is difficult to limit the expressivity of our actions, and their effect upon others. We cannot restrict the way actions are interpreted. And we also act unreflexively in organizing sequences of movements in routine, everyday practices. The production of these unreflexive, embodied forms of practical knowledge engenders patterns of communal association and a sense of dwelling (Seamon, 1979: 58–9). The co-ordinated, collective patterns created by these everyday, habitual gestures and movements are described by David Seamon as 'place ballets', inscribed upon locales to constitute a compendium of regular and routinized dances. Whether the repetitive movements of gendered tasks or the monotonous work undertaken by some workers, such choreographies have been compared to the patterns of migratory birds as 'they trace and retrace the same restricted set of options' (MacDonald, 1997: 153).

Edward Schieffelin (1998) describes performance as an interactive and contingent process which succeeds according to the skill of the performance, the context within which it is performed and the way in which it is interpreted by an audience. The importance of performance as a communicative tool is recognized by authorities, who inculcate and regulate embodied habits through ritualistic performance to reinforce and re-encode hegemonic meanings. Yet even the most delineated performance, which depends upon the elimination of any deviation for its efficacy, must be re-enacted in different conditions. Each separate performance can never be exactly reproduced and fixity of meaning must be continually striven for. Moreover, its reception cannot always be controlled by the performer. However, there are a range of factors which tend to circumscribe performances, which act to constrict the extent of people's physical exploration of the city and their forms of bodily expression.

Most obviously, forms of surveillance and reflexive monitoring may restrict the scope for movement and limit the range of activities. The extent to which directors or choreographers (in the form of police, security guards and group leaders) channel movement impacts upon the range of manoeuvres available. Likewise, where collective movements are enacted, peer pressure to follow a co-ordinated choreography may oblige participants to conform to group norms. Additionally, the skill of the performer in conveying the intended impression may depend upon the level of rehearsal and practice, the development of the right body-image through training and the acquisition of techniques. In any case, the efficacy of the performance is equally reliant upon the cultural context within which it is enacted and the ability of the audience to share the intended meaning.

An equally important factor in setting the parameters of performance is the nature of the stage, for performances are enacted in 'physically and symbolically bounded space' (Chaney, 1993: 18). In the organization of the city, 'architecture functions as a potential stimulus for movement, real or imagined'. For instance, 'a building is an incitement to action, a stage for movement and interaction' (Yudell, 1977: 59) with a wealth of possible niches, paths, stairs, openings, tactile surfaces which invite physical exploration. David Harvey describes the postmodern city, with its proliferating signs and settings, as akin to a 'theatre, a series of stages upon which individuals could work their own distinctive magic while performing a multiplicity of roles' (1989: 5). The nature of the stage varies from the carefully managed arena which contains discretely situated objects around which performance is organized, to those theatres with blurred boundaries, or those cluttered with other actors following incomprehensible scripts, full of shifting scenes, juxtapositions and random movements coming from a range of angles. In fact, urban stages are incorporated into multiple, overlapping spaces, and thus they are fluid entities whose meaning changes and is contested by performers. Accordingly, there may be myriad forms of performance, following distinct roles, scripts, choreographies, group formations, instructions and cues. These moving performative processes ceaselessly reconstitute the symbolic values of sites and reproduce them as dramaturgical spaces. Whilst meaning always overlays the material organization and form, informing practices and assumptions which reproduce space, space also produces action. The influence of spatial form – for instance in spectacular monumental spaces – might channel movement, might suggest permanence in the erasure of other memories, other identities, thus constraining performance and meaning. However, whilst the range of performances executed varies from the disciplined ritual, to the parodic and ironic, to the wholly improvised, most performances are 'regulated' improvisations, as I will now go on to discuss.

Regulating bodies and cities

The movement of humans through cities is very diverse. Moving bodies are motivated by duty, necessity, the maintenance of the body through exercise, a search for the muse, fantasy, *flânerie*, contemplation, collection, drugs, adventure, to produce a wealth of performances and choreographies: group marches, individual wanderings, incursions, trespass, purposive missions. Yet certain kinds of expressive movements – hopscotch, bicycling, loitering – seem to be disappearing from many Western city streets.

The postmodern city is 'constructed as labyrinthine, free-flowing and uncontrollable, but also as containing and trapping' (Briginshaw, 1997: 35). It is a space of fear and alienation, but also of sensual, fantastic and erotic possibility. As a place of comings and goings, the city promises the possibility of multiple connections, and yet it is also mapped, surveyed, colonized, possessed and regulated. In the city, then, there is ambiguity, complexity and multiplicity. In order to examine the ways in which movement through the city may be constrained, the diverse choreographies of the street, the opportunities for moving bodies to feel, express, experience, move towards the other, wallow in urban sensuality and flux, I want to distinguish between the different types of social and spatial ordering, and explore the processes by which urban space is subject to the contradictory desires for the immutable and the indefinite.

To differentiate between exclusionary and inclusionary spatial demarcation, Sibley (1988) distinguishes between 'purified' spaces which are strongly circumscribed and framed, wherein conformity to rules and adherence to centralized regulation hold sway, and weakly classified, heterogeneous spaces with blurred boundaries, in which activities and people mingle, allowing a wide range of encounters and greater self-governance and expressiveness. The delineation of a purified space tacitly identifies the 'outsider', the stranger or the 'other', as 'out of place' (Cresswell, 1996), simultaneously attempting to regulate the movements of these 'others' and constructing spaces to which they are assigned. These marginal spaces are typically represented as chaotic and dirty, the antithesis of 'purified' space. But paradoxically they are also imagined as realms of desire, permitting of interconnection, hybridity and possibility by virtue of their 'weak framing'.

The urge to construct purified spaces stems from modern attempts to accumulate, categorize and exhibit objects and bits of information under one roof, in public sites such as museums. Following ostensibly 'scientific' principles, these forms of ordering different cultures, artefacts and forms of knowledge require the authority of the expert and signify the attempt to banish the randomness of previous forms of display, such as the cabinet of curiosities, which contained heterogeneous objects arranged according to whim. Bennett shows how the organization of these spaces of knowledge was devised to attain 'new norms of public conduct' (1995: 24). Performative conventions and normative choreographies were co-ordinated by attendants and spatially guided by the layout of display cases, special rooms, information boards and room plans. This staging materializes a linear narrative in the design of exhibitions so that visitors not only consume the principles of classification, but also actively perform a unidirectional dance along devised routes. Thus, the museum encourages visitors 'to comply with a programme of organised walking' (Bennett, 1995, 186–7). The replacement of the cabinet of curiosities by the museum can stand as a metaphor for the evolution of modern processes of regulating space and the development of technologies for regulating bodies through stage directions and boundary-making.

In the same era, spaces of public entertainment, such as fairs and carnivals with their 'heteroclite objects', teeming activities (Foucault, 1986: 26), fleeting and surprising momentary sights and juxtapositions, were 'pruned and replanted at the margins of society' (Shields, 1991: 86). However, the removal of gross and vulgar elements is usually accompanied by a corresponding attraction towards them as objects of 'nostalgia, longing and desire' (Stallybrass and White, 1986: 191); they are sought out at a variety of sites and spaces. Capital has been quick to exploit this desire to escape the predictable, and has attempted to construct regulated urban spaces which offer a more designed heterotopia. Popular desires for the contingent, fragmentary, ever-changing and arbitrary aspects of the carnival have been co-opted by advertising and marketing strategists, who produce spaces which appear to promise a cornucopia but only offer a 'controlled diversity' rather than a realm of 'unconstrained social differences' (Mitchell, 1995: 119). Although designed to exclude the quotidian worlds of work and home, such spaces are familiar enough by virtue of their design codes, spatial organization, and the themes which emerge from films, pop music and advertising. For example, the heritage landscape, with carefully devised street furniture, information boards, reconstituted cobbles and so on, follows the imperatives of theatrical design. Producing a nostalgic simulacrum of urban living, refashioned areas conjure up the signs of urban vitality or the atmosphere of a festival. But instead, this imagery and manufactured ambience tends to produce 'sites of ordered disorder' (Featherstone, 1991: 82) satisfying what Chris Rojek calls 'the timid freedom of respectable leisure' (1995: 80). The impact upon the moving body of this spatial containment and commodification has been the encouragement of a 'controlled de-control of the emotions' (Featherstone, 1991: 78), which includes toned-down, self-regulated forms of physical expression.

The 'soft control' (Ritzer and Liska, 1997: 106) that holds sway in these regulated spaces constitutes a range of external techniques such as security guards and CCTV cameras which exclude 'undesirable elements', and internalized rules about what comprise 'appropriate' bodily postures, dress and voice modulation. Thus whilst the visual, the gestural and the sensual are all important modes of identity and expression in the city, order and exclusion have impacted upon such stylistic bodily practices (Squires 1994: 81). Rather than the 'motion and turbulence that makes the city so appealing' (ibid.: 83), unrestricted movement is discouraged in many of the new urban spaces. Congregating in groups is conceived as a threat to public order or as 'causing an obstruction'. Injunctions against sleeping or resting on benches and floors impose restrictions upon the movements and strategies of the homeless. Samira Kawash explains how simultaneously representational and material 'technologies of exclusion' produce a 'legitimate' public who have rights to their space as opposed to homeless 'usurpers'. The denial of any place which homeless bodies may dwell in or pass through generates a condition of 'perpetual movement' borne of placenessness, movement undertaken by the homeless not 'because they are going somewhere, but because they have nowhere to go' (Kawash 1998: 322–9). More broadly, activities such as 'loitering', 'hanging out' and lounging on the pavement are all deterred, considered potentially criminal, and particular activities such as 'cottaging' and 'begging' are particularly scrutinized. Sanctions range from the advice to 'move on' by security guards and police to more draconian stop-and-search police procedures when pedestrians are adjudged to be suspiciously 'out of place'.

Movement around cities is mediated by power relations, blocked by gendered, racialized, sexualized and classed notions of who belongs where. The corresponding production of purified spaces and the entrenching of notions of what bodily actions are permissible, are maintained by racist and sexual violence, and milder procedures which emerge out of suburban Neighbourhood Watch schemes. Mike Davis (1990) has graphically illustrated the restrictions on passage around Los Angeles by security-conscious inhabitants and the forces of order that protect them.

The politics of movement have been extensively contested in battles over the nature and purpose of space and the right to roam, between the police and the state on one side, and contemporary nomads such as 'New Age Travellers' on the other. Such conflicts are also to be found during mass migrations to symbolic sites for the purposes of temporary transformation, events including those where people wish to hold carnivalesque events such as raves and free festivals. Political concern at the explosion of desire for nomadic lifestyles resulted in the passage of the Criminal Justice Act, a catchall proposal. Specific measures designed to restrict movement are enshrined in Part V of the Act: assemblies on land at historic monuments are prohibited; the number of vehicles lawfully gathered in a field has been limited to six; powers have been given to local authorities to deal with raves, tribal gatherings and mass trespasses; and the obligation for local authorities to provide adequate caravan sites for travellers was removed (McKay, 1996). More improbably, particular rhythms were identified as culpable and, by extension, the styles of dancing which accompanied these rhythms were outlawed (subsection 1 (b) of section 63 of the Act): the legislation prohibited music including that which 'sounds wholly or predominantly characterized by the emission of a succession of repetitive beats' (ibid.: 164).

Subaltern social movements depend upon the temporary seizure and transformation of public space in order to transmit alternative symbolic meanings, which the regulation of space and movement denies (Mitchell, 1995). Trevor

Boddy (1992) has pointed out that collective political protests are increasingly carried out in the 'windswept emptiness' of formerly convivial symbolic squares and streets, now emptied out by the evacuation of commercial and civic functions. Whilst the presence of 'mall rats', 'beggars' and shoplifters testifies to certain forms of resistance in the new heterotopias, such opposition seems fleeting and gestural in the face of intensive regulation, which forces rebels to adopt more covert strategies. The kinds of strategies to survive, explore desire or pursue political aims are of necessity carried out on the run, depend on rapid movement. And consequently, practices such as cruising, bill-posting, graffiti-tagging, busking and soliciting are subject to tailored methods of surveillance.

In contradistinction to these furtive tactics, acceptable forms of movement include the organized walking typically performed by tourists (see Edensor, 1998), where established routes are predetermined and participants are guided towards the signs of the city, often a fixed set of signifiers that are replicated in promotional literature and tourist brochures. The imperative for smooth transit between attractions is facilitated by the rigorous maintenance of networks and tourist spaces to maximize the quest for commodities and experiences.

The importance of seamless passage is materialized in the way in which urban space is increasingly organized to facilitate directional movement by both pedestrians and vehicles by reducing points of entry and exit, and minimizing idiosyncratic distractions. Such designs, instead of being 'relational, historical and concerned with identity' (Auge, 1995: 107–8), produce realms of 'transit' as opposed to 'dwelling', sites of 'interchange' rather than a meeting place or 'crossroads', where 'communication (with its codes, images and strategies)' is practised rather than affective and convivial language. As Tim Cresswell (1997: 364) remarks, the state does not oppose mobility *per se*, but 'wishes to control flows – to make them run through conduits'.

Richard Sennett (1994: 15) argues that urban space has largely become 'a mere function of motion', engendering a 'tactile sterility' where the city environment 'pacifies the body'. The imperative to minimize disruption and distraction for pedestrians and drivers means that movement is typified by rapid transit without arousal. In the case of the car, the physical efforts – the 'micro-movements' – used to negotiate space are minimal, producing a desensitized effect. Indeed, although the 'desire to move freely' has been realized by those with enough time and money, this 'has triumphed over the sensory claims of the space through which the body moves'.

This speed of movement is conducive to the formation of an 'image repertoire' through which moving people can quickly distinguish between signs which indicate points of attraction and direction, picking out informative signifiers like traffic signs, shop logos and advertisements. Shops are commonly designed to strategically maximize the behaviour of consumer-pedestrians; retail managers construct elaborate window displays – 'three dimensional billboards' – to attract custom. Once in the shop, techniques such as the 'decompression zone' are applied so that would-be customers continue at 'normal' pedestrian speed for five steps to encourage acclimatization. Spatial layouts are designed to maximize selling potential, for instance by placing the 'best offers' on the right to capitalize on the near ubiquitous tendency of customers to veer rightwards, and by 'fast-tracking', where hard-surfaced pathways lead customers speedily along pre-arranged routes, where places to linger are devised by installing soft carpeting surrounded by goods. Although claims made for these techniques are frequently hyperbolic, the expenditure on the 'scientific' construction of routes, surfaces and distractions is devised to create an environment in which movement is organized.

This facilitation of movement partly captures the grander designs of what Trevor Boddy (1992) calls the 'analogous city', a set of 'new urban prosthetics', a system of smooth and sealed walkways, escalators, bridges, people-conveyors and tunnels. Escalators and moving walkways are designed according to very specific notions of 'efficiency', controlling the direction and partly the pace of pedestrians' movement, leading them to desirable sites or past designated sights, restricting physical communication and adjustment (Sui, 1999). Constituting a comprehensive movement system to link work, recreational and commercial spaces, these systems produce an aesthetic and material form typified by 'incessant whirring', 'mechanical breezes', 'vaguely reassuring icons', 'trickling fountains', and anaesthetic qualities like low murmurings and insensate movements. Simulating urbanity but filtering out 'troubling smells and winds', these staged environments segregate classes and ethnicities, and are the scene of a distinct repertoire of bodily expression: 'never a clenched fist, a passionate kiss, a giddy wink, a fixed-shoulder stride' (Sui, 1999: 123–4).

Another way of directing movement is through an aesthetic monitoring where anything considered to be an 'eyesore' is removed and notions of cleanliness and the 'exotic' concoct the requisite combination of standardization and 'otherness'. These 'themed milieus', designed according to symbols derived from media cultures to constitute a 'sceneography' (Gottdiener, 1997: 73), are conducive to browsing and the passing of shops 'in review'. Such theming imposes a predictable spectacle through the exclusion of 'extraneous chaotic elements [reducing] visual and functional forms to a few key images' (Rojek, 1995: 62) which echo the 'well ordered, pristine and pure' imaginary worlds fostered through advertising (Sibley, 1988: 415).

Moving through these spaces, pedestrians must be content with 'the tactile but not too physical interaction with a crowd, the sense … of something happening beyond the close world of oneself' (Shields, 1992: 102). Thus, forms of movement such as 'browsing' and 'grazing' are encouraged. The narrow scope for improvised or contingent movement minimizes bodily contact and the smooth continuity of the internal and external flooring texture delimits the sensation of touch. These spaces are regulated by stage-managers, and scenery and props rarely change, constituting fixtures around which movement is organized.

This managed diversity cannot serve to fulfil what I believe to be a widespread yearning for immersion in a sensual and unpredictable environment, and recently there has been an upsurge in escape attempts to establish liminal places and occasions: raves (Thornton, 1995; Saunders, 1995), car boot sales (Gregson and Crewe, 1997), music festivals, alternative rituals and political demonstrations often based around new social movements and New Age beliefs (see McKay, 1996; Hetherington, 1996). These realms seem to permit a greater opportunity for a wider range of bodily expressive practices and meanings. However, these events are always of short duration and a movement towards heterogeneity must always come to an end.

Moving against and across regulation

Michel de Certeau (1984) describes how pedestrians escape the carceral power instantiated in the 'concept city'. Walking through the city is identified by de Certeau as creating contingent 'spaces of enunciation' (p. 98) which evade the strategies of the powerful. This creative inscription is necessarily fleeting but defies attempts to fix and rationalize space. The 'rhetoric of walking', through which pedestrians 'compose a path' (p. 100) does not 'cohere with the constructed, written, and prefabricated space through which they move' (p. 34). This clashing of the language of the city – the 'vocabularies' and 'syntax' of linearity, order, procedure – and the performative enunciations of the pedestrian, are confined to 'the chance offerings of the moment', seized on the wing (p. 37). But also in the carceral city, alternative spatial networks evolve 'in the interstitial spaces between dominant orderings' (Stanley, 1996: 37), along the cracks between regulated spaces. The presence of these marginal spaces, which produce a blurring of boundaries and 'constant ruptures in terms of value' (p. 38), provide unfamiliar contexts for more improvised performances which respond to chance meetings and contingent events. Other moving processes undermine dominant spatial meanings and render incoherent the logic of official functions and modes of circulation and exchange. For instance, Kawash describes how the movement of the homeless body 'is mapped according to the exigencies of bodily functions' as it searches for places to sleep, keep warm, eat, rest, beg and excrete (1998: 333).

Part of the process of regulating the city includes the provision of outlets for bodily expression organized by commercial interests or regimes offering public services. Typical here are the proliferating exercise, sports, martial arts and dance training classes which reinstate communitas through collective bodily exertions (Prickett, 1997), albeit in enclavic spaces. Likewise, the entertainment provided at concerts, dance clubs and raves provides particular occasions and spatial contexts for the release of bodily energies and desires. Sometimes, however, the opportunities for physical performance are provided on the city streets themselves, in carnivals and pageants. At certain events, such as the gay Mardi Gras in Manchester, there are opportunities to celebrate those elements of traditional carnival culture, the gross body and excess physicality, through the foregrounding of desires, giant phalluses and sexualized bodies performing mass dances through the streets. Other events such as the Notting Hill Carnival and the Rio Carnival also contain transgressive expressions, although Don Handelman (1997: 396) argues that, typically, such festivals have been 're-taxonomised, reorganised and disciplined through bureaucratic logic'.

Toronto's Caribana festival offers a temporary escape into physical 'abandonment' but this has been counterposed to 'the dignified display of pageantry and style' (Jackson, 1992: 139) wherein celebrations have been marshalled and watered down into a respectable expression along a set parade route. Objections to the 'bacchanalian' aspect of the festivities have not only come from police but the more class-conscious participants in the carnival. These battles have also taken place at the Notting Hill Carnival, where authoritative attempts to delineate the area within which it occurs have been resisted by participants, although the dense labyrinth provided by local streets facilitates the rejection of a linear procession (Jackson, 1992: 139). Even at the world famous Rio Carnival, 'rhythm, spontaneity and satire are being controlled and constricted' by bureaucratic power (Handelman, 1997: 401). Although it remains an occasion at which normally concealed social tensions are revealed and dramatized, and utopian displays of equality and the body are celebrated, Handelman discusses the forms of regulation that are imposed upon the participating 'samba schools' in the climactic great parade, in accordance with the political imperatives of unity and order. The themes chosen by the dancers must not be satirical or critical of national politics, and marks are deducted by the judges if strict conventions of time allotment, and musical and disciplined rhythmic performance are flouted. Seemingly, highly synchronized, spectacular mass dances are replacing improvisational and innovatory forms in an increasingly controlled carnival, which is organized to provide a spectacle for visual consumption as opposed to an occasion for physical experimentation and immersion.

However, as John Tagg (1996: 181) remarks, urban 'regimes of spectacles and discourses do not work ... they are never coherent, exhaustive or closed in the ways they are fantasised as being ... they cannot shed that ambivalence which always invades their fixities and unsettles their gaze'. Instead, they are 'crossed over, graffitied, reworked, picked over like a trash heap ... plagued by unchannelled mobility and unwarranted consumption that feeds unabashed, on excess in the sign values of commodities'. Thus the totalizing power structure, which de Certeau implies, is contingent, since the fabric of order has to be constantly maintained, and its ambivalent and contradictory features disguised. It is this excess, the proliferating forms, styles and images in the city that open up 'a horizontal vista of mobile meanings, shifting connections, temporary encounters', which provides the resources for critical, yet popular, cultural practices by an urban 'community of modern nomads' (Chambers, 1986: 213). This abundance of refuse, a jumbled heap of products continually juxtaposed in new ways, refutes the continual upkeep and policing that is required for the seamless ordering of commodified space. It is the very movement of the city-dweller through the urban landscape, inhabiting and decoding the familiar signs and symbols whilst simultaneously subverting and transforming them, which disrupts dominant meanings.

Whilst materials, meanings and practices are organized according to pre-dictable codes, these elements can be ordered in innumerable alternative ways. A pertinent example of such a performance is recounted by Klugman (1995) in her construction of an 'alternative tour' around Disneyland. In order to escape the highly regulated nature of the stage and the normative expecta-tioris about what performances are appropriate, Klugman imagines her fellow-visitors are actors whom she is directing. Through this subversion of conventional meaning and practice, the scene becomes saturated with other-ness, the site of the weird and fantastic. Moreover, simply by adopting a gaze askance to the norm, Klugman reveals the host of unpredictable oddities that abound in Disneyland. Odd angles, unusual sights and cracks in the fabric are always present, and can be experienced by shifting perspective. However, to avoid any over-optimistic cultural populism, it is important to acknowledge that the ability to see through the arrangements of the city requires a form of cultural capital, one that is perhaps typified by the techniques and disposi-tions of the 'post-tourist' (Feifer, 1985), whose pleasure in the fabricated nature of the produced scene is gained from enacting a playful, detached, ironic performance.

Another figure whose practices can be considered to offer a set of tech-niques and dispositions for experiencing and moving through the city is the *flâneur*. The *flâneur* mobilized distinct modes of strolling and looking, and developed ways of representing the city, using 'a multifaceted method for apprehending and reading the complex and myriad signifiers in the labyrinth of modernity' (Frisby, 1994: 93). The *flâneur*'s approach is not merely con-cerned with employing a detached mental procedure to gather fragments of conversation and other impressions, but also in developing a 'tactile ability' (ibid.) For his desire was to 'merge with the crowd ... to establish his dwelling in the throng, in the ebb and flow, the bustle, the fleeting and the infinite' (Baudelaire, 1972: 403). Continually on the move, the *flâneur* lapped up transient experiences, attempted to catch things 'in flight'.

The *flâneur*'s performance was enabled by a heterotopic ordering of space whereby the city resembled a labyrinth in which he could roam, discovering lost dreams, obscure experiences and residual processes, revealing 'things hidden to those intent on purposive, linear goals' (Savage, 1995: 207). The activity of wandering amidst a sea of unpredictable, varied things and people was crucial. However, to find contemporary *flâneurs*, to identify equivalent ways of moving through the city, seems difficult. Perhaps this is partly due to the call of advocates of commodity capitalism who espoused the slogan '*guerre à la flânerie*', to direct movement to their more instrumental require-ments. Nevertheless, to escape these distractions, the *flâneur* may walk whilst the city sleeps, amongst the denizens of the night, where different perspec-tives and sensations present themselves in a landscape of limited visibility, where an alternative order might exist unseen in alleyways and other unlit spaces (Schlör, 1998).

A more popular recent metaphorical figure has been the nomad, continu-ally on the move, and hence difficult to place and classify. Paying no allegiance to the dominant order of the city, and the illusion of permanence that buildings, institutions and procedures perpetrate, the nomad embodies the idea that there is 'no place in which meaning and identity can rest' (Cresswell, 1997: 362). As I have mentioned above, the nomad is seen as a threat to the norms of home, rootedness, work and structure which official-dom demands. Resisting attempts to reterritorialize, the nomad moves across 'tactile space', transgressing spatial divisions and purified spaces.

Perhaps the most ambitious attempts to resist the inscriptions of the city and to move against the grain were the Situationist projects that aimed to dis-locate the spectacular and banal. To substitute adventure for 'traffic circulation and household comfort' (Constant, 1997), the Situationists introduced the 'psy-chogeographical' technique of the '*dérive*', whereby city-dwellers were urged 'to drift around the whole time', to remove themselves from habitual spaces, refuse to acknowledge private property and bureaucratic barriers as obstacles to the random and improvised excursions that were undertaken. Such rhizo-matic (as opposed to linear) techniques of movement, it was alleged, would reveal the peculiar elements of city life (Bonnett, 1989). Rather than following set routes, pedestrians were advised to be influenced by sudden changes of ambience and other non-purposive sentiments and 'irrational' feelings to 'excavate a network of anti-spectacular spaces' (Sadler, 1998: 92). Accordingly, it was posited that areas within cities could be more appropriately conceptual-ized according to poetic and emotional criteria, so pedestrians would enter the 'Happy Quarter', the 'Bizarre Quarter', the 'Sinister Quarter', or the 'Noble and Tragic Quarter'. This jamming of the messages imparted by the 'mechanis-tic functioning of the city' (Sadler, 1998: 91) would, as a consequence, reveal the absurdities of the paths pedestrians follow (Bonnett, 1989). It was imag-ined that this practical decoding and deconstructing of urban norms of movement and perception would be replaced by 'an infinite variety of envi-ronments, facilitating the casual movement of the inhabitants and their frequent encounters', and that these spaces would be continually changed by 'professional situationists' (Constant, 1997: 112).

But there are more mundane activities which defy the norms of urban reg-ulation through movement. Bonnett (1996) has written about small acts of physical resistance to the regulation of space, depicting the 'till-talkers', who hinder the progress of checkout queues at supermarkets by their dalliance at the till, and 'roadrunners' who perform acts of youthful masculinity by dis-turbing the flow of traffic, running across the road and impeding vehicles. Borden (1998) provides a particularly romantic depiction of skateboarding which he describes as 'nothing less than a sensual, sensory, physical emotion and desire for one's own body in motion and engagement with the architec-tural and social other' (p. 216). The performative display of technique and 'attitude' in skateboard parks, and increasingly deserted urban areas, as well as pavements, roads, walls, stairs and benches, subverts the normative uses of the city. A heightened consciousness of the material textures of the city, and the sounds and tactile experiences they produce with the skater, as well as a transformed relationship of the body with verticality and diagonality, challenges the hegemony of linear, upright bodily positions. Skateboarding thus produces a body-centric space produced dialectically with the built environment. And Borden argues that skateboarding can be distinguished from the proliferating movements of professional spectacular commodification, conscious intellectual and artistic performativity, and narcissistic activities of shopping and body-building (p. 207), and yet attempts to legislate against the practice have proved difficult as the whole city serves as the skateboarders' domain.

Paul Gilroy (1997) provides another example of the appropriation of city-spaces, by the spectacular break-dancing displays of black youth. However, he ruefully remarks that these expressive and politicized performances of black bodies in public space have been replaced by basketball, a more organized, disciplined, heroic, hyper-masculine expression of bodily performance, movements which are 'always purposive, strongly directional and precisely goal-oriented' (p. 24). Gilroy's example is an apt metaphor for the colonization of expressive bodily practices by commerce, as part of the ongoing and uneven processes of ordering and disordering movement in the city.

Moving through the bazaar

In this section, I explore the forms and contingencies of movement through an Indian bazaar in the city of Agra. Other modes of spatial ordering to those discussed above can promote a differently regulated, more sensual realm which facilitates a wider range of movement. The production of highly regulated, commercial and enclavic space is a globally uneven process, and yet Western notions of 'civic consciousness' and 'an order of aesthetics', of health and hygiene, are part of a global process where increasingly 'the thrills of the bazaar are traded in for the conveniences of the sterile supermarket' (Chakrabarty, 1991: 16).

Bazaars constitute an unenclosed realm which provides a meeting point for a variety of people and multiple activities, including forms of 'recreation, social interaction, transport and economic activity' (King, 1976: 56). Not dominated by one economic activity, bazaars mix together small businesses, shops, street vendors, public and private institutions and domestic housing. Hotels co-exist alongside work places, schools, eating places, transport termini, bathing points, political headquarters, offices, administrative centres, places of worship and temporary and permanent dwellings. This multi-functional structure provides an admixture of overlapping spaces that seem to merge Western dichotomous notions of public and private, work and leisure, and holy and profane. And the bazaar contains a host of micro-spaces: corners and niches, awnings and offshoots, providing a labyrinthine structure with numerous openings and passages which enables a flow of different bodies and vehicles which criss-cross the street in multi-directional patterns, veering into courtyards, alleys and cul-de-sacs.

Bazaars are 'centres of social life, of communication, of political and judicial activity, of cultural and religious events and places for the exchange of news, information and gossip' (Buie, 1996: 277). Everyday social activities such as loitering, sitting and observing, and meeting friends are organized around particular micro-spaces such as transport termini and tea stalls. The street is often a site for domestic activities such as collecting water, washing clothes, cooking and child-minding. For some, streets may serve as a temporary home, necessitating the carrying out of bodily maintenance such as washing. Such temporary sites and activities challenge notions of ownership, and the distinction between private and public (Chandhoke, 1993).

The bazaar is occupied by diverse commercial enterprises. Whilst there are fixed shops, the streets may also be the workplace of mobile providers of services of all kinds, including disparate hawkers and beggars. Moreover, the open fronts of most workshops mean that the work of artisans adds to the multifarious street activities. This contributes to a gregarious environment which privileges speech, and facilitates visual and verbal enquiry, most evidently in the practice of barter, which Buie (1996) describes as a sensual and performative, as well as economic activity; an 'art', a 'ritual', and a 'dance of exchange' (p. 227). The bazaar's streets can be theatrically transformed into channels of embodied transmission by collective performances such as political and industrial demonstrations and religious processions. And as a site for entertainment, children and adults play games, and travelling entertainers such as musicians and magicians attract crowds.

Heterogeneous spaces such as bazaars are typically subject to contingent and local forms of planning and regulation, and surveillance is rather low-level. Rather than security guards and video surveillance, local power-holders exercise policies of exclusion and control. Forms of restriction over movement do apply. There may be gendered conventions about who should pass through particular public spaces, and other local struggles over space, shaped around class, ethnic and religious conflicts, may persist. Yet such restrictions are often foiled by intrusions that challenge the spatial ordering of the city. For instance, the 'unintended city' of the 'shanty town' insistently projects into and subverts planned urban spaces. Chandhoke argues that the 'urban poor make and remake space ... seize spaces and reshape in this way the entire urban form':

> They intrude into individual consciousness at traffic crossings ... they inform us that cities are unequally constructed and maintained ... [They] disrupt the coherence of the planned urban landscape, they retaliate and talk back to history and geography by making the homelessness of these people dramatically visible. (Chandhoke, 1993: 64)

Whilst formal traffic rules exist, vehicles pay little heed to them as they jostle for position, creating a competitive and rather chaotic race for road space. Itinerant beggars and workers are rarely advised to 'move on', and while the animals that share the streets may be subject to cruelty, there are few systematic attempts to control their movements or numbers. Aesthetically, the maintenance of a cultivated appearance through control or theming is rarely imposed; an unplanned bricolage of structures is infested with ad hoc signs, contingent and personalized embellishments, and unkempt surfaces and facades.

The pedestrians who move through this protean space are denied the option of seeking refuge in an aloof disposition and are often thwarted in their desire for rapid progress. For it is often difficult for them to move in a straight line. Instead they must weave a path by negotiating obstacles, and remain alert to hazardous traffic and animals (such as monkeys, cows and dogs). The abundant simultaneous cross-cutting journeys mean that pedestrians must take account of others who will cross their path at a variety of speeds. In addition, the miscellaneous collection of vehicles and other diverse forms of transport, all moving at different speeds as they manoeuvre for space, provides an ever-changing dance of traffic which echoes the stop–start choreographies of pedestrians. Walking cannot be a seamless, uninterrupted journey but is rather a sequence of interruptions and encounters that disrupt smooth passage.

This disrupted progress is not only produced through the exigencies of avoiding collisions, but also by the distractions and diversions offered by heterogeneous activities and sights. There is an ever-shifting series of juxtapositions and assemblages of diverse static and moving elements which can provide surprising and unique scenes. Such haphazard features and events dis-order the gaze as the eye continuously shifts, alighting on changing episodes to the left and right, far ahead and close at hand. Moreover, the pleasurable jostling in the crowd engenders a haptic geography wherein there is continuous touching of others and weaving between and amongst bodies. The different textures brushed against and underfoot render the body aware of diverse tactile sensations. These impacts on the haptic system of the body produce distinct forms of feeling and experience (see Lyon and Barbalet, 1994). Additionally, the 'smellscapes' and 'soundscapes' of the bazaar are rich and varied. The jumbled mix of pungent aromas (sweet, sour, acrid and savoury) produces intense 'olfactory geographies', and the combination of noises generated by numerous human activities, animals, forms of transport and performed and recorded music, produces a changing symphony of diverse pitches, volumes and tones.

It seems as if the sensual and social body moving through the bazaar is continually imposed upon and challenged by diverse activities, sensations and sights which render a state at variance to a restrained and aloof distraction. Rather than being a distanced spectator of manufactured spectacle, the pedestrian is part of a heterotopic diversity where the senses are excited by a more variegated set of stimuli.

Conclusion: the value of disruption

> Dominated by overpowering forces, including a variety of brutal techniques and an extreme emphasis on visualisation, the body fragments, abdicates responsibility for itself … Any revolutionary 'project' … must … make the reappropriation of the body, in association with the reappropriation of space, into a non-negotiable part of its agenda. (Lefebvre, 1991: 166–7)

The body can never be mechanically ordered to pass seamlessly through space, for the interruptions of stomach cramps, headaches and injuries, limbs that 'go to sleep', foreground an awareness of the body that can dominate consciousness, to the exclusion of imperatives about efficient movement. Yet I have argued that certain regulatory pressures are restricting physical expressivity in Western cities, which is being displaced to more bounded spatial contexts. In most cities, the balance between the Apollonian and disciplinary, and the heteroclite, baroque dimensions of modernity seems to have tilted decisively in favour of the dream which 'wishes away the aggression, the conflict, and the paranoia that are also part of urban experience' (Donald, 1997: 182). However, these desires are displaced into other realms – into temporary festivals, tourism, raves and dances, car boot sales – where tactile sensations and visual shocks provide contexts for more experimental and expressive forms of movement and bodily display. The seeking out of such experiences attests to the contemporary desire for disorder and unpredictability. Symbolically then, these heterogeneous spaces provide a realm of escape from quotidian order: they are an alternative system of spatial formation where transitional identities may be sought, difference confronted, and sensual and imaginative experimentation indulged.

To counter this tendency, we should celebrate and foster spaces which contain confusion and the energy created by contrasts and clashes. The value of disruption – that which ordering processes try to expunge – lies in its potential to dramatize and reveal the complexities of co-existence, difference and friction that permeate the city. Richard Sennett (1994: 310) writes that the 'body comes to life when coping with difficulty', is roused by the resistance which it experiences. Moments of confrontation, of self-displacement, are vital to preserve openness to stimuli, to awaken the senses. The confrontation with the uncomfortable and the surprising can engender procedures, such as dialogue and negotiation, which help to make sense of the jarring shocks of urban streets, and can encourage a critical approach to conventions. The acceptance of 'impurity, difficulty, and obstruction', according to Sennett, is 'part of the very experience of liberty' (pp. 309–10).

By looking at the Indian bazaar, I have argued that certain forms of urban space are more likely to stimulate sense, feeling, dialogue, excitement, a sense of commonality and diversity. For physical performances and forms of movement are likely to be more improvisational in spaces without stage-directors and strong conventions. In less regulated urban spaces, fluid events, activities and movements arise, random juxtapositions of objects and people occur, and sensory stimulation precludes anything other than a contingent performance. Experience may be akin to the 'vertigo' described by Alex Caillois (1961) wherein perception is temporarily destabilized by a fore-grounding of physical sensation.

A few years ago, Seamon (1979) recommended that diverse and dynamic street choreographies be encouraged by arranging a form of channelling designed to generate face-to-face interaction. He suggested the creation of centres to provide 'stage settings', a mixed space that could bring together

difference and encourage mingling, and called for a reduction in technological forms of transit and the removal of physical and social barriers to
incomers. Failure to attend to these imperatives threatens vitality, motion,
interaction and communality, characteristics epitomized in the pavement ballets described by Jane Jacobs (1961) in her classic text, *The Death and Life of
the Great American Cities*.

I have argued, however, that whatever regime of spatial ordering persists,
the body can never be assumed to passively acquiesce in the performance of
a dance of duty. Bodies are not only written upon but also write their own
meanings and feelings upon space. Bodies belong to places and help to constitute them, whether they stay in place, move through place or move towards
other spaces (Casey, 1996). Bodies cannot be reduced to discursive objects,
but engage in concrete, material practices. These enactions are informed by
conventions of performativity – the 'schemas of acting, perceiving, cognizing
and feeling that make up social representations of the body' – and so whilst
they constitute an active and embodied expression of 'being in the world'
they are also processes of socialization (Turner, 1994: 44).

The body moving through the space of the city evades and conforms to
regulatory norms, impacts upon the city's materiality and is impacted upon.
This relationship is characterized by the ways in which the organization of
space directs, stimulates and disrupts the body's passage. Thus the material
nature of cities impacts upon modes of walking through cities but cannot
determine them, although less regulated urban spaces are more likely to stimulate less normative experiences and readings. A politics which foregrounds a
sensual, expressive revolt against conventions through bodily practices and
movement provides possibilities for transgressing and transcending urban
sterility. The responses of city regulators reveal the constant policing that is
required to maintain an orderly city, showing that whilst power is always
encoded in space, it may be contingent, arbitrary and impermanent.

The relationship between the contrary modern desires for order and its
transcendence is continually being worked out across diverse cultural sites.
The contemporary notion that over-socialized selves and over-determined
communities should be escaped and affective experience sought is manifest
in the evolution of a wide range of cultural sites, from centres for New Age
and new social movements, to raves and drug cultures. The city is a realm
where ordering tendencies are often ascendant but various counter-processes
erupt and are displaced to the night or the margins, or are consigned to
bounded spaces where they are permitted. Attempts to encourage a more
sensual experience of the body through moving against the grain, towards
otherness and into a more sensory awareness can feed into Lefebvre's (1995:
159) declaration that the most important aim is 'to multiply the readings of the
city'.

References

Auge, M. (1995) *Non-Places: Introduction to an Anthropology of Supermodernity*, Verso, London.

Baudelaire, C. (1972) *Selected Writings on Art and Artists*, trans. and ed. Charvet, P.E., Penguin, Harmondsworth.

Bennett, T. (1995) *The Birth of the Museum*, Routledge, London.

Boddy, T. (1992) Underground and overhead: building the analogous city. In Sorkin, M. (ed.) *Variations on a Theme Park*, Hill and Wang, New York.

Bonnett, A. (1989) Situationism, geography and poststructuralism. *Environment and Planning D: Society and Space*, **7**, 131–46.

Bonnett, A. (1996) The transgressive geographies of everyday life. *Transgressions*, **2/3**, 20–37.

Borden, I. (1998) Body architecture: skateboarding and the creation of super-architectural space. In Hill, J. (ed.) *Occupying Architecture: Between the Architect and the User,* Routledge, London.

Briginshaw, V. (1997) 'Keep your great city Paris!' – the lament of the Empress and other women. In Thomas, H. (ed.) *Dance in the City*, Macmillan, London.

Buie, S. (1996) Market as mandala: the erotic space of commerce. *Organisation*, **3**, 225–32.

Caillois, R. (1961) *Man, Play and Games*, Free Press, New York.

Carlson, M. (1996) *Performance: A Critical Introduction*, Routledge, London.

Casey, E. (1996) How to get from space to place in a fairly short stretch of time: phenomenological prologema. In Feld, S. and Basso, K. (eds) *Senses of Place*, School of American Research Press, Santa Fe.

Chakrabarty, D. (1991) Open space/public space: garbage, modernity and India. *South Asia*, **16**, 15–31.

Chambers, I. (1986) *Popular Culture: The Metropolitan Experience*, Routledge, London.

Chandhoke, N. (1993) On the social organisation of urban space – subversions and appropriations. *Social Scientist*, 21, 63–73.

Chaney, D. (1993) *Fictions of Collective Life*, Routledge, London.

Constant (1997) A different city for a different life. Reprinted in *October*, **72**, 109–12 [originally in *Internationale situationniste*, **3**, (December 1959)].

Cresswell, T. (1996) *In Place / Out of Place: Geography, Ideology and Transgression*, University of Minnesota Press, London.

Cresswell, T. (1997) Imagining the nomad: mobility and the postmodern primitive. In Benko, G. and Strohmayer, U. (eds) *Space and Social Theory: Interpreting Modernity and Postmodernity,* Blackwell, Oxford.

Csordas, T. (ed.) (1994) *Embodiment and Experience*, Cambridge University Press, Cambridge.

Davis, M. (1990) *City of Quartz*, Verso, London.

de Certeau, M. (1984) *The Practice of Everyday Life*, University of California, Berkeley.

Desmond, J. (1994) Embodying difference: issues in dance and cultural studies. *Cultural Critique*, Winter, 33–63.

Donald, J. (1997) This, here, now: imagining the modern city. In Westwood, S. and Williams, J. (eds) *Imagining Cities: Scripts, Signs, Memory*, Routledge, London.

Edensor, T. (1998) *Tourists at the Taj*, Routledge, London.

Featherstone, M. (1991) *Consumer Culture and Postmodernism*, Sage, London.

Feifer, W. (1985) *Going Places,* Macmillan, London.

Feld, S. (1996) Waterfalls of song: an acoustemology of place resounding in Bosavi,

Papua New Guinea. In Feld, S. and Basso, K. (eds) *Senses of Place*, School of American Research Press, Santa Fe.

Foucault, M. (1986) Of other spaces. *Diacritics*, Spring, **16** (1), 22–7.

Foucault, M. (1988) The subject and power. In Dreyfus, H. and Rabinow, P., *Michel Foucault: Beyond Structuralism and Hermeneutics*. Harvester, Brighton.

Frisby, D. (1994) The flâneur in social theory. In Tester, K. (ed.) *The Flâneur*, Routledge, London.

Geertz, C. (1993) Deep play: notes on the Balinese Cockfight. In Geertz, C. *The Interpretation of Cultures: Selected Essays*, Fontana, London.

Gilroy, P. (1997) Exer(or)cising power: black bodies in the black public sphere. In Thomas, H. (ed.) *Dance in the City*, Macmillan, London.

Goffman, E. (1959) *The Presentation of Self in Everyday Life*, Doubleday, New York.

Gottdiener, M. (1997) *The Theming of America: Dreams, Visions and Commercial Spaces*, Westview Press, Oxford.

Gregson, N. and Crewe, L. (1997) The bargain, the knowledge, and the spectacle: making sense of consumption in the space of the car boot sale. *Environment and Planning D: Society and Space*, **15**, 87–112.

Handelman, D. (1997) Rituals/Spectacles. *International Social Science Journal*, September, **15**, 387–99.

Harvey, D. (1989) *The Condition of Postmodernity*, Blackwell, Oxford.

Hetherington, K. (1996) The utopics of social ordering: Stonehenge as a musuem without walls. In Macdonald, S. and Fyfe, G. (eds) *Theorising Museums*, Blackwell, Oxford.

Jackson, P. (1992) The politics of the streets: a geography of Caribana. *Political Geography*, **11**, 130–51.

Jacobs, J. (1961) *The Death and Life of the Great American Cities*, Random House, New York.

Kawash, S. (1998) The homeless body. *Public Culture*, **10** (2), 319–39.

King, A. (1976) *Colonial Urban Development: Culture, Social Power and Environment*, Routledge, London.

Klugman, K. (1995) The alternative ride. In Klugman, K., Kuentz, J., Waldrep, S. and Willis, S. *Inside the Mouse: Work and Play at Disney World*, Rivers Oram Press, London.

Lefebvre, H. (1991) *The Production of Space*, Blackwell, Oxford.

Lefebvre, H. (1995) *Writings on Cities*, Blackwell, Oxford.

Lyon, M. and Barbalet, J. (1994) Society's body: emotion and the 'somatisation' of social theory. In Csordas, T. (ed.) *Embodiment and Experience*, Cambridge University Press, Cambridge.

MacDonald, A. (1997) The new beauty of a sum of possibilities. *Law and Critique*, **8**, 141–59.

McKay, G. (1996) *Senseless Acts of Beauty: Cultures of Resistance Since the Sixties*, Verso, London.

Mitchell, D. (1995) The end of public space? People's Park, definitions of the public, and democracy. *Annals of the Association of American Geographers*, **85**, 108–33.

Musil, R. (1954) *The Man Without Qualities*, vol. 1, trans. Wilkins, E. and Kaiser, E., Martin Secker and Warburg, London.

Prickett, S. (1997) Aerobic dance and the city: individual and social space. In Thomas, H. (ed.) *Dance in the City*, Macmillan, London.

Ritzer, G. and Liska, A. (1997) 'McDisneyization' and 'post-tourism': complementary perspectives on contemporary tourism. In Rojek, C. and Urry, J. (eds) *Touring Cultures: Transformations of Travel and Theory*, Routledge, London.

Rojek, C. (1995) *Decentring Leisure*, Sage, London.

Sadler, S. (1998) *The Situationist City*, MIT Press, Cambridge, Mass.

Saunders, N. (1995) *Ecstasy and the Dance Culture*, Nicholas Saunders, London.

Savage, M. (1995) Walter Benjamin's urban thought: a critical analysis. *Environment and Planning D: Society and Space*, **13**, 201–16.

Schieffelin, E. (1998) Problematising performance. In Hughes-Freeland, F. (ed.) *Ritual, Performance, Media*, Routledge, London.

Schlör, J. (1998) *Nights in the Big City*, Reaktion Books, London.

Seamon, D. (1979) *A Geography of the Lifeworld*, Croom Helm, London.

Sennett, R. (1994) *Flesh and Stone*, Faber, London.

Shields, R. (1991) *Places on the Margin*, Routledge, London.

Shields, R. (ed.) (1992) *Lifestyle Shopping*, Routledge, London.

Sibley, D. (1988) Survey 13: purification of space. *Environment and Planning D: Society and Space*, **6**, 409–21.

Squires, J. (1994) Ordering the city. In Weeks, J. (ed.) *The Lesser Evil and the Greater Good*, Rivers Oram Press, London.

Stallybrass, P. and White, A. (1986) *The Politics and Poetics of Transgression*, Methuen, London.

Stanley, C. (1996) Spaces and places of the limit: four strategies in the relationship between law and desire. *Economy and Society*, **25**, 36–63.

Sui, K. (1999) The escalator: a conveyor of Hong Kong's culture. *Human Relations*, **52** (5), 665–81.

Tagg, J. (1996) The city which is not one. In King, A. (ed.) *Re-presenting the City: Ethnicity, Capital and Culture in the 21st Century Metropolis*, Macmillan, London.

Thornton, S. (1995) *Club Cultures: Music, Media and Subcultural Capital*, Polity Press, Cambridge.

Turner, T. (1994) Bodies and anti-bodies: flesh and fetish in contemporary social theory. In Csordas, T. (ed.) *Embodiment and Experience*, Cambridge University Press, Cambridge.

Yudell, R. (1977) Body movement. In Bloomer, K. and Moore, C (eds) *Body, Memory and Architecture*, Yale University Press, London.

Chapter 10

'Finding a place in the street': CCTV surveillance and young people's use of urban public space

Ian Toon

> The human landscape can be read as a landscape of exclusion. Because power is expressed in the monopolization of space and the regulation of weaker groups in society to less desirable environments, any text on the social geography of advanced capitalism should be concerned with the question of exclusion. (Sibley, 1995: ix–x)

Introduction

The above quotation from Sibley (1995) draws attention to the desires inherent in dominant representations of space which are manifest in official attempts to purify the public sphere of disorder and difference through the spatial exclusion of those social groups who are judged to be deviant, imperfect and marginal in public space. Sibley notes how teenagers are often a primary target of such exclusionary processes and points out that young people also contest various forms of spatial restriction that they encounter in their everyday lives. It is difficult to understand the behaviour and activity patterns of teenagers in public spaces without also relating them to issues of social control, spatial exclusion and resistance. However, while Sibley's analysis is insightful in these respects, it also tends to be somewhat over-theorized and empirically weak. In contrast, this chapter attempts to provide some empirical basis to the growing body of work on the 'geographies of exclusion' (Sibley, 1995) by investigating young people's experience of CCTV surveillance and exclusionary youth policing.[1] Little is known about the practices and actions of teenagers in surveilled urban environments or about the kind of spatial practices they employ in order to cope with police removal. Drawing on accounts produced in a study with a group of 40 white, working-class teenagers (girls and boys aged 12–16) in Tamworth, a large town in the British Midlands,[2] the first part of the chapter explores the links between their various social uses of public space and identity formation and the second part goes on to assess the associated spatial outcomes of these new ways of regulating urban public space as they impact on the daily lives of these young people.

What interested me in this study is the relationship between official representations of space constructed from above and the cultural geographies of youth which are lived out at the level of everyday life. Working on the premise that public space is a contested terrain which is open (no matter how panoptic and controlled it may be) to practices of human creativity and resistance (Harvey, 1989), I wanted to find out how these new forms of policing public space are negotiated and contested by youth at street level. The central purpose of this chapter, then, is to show how, set against a backdrop of increasing levels of socio-spatial circumvention and regulation, these young people use public space to create meaning and context in their social existence. In this task I draw on and rework Michel de Certeau's notion of the 'tactic' as it relates to pedestrian movement in urban space, and describe how these youth respond to the constraints and restrictions that are routinely placed on their free access to public space. I show how they reappropriate public space to 'find a place in the street' through the 'tactical inhabitation of spaces' (Ruddick, 1998: 358) within the controlled urban environment. I also offer a consideration of the possibilities and limitations of their spatial resistance, focusing on the positive, as well as addressing the more negative aspects of their social use of urban space. Finally, the chapter concludes with a summary of the findings and a discussion of issues relating to planning spaces for school-age teenagers.[3] I want to begin by discussing the status of the street as an emblematic landscape form for youth.

The geographical constitution of youth culture: school-age youth and the importance of the street

Critical geographers have emphasized the ambiguity and complexity of public space and moved us 'towards an appreciation of the ways in which both the "public" and its activities as well as the city and its spaces are mutually constitutive' (Lees, 1998: 237). In this regard Susan Ruddick (1998: 353) has shown how space plays a crucial role in the construction of youth identities, highlighting the interdependency of the production of 'space' and 'self' for youth, and the ways in which youth both orient themselves to and draw upon space as part of the homological construction of the sub-culture. The starting point of this study was to find out about the daily lives of teenagers (particularly those in early-to-mid adolescence) and how they spend their 'free time', and to develop an effective analysis of the ways in which they construct identity in and through the use of space. In particular, I was interested in relatively under-explored teenage uses of space such as hanging around, doing nothing, and moving around within the built environment – 'ordinary' dimensions of teenage spatial practice that are difficult to explore (see Corrigan, 1979; Leiberg, 1996; Loader, 1995; Watt and Stenson, 1998, for notable exceptions), but are distinctive activities, nonetheless, which play an important role in fulfilling their social needs and interests, condensing many important aspects of their youth identities including the dynamics of class, gender and generation and the cultural politics of space and place.[4]

A large number of youth studies have consistently argued that teenage boys' and girls' youth cultural production takes place at largely different spatial scales – in separate public (boys) and private (girls) domains (see for example, McRobbie and Garber, 1976; Cohen and Robins, 1978). However, these studies offer a narrow consideration of what girls do, where they do it, and what constitutes the distinctive elements of their cultures. Crucially, they fail to take account of girls' uses of public space in the construction of feminine subjectivity and the increasing use of the home as a key site of youth culture by boys (McNamee, 1998). In contrast this study has revealed that outdoor public space is used as a site for self-definition and cultural expression as much by these young women as their male peers. Indoor activities such as watching television and playing computer games (Leiberg, 1995; McNamee, 1997; Willis, 1990); using friends' houses as meeting places when parents are out; and attending semi-public spaces like the youth club and under-18s discos occasionally draw these lower teens in from outdoor social life and activities. However, it is primarily through spatial activities like hanging around and movement in circuit walks on the streets, in parks and in shopping malls that these teenagers (girls and boys) explore and define a sense of self and identity. It is important to emphasize that the reason these teenagers spend so much time in public spaces is that they lack any other choice and thus face daily the increasing 'urban dilemma' (Leiberg 1995: 735) of having nowhere else to go.[5] As Loader (1996: 50, 26) argues:

> Teenagers, perhaps, more than any other social group, make routine use of a whole host of ... public locations, and they are certainly one of the few groups whose use of such space is not merely fleeting, instrumental and organised around consumption ... [Yet] more than any other social group, young people are dependent on a range of public places, especially in relation to the pursuit of leisure. Denied access [as a consequence of age-based and economic exclusion] from a whole host of cultural amenities, young people come to rely on local streets, city centres, shopping malls and the like to build cultural identities away from direct supervision of adult authority.

Numerous empirically situated studies of youth have demonstrated the enduring significance of local neighbourhood as an important and meaningful context for young people in the formation of identity, highlighting the various ways in which youth continue to identify with local places (see, for example Back, 1996, 1999; Cohen, 1999; Loader, 1996; Pearce, 1996; Westwood, 1990; Wulff, 1995; Watt and Stenson, 1998: 252–3). However, these studies tend to lose sight of the ways in which young people could, do, or hope to live out other, wider spatial orientations in important 'elsewheres' beyond neighbourhood environs (for notable exceptions, see Leiberg, 1995; Watt and Stenson, 1998). For example, the symbolic geographies of the youth in this study do not fit easily into a framework which posits the 'local' at the centre of analysis because neighbourhood space has little significance in relation to the ways in which they fashion their identities. They attribute special significance and meaning, instead, to urban space.

These young people live in different social housing estates and suburban-like environments often located a number of miles away from the town centre but these locales fail to arouse any positive identification and emotional attachment for these youth. Instead, they express a strong sense of disconnection from local estate life and community, seeing their neighbourhoods as confining and isolating. In the pursuit of leisure they rarely choose to base their social lives and activities within their neighbourhoods but spend extensive periods of their 'free time', instead, together in the streets and other public places in the town centre. There are a number of important reasons why these teenagers gravitate to the urban centre. It not only constitutes a space in which they feel free from parental jurisdiction but it also provides them with a means of finding and creating a different spatiality, more exciting than that offered by the impoverished landscape of the neighbourhood. This distancing from the 'local' is part of an important struggle for self-definition for these teenagers. However, while they have no embedded territorial ties or local loyalties to 'place', their identities are, nonetheless, premised upon a sense of spatial opposition to particular youth groups associated with specific neighbourhood locales and are operationalized around the collective appropriation of urban public space. As the following statement shows, they are keen to distinguish themselves from these Other youth formations:

> We never shop doss where we live like the H.B.A. [The 'Hoppo Barmy Army'] or the 'Romma' [youth groups with strong place-based identifications to particular local neighbourhoods]. We are different to them and we just go where the action is in town to get away from where we live and they just stay in their areas. (Joley, aged 14)[6]

Their symbolic geographies and identities, then, are condensed explicitly around the social space of the town centre where they go in search of 'bright lights ... of another way of being' (Walkerdine 1997).

The meanings and uses of public space for youth

At this point I want to draw attention to the scope of these young people's uses of urban space – which particular spaces they use; what the characteristics of their patterns of social activity are in these spaces, and what symbolic meaning these sites have in relation to the constitution of their subjectivities.

Like the teenagers in Mats Leiberg's (1995) study, these youth organize the urban landscape into a sequence of places of retreat (backstages) and places of interaction (exposed frontstages). But these sites are more permeable and integrated for these young people and function as interchangeable kinds of hangout for different peer groups at different times. For example, these teenagers enliven the space of the street by developing male, female and mixed informal groups, 'that enable them to explore public places in mundane, unstructured and potentially unpredictable ways' (Loader, 1996: 8).

They place importance on the social dimension of public space – of who and what can be found when they meet up at their hangouts – and 'going downtown' means that they are able to enter a space of 'social centrality' (Shields, 1992), of adventure and expectation. They use the urban as a set of key places to establish important patterns of sociability, as a context to create meaning in their social existence where they spend extensive periods hanging around 'doing nothing', in a manner similar to that described by Corrigan (1979) in his study of 'The Smash Street Kids' in Sunderland. As Corrigan shows, 'doing nothing' is far from a purposeless activity, and in creating their youth culture in these ways the teenagers in this study have evolved a complex set of social practices in places which provide for their everyday needs and interests. In 'doing nothing' and hanging around – activity patterns in which *something* rather than nothing should happen in a spontaneous and unplanned fashion (Leiberg, 1995) – they are provided with clear-cut images of town-centre streets and open public spaces as a distinctive type of space offering particular kinds of opportunities; as a place of interest and excitement for highly visible activity and action; a space where significant events take place and where they can be confronted with unexpected events, chance encounters and intensive sociality – things they feel are missing from their neighbourhoods.

Walking, style creation, looking and being looked at are fundamental aspects of the formation of these young men's and women's subjectivity. Despite their low spending power, like a large number of contemporary youth who find themselves in a moment of postmodernity, hybridity and amalgamation in youth style (Muggleton, 1995), both the girls and the boys create looks and clothing styles by drawing on a diverse range of commercially produced and media-transmitted youth cultural forms, artefacts and ephemera such as fashion, clothes, make-up, hairstyle and music. They adapt these disparate elements and reforge their own looks and meanings from the remnants. This process of 'grounded aesthetics' (Willis, 1990) involves a particular articulation of contacts and influences, ways of dressing which make 'the correct up-to-the-minute references to a wider [transnational] youth culture' (Massey, 1998: 129), but which also follow other cultural trajectories in that their styles and looks are embedded in, and engage with, a host of local cultural elements and particularities.

The public spaces of the street, park and shopping centre are important in the social process of style creation because they provide a context in which the performance of appearance takes place.[7] For example, these teenagers use their looks as distinctive elements in the exploration and expression of identities which they enact spatially through modes of self-display. This 'dramaturgical' (Goffman, 1967) practice involves a considerable amount of walking around within the urban landscape where they search out a range of places which provide a setting within which new styles and self-presentations can be shown off both amongst their admiring peers and for a public audience for pleasure and self-recognition (Langman, 1992). For these youth (girls and boys) being on view and getting noticed is the whole point of personal aesthetic production: 'they are completed through the admiring glances of a stranger [and friend]' (Hebdige, 1997: 402). Through this social practice they transform what is otherwise thought of as mundane shopping space into an arena for the public performance of femininity and masculinity. At the same time, they are both producers and consumers of the scene because public space not only functions as a key site in and through which these youth establish and enact, form and re-form their modes of expression, but it also acts as a 'stylistic monitorium', a unique vantage point where they can scrutinize each other's styles and observe the latest youth styles worn by others (Leiberg, 1995). It is, then, not only being seen but also seeing (the constant monitoring of self and other) that is an important practice of self-definition in these young people's social worlds, where they are simultaneously actors and audience, consuming social space and each other in 'frontstage' situations.

These practices reflect the highly spatialized nature of their identities. It is important to note in this respect that young people's use of public space is a highly gendered process. Due to the constraints of space a full analysis of the different ways in which these youth gender the street is impossible.[8] But I want to give a flavour of these social processes. For example, whilst these young men and women share many hangout sites and use public space in a number of interconnected ways, there are some interesting variations in the use of particular sites by girls and boys. This is reflected in the time spent engaging in certain primary activities, the particular places in which these actions take place, and the different meanings that these spaces and activities have for male and female youth. This is closely linked not only to processes of identity formation but also crucially to perceptions of safety and danger.

The girls in particular expressed a strong concern for personal safety in public space. As a number of other researchers have noted (for example, see Back, Cohen and Keith, 1999; Loader, 1996; Stanko, 1990; Valentine, 1989; Watt and Stenson, 1998) girls' use of, and access to, public space is shaped in large part by the contours of gendered fear. The geography of these young women's fear centres both on anxieties about male behaviour in public space – they reported being threatened, sexually harassed and 'flashed at' by young males hanging around and by older men at night – and on the risks associated with other rival youth groups. Crucially, the girls' identities are ultimately connected with particular public spaces and are organized and performed in and through the use of sites they define as safe: these include the shopping centre, main shopping streets, core open spaces and certain fast-food restaurants. As Pearce (1996: 7) noted in research on young people in the East End of London, young women go to the shops in order 'to see friends, to hang around and in their words, to have somewhere safe to be'. These sites meet important social functions for the girls in that they are a focus and a conveyor of activities like self-display, the cultural practice of 'promenade' (Blackman, 1998: 211), and the social experience of window shopping (Ganetz, 1995). These ways of using public space include walking together in girl-only groups following established circuits, trying on clothes and shoes, experimenting with make-up, messing up displays in shops and other 'dossing pranks' – gendering space through movement and display. The experiential pleasures of 'tactility and togetherness' (Malbon, 1998: 269) – of just 'being' out together in social situations sharing public space – provide these young women with the opportunity to insert themselves in a highly visible way into the urban landscape which allows them to claim a place in the public realm (Breitbart, 1998). These actions enable them to explore space from a feminine perspective and are central practices of self-definition and femininity construction, constituting the primary ways in which they use and appropriate public space.

Whilst the boys spend considerable periods engaged in self-presentation in core spaces, in contrast to the girls their primary hangout sites are concentrated in different 'bits of the street' (Crouch, 1998) and shopping mall, in other sections of parks, in the churchyard, or outside amusement arcades and leisure amenities, sites often located on the inner fringes of the town centre. They spend more time inhabiting 'back' regions and niche spaces in prime space, withdrawing from the public sphere where they can keep track of each other and of what is happening, talk, joke and argue about clothes, appearance, tastes in music and other common interests and shared experiences (Leiberg, 1995). These kinds of gathering offer a different way of promoting group identity and they form the main ways in which they construct masculine subjectivities in public space.

These young people organize their lives, then, in active and creative ways in their use of the built environment, gendering the street by appropriating particular gathering places in which they fashion, locate and give meaning to their emergent social identities. Their social lives are played out in the open, on the street, in the spaces of the everyday and they are deeply dependent on these public spaces for cultural production. However, whilst the town centre is a relatively 'open' space for these youth during daylight hours, finding places in which to gather at night is now increasingly difficult for these young people as a result of a hardening of police response to the presence of teenagers hanging around on the urban streets. Tighter levels of restriction and control are now being placed on their free access to public spaces in the urban centre through the extensive deployment of CCTV surveillance and exclusionary policing strategies.[9]

The installation and operation of the 24-hour CCTV system in Tamworth town centre has been financed through a partnership between public and private sectors, and is maintained by a private security force which operates from a central control room in the main shopping centre.[10] A total of twelve cameras with rotating heads are located on four-metre mast poles and mounted on the sides of buildings to monitor and record public activity in main streets, open spaces, parks, and car parks below, and to oversee all walkways leading to, and including, new commercial recreational developments on 'Leisure Island', a tract of land adjacent to the town centre which, along with the main shopping district, forms a key component of Tamworth's recent urban renewal programme. The building of these new developments raises some interesting questions in the light of CCTV installations in the town centre. What is the connection between urban regeneration, CCTV public space surveillance and the tighter regulation of young people's use of urban public space?

The regulation of public space: CCTV surveillance, urban renewal and youth

> The built environment is built because it's been allowed to be built, it's been allowed to be built because it stands for and reflects an institution or a dominant culture. (Acconci, 1990: 176)

> In societies such as ours, youth is only present [only becomes visible] when its presence is a problem, or rather when its presence is *regarded* as a problem. (Hebdige, 1997: 402)

During the 1980s the economic decline of the town centre became a matter of concern for the local council, who identified as a key issue the public perception of central areas as run-down and lacking adequate shopping and leisure facilities, and as being unsafe and hostile after dark. The response by local politicians in the 1990s has been to undertake an ongoing urban regeneration programme in a public–private partnership bid to revive the economic role of the town centre (Bulos and Sarno, 1995).[11] In the drive for urban renewal policy makers, planners, architects, private developers and other powerful commercial and civic interests have embraced shopping and leisure as a key economic development strategy. On the one hand, this public–private coalition has sought to restore the commercial viability of the town centre through the construction of a new and progressive urban 'place-image'. On the other hand, the entrepreneurial 'selling' of the town centre through the manipulation of place-imagery has gone in hand with the creation of a transformed, physical urban landscape and the spatial reorganization of economic activity. Key areas of the town centre have been re-aestheticized through the development and redesign of new urban forms such as shopping centres, heritage sites, tourist attractions, and entertainment and leisure facilities in an attempt to cultivate a new urban aesthetic which blends the past, present and future to create a lively and vibrant urban environment of (and for) shopping, leisure and 'place-history' tourism. The invention of a new local economy, through the activities of a public–private partnership of speculative image-making, is designed to create a post-industrial urban setting geared towards consumption rather than production, and is intended to secure a new economic role for the locale in a network of inter- and intra-regional competition. However, as Sibley (1995: ix–x) argues, there are assumptions about inclusion and exclusion which are implicit in the design of spaces and places. Thus, the crucial questions that need to be asked here are: who is this urban space for, who does it exclude and how are prohibitions maintained in practice?

In her analysis of urban regeneration in Manchester city centre, Rosemary Mellor (1996: 67–8) has argued that

The regional centre is socially eclectic – extensively used by the local poor as well as the young ... It is the shopping centre, marketplace, source of entertainment and recreation for the inner-city populations. British cities are relatively permeable and current policy is to retain that characteristic.

The major problem with Mellor's analysis, however, is that it promotes a kind of myth of inclusion which is central in securing capitalist hegemony. From this perspective the city becomes the perpetually acceptable face of capitalism because the perspective fails to expose the processes of exclusion and spatial prohibition that are informally, routinely and selectively enforced by formal agencies of control in public space (Pile, 1996). As Wardhaugh (1996: 703–4) argues, 'Spatial boundaries are used across cultures to assert a moral order. In the postmodern, post-industrial city boundaries may not be as clearly delimited as in pre-industrial societies, but they are no less vigorously enforced.' And crucially, as a number of critical geographers have shown (see Ruddick, 1996: Valentine, 1996) the exclusion of children/youth is built into so-called universal public spaces (Lees, 1998: 237).

But why are youth (and younger teenagers in particular) one of the primary targets of police removal in the 'aestheticized' public space of Tamworth town centre? There are a whole set of assumptions mobilized in official and public discourses about the potential 'delinquent' behaviour of teenagers in public places. In the dominant imagination the category 'youth' occupies a central position in the tension between order and control and disorder and deviance. Media panics which frame 'youth as trouble' and a potential threat to public order play a large part in the social construction of youth and youth cultures (Valentine *et al.*, 1997: 4). What is significant about recent youth panics, however, is that they are strongly 'underclassed'.[12] For example, in Britain a series of new and dominant images have recently appeared on the horizon around which various ideas about youth and youthful misbehaviour have been ordered. Official and public anxieties have recently centred on the theme of 'dangerous masculinities' (Cambell, 1993), 'yob culture' and 'lawless girls gangs' amongst a 'new underclass' (Murray, 1993) of contemporary youth (Back, 1999). Orchestrated media panics have condensed explicitly around the presence and practices of teenagers hanging around on the urban streets (and especially upon groups of youths under 16 who gather without adult supervision in public space). The street is commonly represented within media panics as a place where deviance is so often located and in relation to youth is cast as an emblematic site for particular forms of criminal activity, gang behaviour and drug dealing (Keith, 1999). This logic positions youth as a particular kind of social imaginary – as threatening, dangerous and 'undesirable' occupants of public space. Importantly, as Breitbart (1998: 307) points out, the demonization of youth by the media is often used to justify the restrictions that are placed on young people's free access to public spaces in the city. Media panics combined with the increased privatization and revitalization of public space result in 'public policies that seek to remove young people from public places, delimit their geography and enforce their invisibility' (ibid.). This 'narrative harmony' (Back, 1999) between media and dominant political discourses is especially important in relation to recent government legislation affecting youth in the UK. As Keith (1999: 5) points out, young people have been rendered a 'problem' in government discourse because they are often taken to personify the risks of the city, '"youth" has become a subject of [punitive] policy intervention across a range of related pieces of legislative change. A whole series of policy initiatives have targeted Britain's young people as the objects of policy concern.' These principles and punitive policies are enshrined in New Labour's recent adoption of 'zero tolerance' youth policing which forms a core component of the Crime and Disorder Act 1998 (Home Office, 1998).[13] 'Zero tolerance' is based on a US model of law enforcement and translates in the UK into a more authoritarian approach to 'non-chargeable' youth crime by giving police more powers to clear teenagers off the streets.[14] The series of disciplinary legislative measures contained in the Crime and Disorder Act are designed to enhance the powers of the local state to regulate and control young people's activities in public space (Fyfe and Bannister, 1998). This project is deeply bound up with and a moralizing reinvention of youth identity. As Massey (1998: 128) argues, the control of spatial behaviour is part of the social definition of what are/are not acceptable forms of youth culture and identity. Thus the attempt to control and define 'youth' hinges centrally on spatiality: 'The control of spatiality is part of the process of defining the social category of "youth" itself. It is also part of the process of defining what is deemed as acceptable behavior on the part of the group.'

The deep-rooted changes that are taking place in the policing of young people in public space as part of a national 'law and order' political agenda are also mirrored at the local political level (Fyfe and Bannister, 1998: 259). As Keith (1999: 6–7) points out, the local is the level on which the conduct of young people is to be mapped, and, importantly, CCTV has a central role to play in this process. For example, in the remaking of the townscape in Tamworth, the local authority has taken excessive steps in the name of public safety by deploying 'rational strategies of social control' (Goss, 1998: 238) and regulation in the form of CCTV and exclusionary policing measures. Local geographies of intolerance have revised codes of public conduct in an attempt to reinvent public space, promoting the commercial functions of the street over all other uses by accommodating shoppers, tourists and office workers during daylight hours and those at play in the pubs, clubs and restaurants at night-time. In a space dedicated to consumption by the family and others (including consuming youth) it is non-consuming school-age youth (and others) with low spending power who are defined as a central 'suspect category' (Cohen, 1999) and a 'polluting presence' on the street, and who thus constitute a primary designated target of new spatial injustice strategies.[15] This is a strategic attempt to circumvent the streets as a socially secured space through the exclusion of youth, reducing diversity on the streets by domesticating urban space in order to make it 'safe' for reinvestment and use by those who qualify as consumer citizens (Lees, 1998: 237).

In the contemporary economic and political climate in which the role of the local state is increasingly seen as an organizer of 'risk management' and the creator of officially sanctioned geographies, the evident rationale in hegemonic cartographies of place in Tamworth, then, is to create a kind of geographically enclosed 'security bubble' for shoppers and the consuming public, maximizing safety by creating time-spaces in the urban public sphere where young people are not allowed (Keith, 1999). It is the visible presence of youth hanging around on the street that represents a threat to the sense of place, public order and orderliness enforced by the urban regime and this has led to the application of consistent police pressure to limit teenagers' free access to the urban outdoors (particularly at night) by enforcing their eviction from all street, park, wall and bench space in areas under electronic control.[16] What does it mean to be on the receiving end of constant police efforts to (re)gain control of the street?

'It's like being sent to another country': the problem of CCTV surveillance and spatial closure for youth

The introduction of these new kinds of disciplinary measures has meant that 'fortress city' is now increasingly a daily reality for these youth, as this statement from John (aged 14), who recently moved to the area from Glasgow, shows:

> The cops are really strict here and they get you all the time for anything. In Glasgow like you could get away with hanging around there in the city centre. They stop you here though all the time and search you and tell you to move on out of the town.

This is confirmed by Rhia (aged 15) who claims that the group has become a primary target of excessive police control because the local council wants to 'keep up appearances':

> They try to stop you doing anything in town now ... it's like being sent to another country. They just don't want you on the streets so they get you for anything now even though we don't do nothing 'cos they're trying to make the place look posh ... and they don't want us kids being there now 'cos it don't look good. So the cops just push us out of the town centre all the time.

Importantly, almost all of the girls in the sample expressed concern about being watched and recorded by CCTV cameras and were conscious of being under the scrutiny of an objectifying male gaze:

> I bet they get a thrill out of it like ... it's pervy 'cos they have just got to sit there and watch you ... you don't know who they are all those men just watching us and looking at the screens all day ... 'cos they might think 'cor look at her' ... you don't know what they might do with the tapes and that ... 'cos some people prey on us young girls, like.

These youth rely on public space as a means to build cultural identity. Their subjectivities are intimately bound up with, and dependent upon, access to and control over particular material and symbolic spaces within the town centre, but concerted police efforts to clear them off urban streets are jeopardizing the survival of their youth cultures. Whilst a number of these teenagers have been discouraged from inhabiting central urban spaces as a direct result of CCTV and police removal strategies,[17] the majority of these young people are determined not to be displaced from urban space because they continue to see creative and meaningful possibilities for activity in this secured urban setting. Here is a typical statement on this theme:

> We still hang around in town 'cos it's just the place that we like being in ... It is where everything happens. Anyway all the new leisure stuff costs too much ... We're not going to go away just 'cos they [the police] try to push us out and the cameras watch us all the time.

This highlights that dominant topographies of discipline and spatial closure are repeatedly contested and resisted by these young people. In this way, the town centre is simultaneously a regulated environment infused with new configurations of power and control and a contested place, a site of countervailing youth resistance to spatial authority. As Hebdige (1997: 403) argues, 'As power is deployed in new ways, so new forms of powerlessness are produced and new types of resistance become possible.' Crucially, as Herbert (1998: 226) has pointed out, police efforts to claim sovereignty over the street 'are always subject to contestation ... But resistance to the police, like resistance more generally, is rarely ... cataclysmic; it typically involves more subtle evasions and resistances' and it is in this manner that these teenagers resist the spatial hegemony of the dominant city. The work of Michel de Certeau (1984) provides a good starting point from which to map the dynamics of this resistance.

'Hide and seek': teenage transit points and temporary abodes[18]

> All I need is a place to find and there I'll celebrate (Air, 'All I Need').

In de Certeau's terms public spaces of commerce and power are sites of everyday resistance where, he argues, there has been too much emphasis on strategies of control and too little emphasis on the 'tactics of the weak'.[19] For de Certeau, walking in public space is a central resistant tactic. It is associated with what de Certeau (1984: 190) calls 'delinquency' because it undoes the planned city and creates within it a mobile city; it defines a 'space of enunciation', a space from which to articulate a 'spatial story' or narrative of resistance:

> [t]he act of walking is to the urban system what the speech act is to language or to the statements uttered. At the most elementary level, it has a[n] ... 'enunciative' function; it is a process of *appropriation* of the topographical system on the part of the pedestrian. (ibid.: 97)

For de Certeau, then, pedestrian practices can be analysed as a fleeting, though ineradicable, space of resistance. However, while walking is a transient and evanescent practice which 'always involves a lack of place (not being somewhere)' (Pile, 1996: 226–7), at the same time, according to de Certeau, the pedestrian is unaware that their movement produces unmappable space: 'These practitioners make use of spaces that cannot be seen [but] their knowledge of them is as blind as that of lovers in each other's arms' (de Certeau, 1984: 93). Moreover, for de Certeau (1984: 101) the 'long poem' of walking 'manipulates spatial organisations, no matter how panoptic they may be: it is neither foreign to them (it can take place only within them) nor in conformity with them (it does not receive its identity from them)'. However, de Certeau's 'dual vision' of the city (the idea that the city 'from down below' remains unseen when seen in panorama 'from above') may start to look problematic when we consider the extensive use of various forms of 'techno-policing' in urban centres. It might be argued, for example, that the introduction of CCTV surveillance cameras in street policing mediates and nullifies de Certeau's division between the 'static city' and the 'mobile city', offering regimes of place the means with which to impose order both 'from above' and 'from down below', to unveil and intervene in pedestrian territories and thus make all those 'invisible', 'polluting' and 'illicit' spatial practices and locations, that for de Certeau (1984: 93) exist 'below the thresholds at which visibility begins', arrested by vision, immobilized, repressed and controlled. I want to suggest that the concept of pedestrian 'tactics' as envisioned by de Certeau cannot help us at this point but requires further refinement in order to explain the various ways in which space is seized by these youth within secured urban space. I want to argue that the practised place of the city is a space still capable of manipulation at street level despite the presence of CCTV and police control. CCTV cannot command total surveillance and displace all 'improper' productions of space because there are cracks in the vision of this scopic technology, pockets of space within monitored environments which remain invisible.[20] And because, as Herbert (1998: 234) contends, 'the police can only ensure control of part of public space and only for limited periods of time', this means that there is room for subversion of all kinds within public space.[21]

How do these young people escape the gaze of social authority and re-find a place in the street? What kind of steps do they take to manipulate surveilled space and how does their movement differ from de Certeau's ordinary practitioners? These teenagers mobilize particular modes of movement within controlled urban space in ways that are neglected in de Certeau's analysis. For example, the essential problematic they faced from the onset of CCTV and police control of the street was the fact that, in order to remain in the town centre, they needed to re-organize their hangout sites by establishing new gathering places where they could not be seen by cameras and easily moved on by the police. In direct response to the particular conditions of disciplinarity and exclusion to which they are subject, these youth have developed deliberate 'narrative footsteps' – a specific evanescent kind of transit – enacted as a series of countermoves in order to make it hard for police and CCTV operatives to co-ordinate their efforts to encircle and capture. They have discovered a way around the system of control in which they find themselves by finding concealed interstitial spaces within, and 'invisible' routeways through, the town centre, employing tactical 'choreographies' in order to navigate the controlled landscape and re-appropriate space for themselves. Unlike de Certeau's everyday practitioners, then, these youth have a *conscious* relationship with the urban landscape, intentionally producing unmappable spaces within surveilled space which are plotted out within their everyday walked landscape beyond CCTV's controlling vision, as this statement shows:

> We can go and hang around in town ... but we just have to make sure we get away from the cameras and police all the time, like ... We just keep moving about and go to places where they don't know where we are and can't see you. It's like 'hide and seek' now in town with all those cameras. (Debbie, aged 14)

Time, as was noted earlier, is a significant dimension both in these young people's experience of spatial exclusion and in their resistance. For example, these youth are routinely evicted from public space by the police in 'respectable' daylight hours, but report that they experience sharper levels of police intolerance at night. Some commentators have argued that 'increasingly the spaces of the street are far from secretive at night ... this time/space is being progressively eroded by the ever-widening gaze of closed-circuit television with its disciplinary function' (Cresswell, 1998). However, these teenagers have evolved mobile tactics which enable them to remain invisible to the forces of order within this landscape of exclusion – a place in which it is seemingly impossible to hang around by virtue of the sheer number of cameras overlooking the street – because they have a highly sophisticated degree of situated knowledge of this urban locale. They know the urban landscape intimately and have interpreted information about CCTV – how it works and its weaknesses – to their advantage. They have discovered the location of an extensive network of 'blindspots' in CCTV's viewmap which they use as signposts to re-map, traverse and organize surveyed space; to pattern their surroundings in ways that ensure that their activities can pass unnoticed. As well as inhabiting existing unmonitored spaces within core surveilled space, one of the main ways in which they evade visual capture is to split into small groups as a way of dividing the regulatory gaze and then orchestrate moves in the opposite direction to which camera heads are pointing, using vision and movement to walk the blindside of cameras and exploit 'fluid' blindspots and gaps within the disciplinary field, weaving and wefting around the scopic regime to clear unmappable paths from one hidden hangout to another. At other times they take advantage of their knowledge of local geography to successfully flee approaching officers and retreat into hidden spaces to escape detection (Herbert, 1998). As Alex (aged 14) puts it:

> There are like these blindspots everywhere. We sort of dodge the cameras when they follow us you can hear 'em clicking like mad trying to spot us. We know where to go where they can't see you and we just hang around there 'cos between

the cameras there are gaps where they can't see you and we move around so they can't follow you all the time. Sometimes we just run from the police and hide where they don't know.

According to these teenagers, the police and CCTV security staff are aware of some of these various blindspots, but on the whole tend to assume total panoptic surveillance:[22]

They see us by McDonalds and then we disappear and they must think that we have gone out of town but we don't we go all over the town. It's going on under their noses but they don't know where 'cos they can't see you.

In playing 'hide and seek' with CCTV and police patrols these youth work the unmonitored angles, vanishing points and fade-out spots within CCTV's horizons, cutting the totality of ordered space into a multiplicity of concealed spatial fragments and stitching places of concealment together to establish their own mobile map of hidden transit points and temporary abodes beyond the regulatory gaze of social authority.

The limits and possibilities of evasive resistance in surveilled space

The question of whether youthful opposition 'makes a difference' and constitutes '"real" acts of resistance or the extent to which, conversely they [are] simply symbolic acts serving to reproduce the very structures of inequality that they challenge ...' (Ruddick, 1998: 343) has been a matter of theoretical debate for many decades. Recently, Edensor (1998: 214), echoing Davis' thesis (1990) on the end of the 'authentic' public space of the street, has argued that the commodification of most Western high streets has resulted in the

triumph of 'non-space' [which] depoliticises the street, forcing forms of resistance to adopt more covert strategies. Subaltern social movements depend upon the temporary seizure and transformation of public space in order to transform alternative symbolic meanings, which the regulation of spaces for representation denies ... Whilst the presence of 'mall rats', beggars and shoplifters testifies to certain forms of resistance, such opposition seems fleeting and gestural in the face of intensive surveillance.

The conclusion drawn by Edensor (1998) certainly captures something about the experience and control of urban space in late modernity, particularly in relation to the ways in which certain forms of sociability are facilitated by particular kinds of space. However, there are a number of problems with Edensor's (1998) deterministic analysis, not least of which is the undertheorization of the possibilities of youth resistance in urban space. He frames 'successful' youth resistance in highly limited terms, citing 'mall rats' as evidence of a circumscribed form of resistance. Yet, as I have shown, the shopping mall is a fundamentally different kind of space for youth – in terms of its gendered use, the particular opportunities it affords teenagers and its perceived level of safety – in comparison to the space of the street however panoptic it may be (Edensor, 1998). His over-riding concern with control and over-emphasis of the process of being rather than *becoming* in space and place (Harvey, 1989) disqualifies the tactical ways in which youth deflect disciplinary power, and thus marginalizes the ability of youth to open up and re-appropriate meaningful space for themselves within the seemingly 'dead' space (Berman, 1988) of tightly controlled urban environments. Crucially, we need to think more contextually about youth resistance and recognize that CCTV and police control cannot be universalized as always all-powerful and all-encompassing across time, space and context because any attempt to control spatiality is always mediated by the particularities of place and the social and psychic forces of contestation. As Lees (1998: 238) argues, public space is as much a cultural space which embodies struggle and resistance as it is a controlled milieu: 'Whilst there may have been an increase in the control and surveillance of public space ... public space is being opened up in new and complex ways ... the control of public space can always be countered, subverted and resisted.' This observation is important in light of these teenagers' evasive choreographies, practices which allow them to remain shielded from the watchful eye of town centre security apparatus and escape, however momentarily, particular forms of disciplined selves which might only be a product of time and space but are no less powerful for being so (Keith, 1999).

It is important to emphasize that these teenagers have re-constituted their youth cultures in surveilled space by re-configuring their sense of gathering place on a smaller scale. They inhabit a city of bits, a fractured landscape that CCTV will never see. This is an important alternative space of youth interaction which has hitherto been neglected in much cultural and sociological theory (however, for notable exceptions see Back, Cohen and Keith, 1999; Ganetz, 1995). As Keith (1999: 11) argues:

> Much sociological analysis of public space has tended to focus on notable sites in which particular forms of social interaction have occurred; Sennett's rendition of the parks and cafes of London and Paris or Berman's invocations of the streets of New York [but] more mundane places emerge at the interface between public spaces ... that highlight the existence of a micro-public sphere.

The important question remains, however, about the nature of their access to public space in terms of the quality of the micro-spaces they inhabit. For example, it could be argued that by concentrating their social activities into unmonitored gathering places these youth are in effect policing themselves out of prime public space, that the ironic consequence of their spatial resistance is merely 'self-surveillance' (Foucault, 1980). Yet this misses the important point that whilst CCTV and police control surrounds these youth and has re-defined their relation to the urban by reducing the size of their environment, inhibiting certain forms of urban sociability in certain prime time–space contexts, it is not constitutive of their social experience in public space; it does not enclose them within marginal space. For example, these youth have arrived at a new sense of their landscape by shifting their social centre from one set of spaces to another (but still within prime core spaces), re-defining the purpose and meaning of public space in the formation of their subjectivities by creating a new 'spatial bricolage' of 'hybrid' sites which link all their activities. These sites represent resistant spaces or 'free zones', places within which vital spatial practices can develop and spontaneously unfold without being captured or determined by CCTV and police control. They are organized in relation to multiple ends and are crammed with all the important experiences and established relations they seek from the urban realm; these are also highly gendered spaces where single-sex or mixed friendship cluster groups can converge and meet (or where they can be alone in their primary peer groups) depending upon the particular times they are inhabited.

Importantly, it is not simply the case that these youth are hidden away in compromised solitary spaces or that public space is only a space of retreat for these youth because a large number of these interstitial hangout spaces are highly *public* spaces which are located within or overlap main shopping, leisure and business districts, or are situated on and next to the busiest pedestrian thoroughfares in the town centre. These are spaces of both high and low visibility where these youth can both withdraw and show themselves. In this way, they are able to shield their activities and whereabouts from CCTV and police patrols but remain a sustained and highly visible presence in public space.

These youth have produced a new 'stealth' youth culture, then, which slips beneath authoritative power and control and remains invisible to, but in the midst of, the dominant city. They express a social space where the ordering of front and back region are inverted. They make a tactical appropriation of hidden space in ways that change the value of these spaces so that these spaces do not confirm for them their own marginality but have become their privileged front region (Ruddick, 1998: 358). It can be argued, therefore, that they occupy *space on the margins* but not marginal space. However, there is a downside to the ways in which they make themselves a home in this prohibited location which is evidenced in the ways in which these teenagers (girls and boys) set up their own spatial prohibitions against other youth in an attempt to hold the ground they have won.

The 'race for space': youth, territoriality and violence

Surveilled space is, of course, a space of power, tension and violence enacted 'from above', but it is also contested 'down below' both across and within gender and generation. In town centre Tamworth there is a complex play of different modalities and relations of power, territoriality and control enacted on the street, and youth cultures are central in these processes. The urban centre has become a crowded space for youth in the sense that CCTV has reduced the number of public places in which teenagers are able to gather and this has led to a situation in which there is a high degree of contestation over space amongst youth.[23] The unmonitored interspaces these youth inhabit are, of course, also attractive to other young people who insist on their rights to these 'liminal' places. In the intense competition amongst youth for a place in the street these teenagers (girls and boys) collaborate as a large network group and claim these spaces through violence. The claiming and closure of youth territory is integral to the identity that these youth are striving to establish and is premised upon a rigid insider and outsider dichotomy. However, this is also a highly selective process because they conceptualize these spaces as thresholds over which they have control and possession (where they rule first) but include some youth who are known to them and exclude others who are either unknown or who, in particular, are members of rival neighbourhood-based youth groups. In this way, these 'syncretic' spaces are simultaneously 'frontlines' and 'backyards' (Cohen, 1997) linking these youth with other teenagers in the youth public sphere, but in a highly selective way as the crucial difference between being known and not being known to this dominant youth cultural formation can mean the difference for many young people of their age between safety and danger in public space.

Importantly, their attempts to assert their own rules and rituals of territoriality and monopolize hidden areas for their own use have a significant adverse impact on other young people's social experience in urban space (Loader, 1996). For example, many young people who took part in this study but who had no connection with the primary focus group identified certain places in and around the town centre as sites fraught with danger and 'trouble' because they associate these spaces primarily with the teenage friendship group. The urban centre represents for a large number of young people a landscape of risk rather than chances, and this is intensified at different times and in particular places: i.e. many are extremely cautious when using the urban environment and avoid certain 'phobic' spaces (Cohen, 1999) or simply avoid visiting the town centre (especially at night). There are marked similarities, therefore, between the efforts of social control agencies and the practices of the dominant teenage group who also attempt to establish territory and exercise their own modalities of power and spatial control. This intersection of power, territoriality and exclusion, then, highlights some of the more destructive and negative aspects of these young people's spatial practices.

Conclusion

Throughout this chapter I have attempted to highlight the ways in which these young people's uses of public space intersect with important aspects of their youth identities such as class, gender and generation, whilst also drawing attention to their ability to subvert and resist the production and protection of adult/consumer-oriented public space and carve out meaningful spaces for themselves within an over-determined environment. The dynamics of the formation of their youth culture has much to do with the particularities of spatial restriction in this urban context. As de Certeau (1984: 37) argues, 'The space of the tactic is the space of the Other. Thus it must play on and with a terrain imposed upon it and organized by the law of a foreign power.' It is in this field of tension, in the space between controlled and resistant spaces, that these teenagers live out their youth cultures. Their 'tactics of invisibility' (Ruddick, 1998) provide them with a degree of imperceptibility which enables them to release micro-spaces from monitored 'non-space', often only temporarily and partially but at times for as long as they need them, to create an important 'elsewhere' – their own spaces for representation, or places out of place. They have disrupted the strategies of control which deny them access to public space and manipulated and diverted the meaning of dominant space, re-inscribing it as a site of possibility and meaning, creating the ground for the re-location of their identities through a 'hidden' mode of resistance. Their 'enunciative' acts shatter the illusion of police and CCTV pre-eminence. As Herbert (1998: 236) notes, the police 'are engaged in an ongoing game of cat-and-mouse, an ever-shifting struggle [with youth on the street] where they do not always win.' CCTV and police removal strategies, then, have not resulted in the fundamental dissolution of their spatial access and have failed to corrode their youth cultures because these youth have not let disciplinary power work on its own terms. However, a note of caution is necessary here because, while these youth are active in myriad ways in the production of their own space, it would be wrong to equate 'active' with 'powerful'. We need to be aware here that these young people's geographies of dissent have an ambiguous outcome because they display both emancipatory and brutalizing features. They are simultaneously positive expressions of freedom and opposition which also have a foot in terror and violence evidenced in their attempts to territorialize and claim space, which can have a potentially negative impact on other young people's free access to public space.

Finally, Susan Ruddick (1998: 358) has drawn attention to the dangers of speaking exclusively of tactical forms of resistance, suggesting that this risks romanticizing the condition of marginalized people in ways that demand no further action. This is crucial because, whilst these youth have found a place in the street, it is not the only place they want to be. For example, all of these young people express a desire for an increase in adequate and appealing leisure spaces and for free access to public space, seeing a clear relationship between leisure improvement, spatial inclusion and a positive impact on the overall quality of their lives (Breitbart, 1998). As Sally (aged 13) highlights, these teenagers have their own ideas about how the local council should go about revisioning urban life:

> I think the council could do something about where we want to go ... 'cos they moan about us hanging around and they moan about lack of money and everything and then they go and put those surveillance cameras up everywhere ... They are stupid! If they want us off the streets they should spend the money on things for people our age that are cheap to get in like an ice skating rink and skate park and they should let us hang around town at night time as well without moving us on all the time. It would be better for everyone really.

This highlights the needs of these school-age teenagers for viable places they want to use and where their free access to public places in the urban centre is a priority. Interestingly, these teenagers' resistant spatial practices have forced a change of direction, albeit partially, in official thinking, evidenced in recent proposals put forward by Kevin O'Leary, the Chief Inspector of Tamworth Police, to develop a 'free zone' for teenagers in Tamworth in the form of a self-contained 'teen village'. The idea is to give youth of school age some space of their own in order to eliminate the problem of 'nuisance' teenagers hanging around the urban streets. Essentially, of course, the main function of a so-called 'teen village' is to displace and contain the 'problem' of teenagers' spatial activities into socially controlled sites which allow youth to be more 'effectively' policed and this is the main problem with such a development in respect of teenagers. It is highly unlikely that young people themselves would give their support to a place which is intended to keep them off the streets and ensure the police even closer control and surveillance of their activities. Such proposals for special locales for youth miss the important point that teenagers, as this research has shown, do not want to disappear from the space of urban streets – the very places that provide autonomy from adult supervision. There is a need for a new and different vision of the urban which meets teenagers' needs for appealing leisure spaces and which also protects their free access to public space. However, recent policy initiatives which seek to deny young people the freedom of the city through the reinvention of its public spaces point to an uncertain future for youth. It looks increasingly as if it is going to be a long game of 'hide and seek'.

References

Acconci, V. (1990) Public space in private time. *Critical Enquiry*, **16** (Summer), 134–57.

Anderson, S., Loader, I. and Kinsey, R. (1994) *Cautionary Tales*, Avebury, Aldershot.

Auge, M. (1995) *Non-Places: Introduction to an Anthropology of Supermodernity*, Verso, London.

Back, L. (1996) *New Ethnicities and Urban Culture: Racism and Multiculture in Young Lives*, UCL Press, London.

Back, L. (1999) *Rights and Wrongs: Youth Community and Narratives of Racial Violence*, Centre for New Ethnicities Research, University of East London, London (Finding the Way Home Working Papers 5).

Back, L., Cohen, P. and Keith, M. (1999) *Between Home and Belonging: Critical Ethnographies of Race, Place and Identity*, Centre for New Ethnicities Research, University of East London, London. (Finding the Way Home Working Papers, 2).

Bell, D. and Binnie, J. (1998) Theatres of cruelty, rivers of desire. In Fyfe, N. (ed.) (1998) *Images of the Street: Planning, Identity and Control in Public Space*, Routledge, London.

Berman, M. (1988) *All that is Solid Melts into Air*, Penguin, Harmondsworth.

Blackman, S. J. (1998) The school: 'Poxy Cupid!'. In Skelton, T. and Valentine, G. (eds) (1998) *Cool Places: Geographies of Youth Cultures*. Routledge, London.

Boyle, M. and Hughes, G. (1995) The politics of urban entrepreneurialism in Glasgow. *Geoforum*, **25**, (4), 453–70.

Breitbart, M. (1998) Dana's mystical tunnel: young people's designs for survival and change in the city. In Skelton, T. and Valentine, G. (eds) (1998) *Cool Places: Geographies of Youth Cultures*, Routledge, London.

Bulos, M. and Sarno, C. (1995) *Closed Circuit Television and Local Authority Initiatives: the First National Survey*, HMSO, London.

Cambell, B. (1993) *Goliath: Britain's dangerous places*, Lawrence and Wishart, London.

Cohen, P. (1972) *Subcultural Conflict and Working Class Community*, Centre for Contemporary Cultural Studies, University of Birmingham (Working Papers in Cultural Studies No. 2).

Cohen, P. (1997) Out of the melting pot into the fire next time: imagining the East End as city, body, text. In Westwood, S. and Williams, J. (eds) *Imagining Cities*, Routledge, London.

Cohen, P. (1999) *Strange Encounters: Adolescent Geographies of Risk and the Urban Uncanny*, Centre for New Ethnicities Research, University of East London, London. (Finding the Way Home Working Papers 3).

Cohen, P. and Robins, D. (1978) *Knuckle Sandwich: Growing up in the Working Class City*, Penguin, Harmondsworth.

Corrigan, P. (1979) *The Smash Street Kids*, Paladin, London.

Cresswell, T. (1998) Night discourse. In Fyfe, N. (ed.) (1998) *Images of the Street: Planning, Identity and Control in Public Space*, Routledge, London.

Crouch, D. (1998) The street in the making of popular geographical knowledge. In Fyfe, N. (ed.) *Images of the Street*, Routledge, London.

Davis, M. (1990) *City of Quartz*, Vintage Books, New York.

Davis, M. (1992) Fortress Los Angeles: the militarization of urban space. In Sorkin, M. (ed.) *Variations on a Theme Park: The New American City and the End of Public Space*, Hill & Wang, New York.

de Certeau, M. (1984) *The Practice of Everyday Life*, University of California, Berkeley, Ca.

Edensor, T. (1998) The culture of the Indian street. In Fyfe, N. (ed.) (1998) *Images of the Street: Planning, Identity and Control in Public Space*, Routledge, London.

Foucault, M. (1980) *Discipline and Punish*, Penguin, Harmondsworth.

Fyfe, N. and Bannister, J. (1998) The eyes upon the street: closed-circuit television surveillance and the city. In Fyfe, N. (ed.) (1998) *Images of the Street: Planning, Identity and Control in Public Space*, Routledge, London.

Ganetz, H. (1995) The shop, the home and femininity as a masquerade. In Fornas, J. and Bolin, G. (eds) *Youth Culture in Late-Modernity*, Sage, London.

Goffman, E. (1967) *The Presentation of Self in Everyday Life*, Penguin, Harmondsworth.

Goss, J. (1998) Disquiet on the waterfront: reflections on nostalgia and utopia in the urban archetypes of festival marketplaces. *Urban Geography*, **17**, (3), 221–47.

Griffin, C. (1985) *Typical Girls? Young Women from School to Job Market*, Routledge & Kegan Paul, London.

Harvey, D. (1989) *The Condition of Postmodernity*, Blackwell, Oxford.

Hebdige, D. (1997) Posing … threats, striking … poses: youth surveillance and display. In Gelder, K. and Thornton, S. (1997) *The Subcultures Reader*, Routledge, London.

Herbert, S. (1998) Policing contested space: on patrol at Smiley and Hauser. In Fyfe, N. (ed.) (1998) *Images of the Street: Planning, Identity and Control in Public Space*, Routledge, London.

Hollands, R. G. (1990) *The Long Transition: Class, Culture and Youth Training*, Macmillan, Basingstoke.

Keith, (1999) *Making Safe*, Centre for New Ethnicities Research, University of East London, London.

Langman, L. (1992) Neon cages: shopping for subjectivity. In Shields, R. (ed.) *Lifestyle Shopping: The Subject of Consumption*, Routledge, London.

Lees, L. (1998) Urban renaissance and the street: spaces of control and contestation. In Fyfe, N. (ed.) (1998) *Images of the Street: Planning, Identity and Control in Public Space*, Routledge, London.

Leiberg, M. (1995) Teenagers and public space. *Communications Research*, **22**, 206–31.

Loader, I. (1996) *Youth, Policing and Democracy*, Macmillan, Basingstoke.

Malbon, B. (1998) Clubbing: consumption, identity and the spatial practices of everynight life. In Skelton, T. and Valentine, G. (eds) *Cool Places: Geographies of Youth Cultures*, Routledge, London.

Massey, D. (1998) The spatial construction of youth cultures. In Skelton, T. and Valentine, G. (eds) (1998) *Cool Places: Geographies of Youth Cultures*, Routledge, London.

McNamee, S. (1998) The home: youth, gender and video games. In Skelton, T. and Valentine, G. (eds) (1998) *Cool Places: Geographies of Youth Cultures*, Routledge, London.

McRobbie, A. and Garber, J. (1976) Girls and subcultures – an exploration, in Hall, S. and Jefferson, T. (eds) *Resistance Through Rituals*, Routledge, London.

Muggleton, D. (1995) From 'subculture' to 'neo-tribe': identity, paradox and postmodernism in 'alternative style'. Paper presented at Shouts From the Street Conference, Manchester Metropolitan University.

Murray, C. (1993) *The Underclass: The Crisis Deepens*, Health and Welfare Unit, ILEA, London.

Pearce, J. (1996) Urban youth cultures: gender and spatial forms. *Youth and Policy*, **52**, 1–11.

Pile, S. (1996) *The Body and the City*, Routledge, London.

Ruddick, S. (1996) Constructing difference in public spaces: race, class and gender as interlocking systems. *Urban Geography*, **17**, (2), 132–51.

Ruddick, S. (1998) Modernism and resistance: how 'homeless' subcultures make a difference. In Skelton, T. and Valentine, G. (eds) (1998) *Cool Places: Geographies of Youth Cultures*, Routledge, London.

Scott, J. (1985) *Weapons of the Weak: Everyday Forms of Peasant Resistance*, Yale University Press, New Haven.

Sennett, R. (1990) *The Conscience of the Eye*, Faber and Faber, London.

Shields, R. (1992) (ed.) *Lifestyle Shopping: The Subject of Consumption*, Routledge, London.

Sibley, D. (1995) *Geographies of Exclusion*, Routledge, London.

Smith, N. (1992) New city, new frontier: the Lower East Side as Wild, Wild West. In Sorkin, M. (ed.) *Variations on a Theme Park: The New American City and the End of Public Space*, Hill & Wang, New York.

Stanko, E. (1990) *Everyday Violence: How Women and Men Experience Everyday Sexual and Physical Danger*, Pandora, London.

Tagg, J. (1996) The city which is not one. In King, A. (ed.) *Re-Presenting the City*, Macmillan, Basingstoke.

Valentine, G. (1989) The geography of women's fear. *Area*, **21**, 385–90.

Valentine, G. (1996) Children should be seen and not heard: the production and transgression of adults' public space. *Urban Geography*, **17**, (2), 205–20.

Valentine, G., Skelton, T. and Chambers, D. (1998) Cool places: an introduction to youth and youth cultures. In Skelton, T. and Valentine, G. (eds) *Cool Places: Geographies of Youth Cultures*, Routledge, London.

Walkerdine, V. (1997) *Daddy's Girl*, Macmillan, Basingstoke.

Watt, P. and Stenson, K. (1998) The street: 'It's a bit dodgy around there': safety, danger, ethnicity and young people's use of public space. In Skelton, T. and Valentine, G. (eds) (1998) *Cool Places: Geographies of Youth Cultures*, Routledge, London.

Westwood, S. (1990) Racism, black masculinity and the politics of space. In Hearn, J. and Morgan, D. (eds) *Men, Masculinities and Social Theory*, Unwin Hyman, London.

Willis, P. (1990) *Common Culture*, Open University Press, Milton Keynes.

Wulff, H. (1995) Inter-racial friendship: consuming youth styles, ethnicity and teenage femininity in South London. In Amit-Talai, V. and Wulff, H. (eds) *Youth Cultures: A Cross-Cultural Perspective*, Routledge, London.

Notes

1   This chapter is based on accounts collected as part of a research project on 'Youth, White Ethnicity, Shopping and Space' for a Ph.D. thesis. The field-work was carried out over a two-year period between January 1995 and January 1997 with 20 girls and 20 boys (aged 12–16) at a youth club in North Tamworth. The research data was collected in a variety of ways including semi-structured, taped individual and group interviews with informants to solicit stories about their leisure time and use of public space, 'mental map' marking exercises focusing on their perception of safe and dangerous areas in the town centre, photographic diaries and direct observation of their patterns of use of public and semi-public space in shopping centres, precincts, parks and on the streets at weekends in day-time and night-time contexts. Interviews were also carried out with a smaller sample of teenagers, 10 boys and 10 girls (aged 12–16) who were unconnected with the main focus group on the theme of 'safety' and 'danger' in public space.

2   The young people that make up the focus group studied here associate closely as a loosely structured friendship network group which is divided into smaller 'cluster groups' of 4 to 6 'best friends' of girls-only, boys-only and mixed subsets that alter and change members.

3   Two points of qualification should be made here before we can begin this discussion. Firstly, this chapter reports on the social activities of one loosely defined friendship group of ordinary youth in their lower teens who regularly go out and use public space together. Therefore, their practices and actions should not be taken to be representative of the social practices of all young people in this locale. Secondly, the intention in this study was not to collect the accounts of police officers and CCTV monitoring staff themselves on how they police teenagers in public environments, but rather to provide an account of prohibitions and constraints on activities from the point of view of the excluded, and thus allow these marginalized and disempowered young people a voice which would enable them to tell their own stories and provide a critical commentary on their own sense of spatial oppression and what they do to enable themselves to escape it.

4   These young men and women's use of urban space also intersects with their ideas about 'white ethnicity', hybridity and 'race' which are dealt with in the larger thesis.

5   Public space occupies a contradictory position in these young people's everyday lives. It is experienced as both an exciting arena and as a tedious and predictable site. They move between the two senses of space continually and their aim is to substitute the latter with experiential pleasure.

6 All real names have been changed in order to protect these young people's rights of confidentiality.

7 Large-scale, out-of-town retail parks are less important in these respects.

8 It is important to emphasize that the girls are not marginalized in the group and the girls and the boys do not exist autonomously from each other; their patterns of spatial activity do not take place in isolation but are shared, alternate and cross over and thus their cultures are constituted through direct interaction, negotiation and 'consent' to each other's wishes and demands. At the same time their social use of public space allows them to generate and explore particular ideas about identity where feminine subjectivities and teenage masculinity are enacted differently in public space.

9 Urban regeneration in Tamworth has been implemented in several consecutive phases throughout the 1990s and the latest improvement plan, proposed as part of the local Labour-controlled council's economic development strategy, is intended to build on earlier redevelopment initiatives by increasing the quality and quantity of shopping and leisure facilities in the town centre.

10 Local authorities are increasingly relying on CCTV as a key means to police town and city centres. Tamworth was one of the first towns in the UK to turn part of the policing of public space over to private company management by introducing CCTV in crime prevention.

11 For example, the *Sunday Times* (13 October 1998) talked of the emergence of a 'pan-European youth underclass', which was also described as a 'teenage time-bomb' (Ian Burrell, *Independent on Sunday*, 18 October 1998, p. 22).

12 Interestingly, according to the young people who took part in this study 'zero tolerance' has been in place as an informal policing strategy for a number of years in Tamworth town centre, and translates in this context into the constant police repossession of urban public space from youth.

13 Importantly, as part of a package of measures, the Crime and Disorder Act includes proposals to introduce night curfews (initially to those under 10) on youth access to public space as a way of cutting youth offending. 'Curfews' are legal time curbs on the free access of young people below the age of 18 to the out of doors at night – 'house arrest for chronological age' (American Civil Liberties Union, quoted in Breitbart 1998: 307). However, young people themselves have had no input in these policies which impact directly on their lives. Moreover, in Hamilton, Scotland (which is often the testbed of future social policy in the UK) curfews are already being enforced by the local police against youth under 16, making it a crime to be out on the streets after dark.

14 The selective deployment of spatial exclusions in urban environments is also disproportionately enforced against black youth (Breitbart, 1998).

15 The urban regime aims to order space and create a coherent disciplinary field by maintaining a continuity partnership between camera detectors and police patrols on the ground.

16 Some youth have decided to fall back to neighbourhood spaces or areas immediately outside the town centre, but the numbers are small. A further problem in this respect is that in July 1997 four neighbourhood shopping and residential centres were equipped with CCTV surveillance cameras and more are planned for a large number of other public housing estates in the town.

17 For de Certeau the term 'tactic' refers to 'a mode of action determined by not having a place of one's own', while the 'strategy' is, 'a mode of action specific to regimes of place' (de Certeau, 1984: 34–9, xix–xx; Pile, 1996).

18 The second part of this subheading is taken from a description of de Certeau's work by Marc Auge (1995: 78).

19 Significant here is that in city watching, regardless of the positioning and amount of cameras in operation, CCTV's gaze is not always constant. For example, at the human level detector staff have a maximum twenty-minute attention span, and without a rest after this period often suffer reduced concentration and a loss of focus (*CCTV Today* [trade magazine], April 1994). This margin of human error calls into question the effective working of close supervision of urban populations. There is also the important issue, of course, that some CCTV staff may not be committed fully to the task, and thus a discrepancy arises between the quality of monitoring service offered by these operatives and that relayed by 'dedicated' detector staff.

20 The major problem concerning the effective operation of CCTV and police part- nerships in crime prevention is the inherent contradiction that exists concerning the level of commitment to law enforcement between the two social control agen- cies. For example, security staff working for private policing franchises are private citizens and answer only to their bosses, being bound only to their working contracts, whereas police officers answer to the law of the land and enforce state legislation. The result is not always a successful integrated assault on youth gatherings.

21 Indeed, Tamworth council leader, Brian Jenkins has claimed that the CCTV scheme 'not only reduces crime but reduces the fear of crime' (Labour news and campaign leaflet, February 1996).

22 Panic rumours about the presence of secret cameras planted by the police in their hidden locations and in other parts of the town centre are common amongst these youth.

23 CCTV surveillance and police removal strategies have undoubtedly intensified the contestation over space amongst youth, creating a more dangerous urban space for teenagers by concentrating the competition for hangout sites. CCTV, then, has perhaps introduced considerably more risk and anxiety into young people's social worlds rather than reducing their psychic maps of fear and their daily experiences of risk and danger.

Chapter 11

Cosmopolitanism and the sexed city

Jon Binnie

Cities have always been stages for politics of a different sort than their hinterlands. But in the era of mass migration, globalization of the economy, and rapid circulation of rights discourse, cities represent the localization of global forces as much as they do the dense articulation of national resources, persons, and projects. (Holston and Appadurai, 1996: 189)

In this chapter I argue that the increased visibility of lesbians and gay men within the commercial hearts of British and North American cities has paradoxically only served to reinforce the marginalization of public sex. This increased visibility has meant that the most threatening components of queer sexual cultures, such as the leather scene (which coincidentally are also the least easily assimilated by corporate interests), have been further marginalized. For example, it is hard to envisage brewing, leisure and retail industries opening up a chain of leather bars, whereas coffee shops and café bars can be marketed at both straight and queer markets.

In current conflicts over sexualized urban space, political economic realities play a major part in determining the organization of space. Disputes over uses of territories within the city are marked by class and social polarization. As cities compete for mobile capital in the global market, they strive to present themselves as safe, business-friendly controlled environments. In this context, the most visible aspects of public sex are zoned out of existence, or are forced out by corporate interests. For some gay men gentrification based on respectability means support for initiatives to cleanse a neighbourhood of businesses involved in public sex. The growth of visibility in global cities should not mask the general invisibility in rural and non-metropolitan areas. Because of this visibility in global cities, queer culture is particularly marked by cosmopolitanism. Queer cosmopolitanism is based on knowingness and sophistication, a rejection of the provincial and rural. Gay men have traditionally been identified as cosmopolitans because they have been excluded from discourses of nationhood, and the restless desire to escape the confines of social control has meant an enforced mobility. However cosmopolitanism masks awareness of provincial gay life, which is the norm rather than exception.

Here I argue that the distinction between cosmopolitanism and provincialism is articulated through discourses of sophistication. Working-class and 'provincial' sexualities are marked as being unsophisticated, and 'less developed'. I will examine the distinction between cosmopolitanism and provincialism through discussion of the 'Bolton 7 case' when seven working-

class men in Bolton, a town in North-West England, were prosecuted for consensual gay sex. My discussion of the trial will demonstrate that these men
were on trial as much for their supposed 'unsophisticated' backgrounds as for
any alleged sexual misdemeanour.

Globalization and the transformation of the urban sexual economy

American urban history (e.g. the work of George Chauncey, 1994; Gayle
Rubin, 1993; John D'Emilio, 1993, and Estelle Freedman (in D'Emilio and
Freedman, 1988) has demonstrated that capitalist development in the United
States led to the formation of the modern gay consciousness; that, after Bech
(1997), homosexual identity in the West is a product of modernity; and that
the city is the key site in the production of that identity and modern gay consciousness. However we need to be careful in not essentializing a global gay
identity. As Dennis Altman notes, 'It has become fashionable to point to the
emergence of "the global gay", the apparent internationalization of a certain
form of social and cultural identity based upon homosexuality' (Altman, 1996:
77). While a 'global gay identity' must be treated with suspicion, I argue that
globalization does impact on the use of urban space and sexual cultures.
Lesbians and gay men are both agents and victims of globalization. It is
important to remember that lesbians and gay men are workers as well as consumers within the global economy. Recent transformations in the urban
sexual geography of two global cities – New York and San Francisco –
demonstrate that urban policies geared to making these cities more attractive
as centres of global corporate headquarters are having major consequences in
the use of space in specific localities within these cities.

In New York City, Mayor Giuliani is promoting new zoning laws which
have radically altered the sexual geography of the city. The new laws are an
attempt to regulate and purify space by forcing out businesses devoted to
public sex. By forbidding these businesses from operating within 500 feet of
one another, but also from operating within a similar distance from residential
buildings, schools and churches, the new laws are calculated to drive sex
establishments out of Manhattan and to the outskirts of the city. As Lauren
Berlant and Michael Warner argue in their essay on intimacy, the new zoning
laws have major consequences for public spaces of intimacy within the city:

> Because the heteronormative culture of intimacy leaves queer culture especially
> dependent on ephemeral elaborations in urban space and print culture, queer
> publics are also peculiarly vulnerable to initiatives such as Mayor Rudolph Giuliani's
> new zoning law. The law aims to restrict any counterpublic sexual culture by regu-
> lating its economic conditions. (Berlant and Warner, 1998: 562)

The local consumption spaces of Manhattan are linked into chains of global
connectedness. Thus changing the sexual geography of Christopher Street
would have a global impact as the large numbers of consumers of this locality
reside outside of the locality, city, or state. It is a popular site for international

gay tourism. As Berlant and Warner note, the new zoning laws assume the desirability of a neat coherent neighbourhood as the basis for a stable thriving city, and this view dangerously misrecognizes who uses such spaces:

> The point here is not that queer politics needs more free-market ideology, but that heteronormative forms, so central to the accumulation and reproduction of capital, also depend on heavy interventions in the regulation of capital. One of the most disturbing fantasies of the zoning scheme, for example, is the idea that an urban locale is a community of shared interest based on residence and property. The ideology of the neighborhood is politically unchallengeable in the current debate, which is dominated by a fantasy that sexual subjects only reside, that the space relevant to sexual politics is the neighbourhood. But a district like Christopher Street is not just a neighbourhood affair. The local character of the neighbourhood depends on the daily presence of thousands of non-residents. Those who actually live in the West Village should not forget their debt to these mostly queer pilgrims. And we should not make the mistake of confusing the class of citizens with the class of property owners. Many of those who hang out on Christopher Street – typically young, queer, and African American – couldn't possibly afford to live there. Urban space is always a host space. The right to the city extends to those who use the city. It is not limited to property owners. It is not because of a fluke in the politics of zoning that urban space is so deeply misrecognized; normal sexuality requires such misrecognitions, including their economic and legal enforcement, in order to sustain its illusion of humanity. (Berlant and Warner, 1998: 563–4)

In *Policing Public Sex* David Serlin comments that 'The zoning ordinance may persuade its advocates that it is preserving New York's reputation as a "world-class" international city, but their regressive notions of culture evoke the safe, sanitized world of the suburbs – itself a mirage (Serlin, 1996: 47). The suburbs are as much sexualized spaces as any 'sex zone'. Pat Califia (1994) and David Bell (forthcoming) note that the production of the suburban myth is dependant upon the containment of threats to its cleanliness in specific localized places within the city. As David Serlin (1996) notes, the current attempt at cleansing the sex zone in Manhattan has its parallel in the cleansing of unsightly blue collar industries out of Manhattan and into the less respectable suburbs:

> As anyone who has followed policy decisions in the city for the past few decades will affirm, New York markets itself as the 'Global City' *par excellence*, and has become an information-based economy that pushes the atavistic *materialismo* of industry and manufacturing out of Manhattan and into the lowly, ethnic, blue-collar outer boroughs and suburbs. (Serlin, 1996: 50)

While there is a rush to proclaim New York's new-found respectability as a totem of its position as key global player and place to do business, it is evident that the new zoning laws are merely the latest stage in the attempt to contain unruly or deviant behaviour. As David Serlin observes, there are clear parallels between the treatment of the sex industry and manufacturing industries, as both are seen as polluting and unsightly, and thereby warrant containment:

[T]he rezoning of commercial sex resonates loudly with the larger goals of New York City politicians. Both strategies – rezoning the commercial sex industry and rezoning industrial and manufacturing operations – tend to target areas where minority groups live and work, converting these neighbourhoods into purified, gentrified zones. (Serlin, 1996: 51)

Similar processes of displacement of spaces of public sex are at work in San Francisco. The South of Market (SOMA) neighbourhood became the main focus of gay male leather culture in the city in the 1970s. It is currently undergoing rapid change which has brought more straight businesses into the area which threaten the future of the distinctive queer leather culture. Gayle Rubin (1998) notes that notions of cleansing and improvement have been mobilized by the city authorities to develop the area as a convenient location for headquarters of global corporations. Rubin is quick to point out that AIDS was seen as being the reason for the decline of the SOMA neighbourhood as the hub of the gay male leather scene, but that the area was under pressure prior to the onset of AIDS due to pressure for development projects such as the Moscone Centre. Bathhouse closure, led by people such as Randy Shilts, has a major impact on the political economy of the neighbourhood as businesses devoted to leather sex closed down (see Berube, 1996a on the politics of bathhouse closure). As Gayle Rubin argues: 'The displacement of gay leather South of Market resulted from geographic competition for the area that long preceded AIDS, and from public policy decisions about disease control, as much as it did from AIDS itself' (Rubin, 1998: 260).

Gay male leather culture is inextricably linked to AIDS as leathersexuality is seen as the most visibly and publicly sexual of cultures. This visibility made leathermen easy targets of blame for the onset of AIDS, even though leathersex and SM, in eroticizing other parts of the body, is widely seen as a part of the safer sex repertoire of many gay men: 'since leathermen were often characterized as more "sexual" than other gay men, it was easy to consider them more prone to exposure to a sexually transmitted disease. Even the South of Market neighborhood became a geographic magnet for AIDS-related apprehensions' (Rubin, 1998: 261). Rubin's study of the SOMA gay male leather scene demonstrates considerable awareness of the global/local interconnections, showing how corporate interests and the re-positioning and reinforcement of San Francisco as a global city through redevelopment of the SOMA neighbourhood has called into question the survival of the local sexual culture of gay leather in the neighbourhood. It is ironic that in some recent essays on gay SM culture (such as that by Donald Morton) – SMers and their culture are equated with forces of global capital, rather than being in opposition or resistance to it. Businesses serving the gay male leather culture in SOMA are being forced out of the area by more powerful agents in market economy. As Rubin notes, despite attempts to safeguard this territory as a safe space for sexual experimentation and play, these more powerful corporate interests are more likely to win in the long term:

> Redevelopment is now rapidly invading and encircling the Folsom ... An Office
> Max store has recently opened just behind the San Francisco Eagle, one of the
> remaining leather bars. The back of the Costco parking lot faces the Eagle on one
> corner and Lone Star, another leather bar, on the other. Shoppers laden with carts
> of paper towels and a year's supply of Windex are not a promising mix with gay
> men dressed in leather. The potential for conflict and violence along these ruptured
> territorial membranes is immense ... The differences in scale between Costco and
> the leather bars in size, capital investment, and mayoral benediction are extreme. It
> is quite evident that if anything gives, it will not be Costco. (Rubin, 1998: 266)

It is significant that political economic conflicts over territory have led to a
paradoxical situation whereby the 'improvement' of the neighbourhood has
led to greater crime and fear of crime:

> It is deeply ironic that, contrary to stereotyped expectations, the displacement of
> those 'threatening men in black motorcycle outfits' by a mostly heterosexual street
> population has made this neighbourhood considerably less safe than it used to be.
> (Rubin, 1998: 267)

Sexuality and class

From the discussion so far, it is worth stressing here that these conflicts over
the sexualized nature of urban space have a class basis. In addition conflicts
over specific localities are framed by political economic realities, namely the
strive to create new business-friendly images for these cities. In New York we
see conservative middle-class respectable gay men pathologizing the sex
zone. Strategies for occupying city space such as gentrification are not open
to all. It is only a minority of gay men that can afford to live in the (gay)
neighbourhoods that others may aspire to live in. For those who do not possess the capital, the ability to take up space is constrained. In their discussion
of public sex, Lauren Berlant and Michael Warner note that the majority of
those queers using and making the city's queer streets are themselves socially
marginal (African Americans, Latinos). Queer urban consumption spaces are
commonly misrecognized as sites of untrammelled affluence. Poverty, debt,
displacement and dislocation are present. Here we see a contrast between
those who are less assertive, have less power, and those who traditionally
dominate activist movements. Peter Cohen's carefully positioned, thoughtful
essay on AIDS activism suggests that the politics of ACT UP were as much
about the politics of entitlement. He argues that ACT UP was dominated by
educated, articulate, middle-class white gay men who had much invested in
maintaining the status quo. This inevitably means that other voices, less loud
and assertive of their entitlements, went unheard:

> ACT UP must ... be understood as something of a paradox: a movement for social
> change that was created in large part by individuals with a significant investment in
> the status quo. In ACT UP/New York and ACT UP chapters around the country, the
> majority of activists were white men, many of whom were quite well off financially,
> and most of whom had little (if any) previous interest in or experience with grass-
> roots activism. (Cohen, 1997: 90)

Those who do not share this sense of entitlement, and are not able to articulate their experiences, do not occupy the same spaces as the more affluent. For instance in his essay 'Intellectual Desire', Allan Berube notes a certain discomfort he experienced in his own participation in activism within the gay community of San Francisco. This led him to reflect on questions of class and ethnicity and how these informed his public role as a gay intellectual: 'What I experienced most directly as a white gay man with little money and no college degree was how the gay community reproduced class hierarchies. There were many gay restaurants, disco parties, conferences, resorts, and bathhouses I couldn't afford. And I didn't have the income to live in the Castro' (Berube, 1996b: 152). Not having the financial capability seriously compromises one's attempts to lead a 'modern gay lifestyle'. It directly impacts upon one's ability to take up space within the city. It means that one does not have the disposable income to participate in, for example, international tourism, thereby impacting on one's ability to become a cosmopolitan queer. While place is seen as an enabler and facilitator of modern gay identities – in the sense of creation of gay neighbourhoods, and commercial spaces within global cities – place can also confine, stifle, control gay identities.

Cosmopolitanism

Modern gay lifestyle and culture is dominated by cosmopolitanism reflected both in a desire to live in the perceived centre of gay culture and commerce, and equally in a certain knowingness and sophistication. 'Provincial' gay life outside the global cities is markedly different. What is mean by cosmopolitanism? One definition is provided by Ulf Hannerz, who argues that

> A more genuine cosmopolitanism is first of all an orientation, a willingness to engage with the Other. It entails an intellectual and aesthetic openness toward divergent cultural experiences, a search for contrasts rather than uniformity. To become acquainted with more cultures is to turn into an *aficionado*, to view them as artworks. At the same time, however, cosmopolitanism can be a matter of competence, and competence of both a generalized and a more specialized kind. There is the aspect of a state of readiness, a persona; an ability to make one's way into other cultures, through listening, looking, intuiting, and reflecting. And there is cultural competence in the stricter sense of the term, a built-up skill in manoeuvering more or less expertly with a particular system of meanings. (Hannerz, 1996: 103)

In homophobic discourse it has not been uncommon for gay men to be represented as exerting a disproportionate global power, and as body invaders constituting a threat to the nation from outside (see Smith, 1994). In his essay on homophobic anxiety and American nationalism since the Second World War, Lee Edelman refers to an editorial from the journal *America* in 1962 in which gay men, like Jews, are seen as a troubling source of internationalism threatening the stability and foundations of the American state:

> In this virulently homophobic article the editors endorse an assertion by Eric Sevareid that homosexuals exercise pernicious international control over the worlds

of fashion, theater, film, and design. The implications of a conspiracy are reinforced by Sevareid's description of homosexual power imposed through 'loose but effective combines'. (Edelman, 1992: 283)

It is significant that homosexuals are seen as 'traditional' cosmopolitans, alongside merchants and Jews. Bruce Robbins (1998: 1) writes:

> Something has happened to cosmopolitanism. It has a new cast of characters. In the past the term has been applied, often venomously, 'to Christians, aristocrats, merchants, Jews, homosexuals, and intellectuals'. Now it is attributed, more charitably, to North Atlantic merchant sailors, Caribbean au pairs in the United States, Egyptian guest workers in Iraq, Japanese women who take *gaijin* lovers.

Cosmopolitanism is a trope, a common theme in modern gay fiction. If cosmopolitanism is, after Hannerz, a question of competence in the global city – the ability to manage new situations – it seems remarkably similar to the processes of moulding and managing a modern urban gay identity. As Kath Weston (1995) has eloquently shown in her essay 'Get Thee To a Big City', coming out and developing a gay identity has commonly gone hand in hand with becoming a sophisticated urban dweller at ease with urban life.

Universalism

> Understood as a fundamental devotion to the interests of humanity as a whole, cosmopolitanism has often seemed to claim universality by virtue of its independence, its detachment from the bonds, commitments, and affiliations that constrain ordinary nation-bound lives. It has seemed to be a luxuriously free-floating view from above. But many voices now insist ... that the term should be extended to transnational experiences that are particular rather than universal and that are unprivileged – indeed, often coerced. (Robbins, 1998: 1)

In one of the few discussions on representations of transnationalism in urban sexual cultures, Martin Manalansan IV critiques dominant gay urban narratives. He challenges the idea that urban life facilitates queer visibility by stressing how, in his interviews with Filipino gay New Yorkers, they argued that visibility was a particularly problematic concept for them:

> For many of my informants, the closet is not central to their personal narrative. They see 'coming out' as the primary preoccupation of gay men from other ethnic and racial groups. In fact, visibility can be dangerous for gay Filipinos. Until the late eighties, U.S. immigration laws both criminalized homosexuality and categorized it with Communist Party membership. And not all gay venues are open to these immigrants. (Manalansan, 1995: 434)

Manalansan also challenges the centrality of the Stonewall metaphor in organizing urban narratives of coming out. Stonewall has come to represent a privileged site of gay liberation, thus ignoring struggles elsewhere. Activists outside of the United States have also sometimes expressed resentment that Stonewall was not the most salient watershed in the struggle for gay liberation in their respective states. Manalansan's paper raises significant questions about the links between race, sexuality and the production of cultural authority. In his discussion of Andrew Sullivan's writing on AIDS and gay male identity – in particular Sullivan's widely publicized book *Virtually Normal* – Philip Brian Harper tackles issues of entitlement and privilege in affluent gay white men. In *Virtually Normal* Sullivan promulgates his version of gay conservatism and assimilation into the white mainstream of US society, as spokesman of an affluent urban gay male élite:

> Identified with these men by mere virtue of his acquaintance with them – which itself already connotes his similarity *to* them – Sullivan emerges through these references as both subject and object of his own representational undertaking, the masculinized white normativity that he projects comprising nothing other than his own subjective status, officialized as 'authentic' homosexuality by the journalistic medium through whose public engagement it is widely registered. Nor is this simply a one-directional development, for not only does the journalistic promulgation of Sullivan's experiential subjectivity as somehow normatively authentic enhance Sullivan's own power to construe the significances of various social situations, but his 'authenticity' itself provides the cultural *authority* on which the press depends for its continued social legitimacy. (Harper, 1997: 14)

In contrast to Manalansan, and Berube – who found transnational space one of profound dislocation, trouble – we see the cosmopolitanism of the urban homo élite, whose knowledge (equated to power), and knowingness about the hippest destinations and urban sites for queer consumption becomes a self-reflective marker of their own sophistication: the ability to travel to global gay destinations, and return with identity unchallenged. Reality for many gay men is somewhere in between – there is a sense of restlessness, a search for somewhere where things could be better (see Weston, 1995). As Dennis Altman notes, 'there is a constant danger of romanticizing "primitive" homosexuality', and he argues that this desire does have colonizing tendencies in certain circumstances where: 'Western romanticism about the apparent tolerance of homoeroticism in many non-Western cultures disguises the reality of persecution, discrimination, and violence, which sometimes occurs in unfamiliar forms' (Altman, 1996: 80).

Opening up discussions about transnational space, and how sexual dissidence is differently configured across space, means a resistance to any attempt to universalize or globalize sexual dissidence, as Martin Manalansan IV argues: 'In the shadows of Stonewall lurk multiple engagements and negotiations. Conversations about globalizing tendencies of gay identity, politics, and culture are disrupted by local dialogues of people who speak from the margins. These disruptions need to be heard' (Manalansan, 1995: 436).

Questions of power, authority and representation loom large as a result of this discussion of discourses on sexuality, body, city – specifically race and class in urban sexual geographies (Elder, 1995, 1998). This is also true for societies outside of the West, as Dennis Altman argues:

The romantic myth of homosexual identity cutting across class, race, and so on doesn't work in practice any more than it does in the West. The experience of sexuality in everyday life is shaped by such variables as the gay between city and country; ethnic and religious differences; and hierarchies of wealth, education, and age. The idea of a gay or lesbian/gay community assumed that such differences can be subordinated to an overarching sense of sexual identity, a myth that is barely sustainable in comparatively rich and affluent societies. (Altman, 1996: 89)

It is the romantic myth of a universal homosexual identity and experience which helps legitimate Western anthropological excursions. However, such a process creates a sophisticated cosmopolitan identity for the writer or researcher, in the construction of the 'native' as unsophisticated 'Other'. Within writing on the gay male experience, the production of sophistication is a major trope. For example, Gilbert Herdt even uses the term 'unsophisticated' to describe non-metropolitan USA. Herdt is the archetypal cosmopolitan gay intellectual: we are told in the blurb for his book *Same Sex, Different Cultures* (1997) that 'He resides in Chicago and Amsterdam', as if to add credentials as an authority on global gay culture. Discussing passing and difficulties of coming out, he notes: 'In many towns and cities, especially unsophisticated and traditionally conservative areas of the country, the possibilities are only now emerging for gay/lesbian identification and social action' (p.126). It is his use of the term 'unsophisticated' that is particularly worthy of comment. Ruth Holliday (1998: 213) writes of Herdt: 'Given the intense scrutiny and self-scrutiny which has questioned the anthropological imagination, what is most remarkable about Herdt's compilation is its lack of reflectivity and theoretical (self-)criticism.'

Spaces of unsophistication: the case of the 'Bolton 7'

Bolton is a large, predominantly working-class, town of approximately 250,000 inhabitants in Lancashire, North-West England, close to the regional centre of Manchester. In contrast to Bolton, Manchester has a large and reasonably diverse lesbian and gay commercial scene focused on the Canal Street area, whose spectacular growth and development has been much studied in recent years (e.g. Whittle, 1994; Quilley, 1995; Ryan and Fitzpatrick, 1996; Taylor, Evans and Fraser, 1996). While it is part of the Greater Manchester conurbation, and only twenty minutes by road and rail from Manchester city centre, Bolton retains a strong local identity as a town in its own right, with its distinctive history and sense of place. This is reflected, for instance, in local outrage when it was rumoured that the town would not be signposted on the renumbered M60 motorway (see Taylor *et al.*, 1996: 327–8):

In November 1994, local councillors and MPs alike in Bolton were up in arms against the Highways Agency of England and Wales at the news that directions to the town would disappear from signs to be erected on the new M60 motorway. The strength of local identity in Bolton was evident in the protests mounted, in particular, against the idea that the signs on the motorway would subsume Bolton into what the Conservative MP for North West Bolton chose to call the 'anonymous mass' of 'North Manchester' (*Guardian*, 24 November 1994).

Bolton is the archetypal Northern town – the 'worktown' of the Mass Observation project of the 1930s. Peter Gurney has argued that the Mass Observation project asserted a middle-class gaze on working-class sexuality. Working-class women were commonly represented as depraved, lacking in morality and infantilized. Gurney argues that the Bolton women resisted, or returned the gaze of the observers. From Peter Gurney's essay we can see that historically Bolton has occupied a special place in the imagination and representation of the northern-ness of the North of England, and that central to that representation is the construction of working-class sexualities as Other.

> Harrison wanted to study the English working class in the same way that he had recently studied cannibals in the New Hebrides, because he reckoned that the customs, habits, and social life of English 'natives' were as obscure as those of the most exotic tribespeople. ... Harrison urged the organization to shift its focus to Bolton, the Lancashire mill town that, he believed, constituted the classic location of working-class culture, a mythical 'Worktown'. (Gurney, 1997: 257)

Gurney notes how anthropological discourses of difference and exoticness were mobilized to examine the morality of Boltonians. In this way working-class Boltonians were approached in the same manner as an exotic tribe: 'The sexual habits and activities of the Blackpool holiday crowd were major preoccupations for Mass-Observation' (p. 268).

Blackpool was of particular interest to Mass Observation as this was the resort where Boltonians went on holiday. The resort had a particular reputation for sexual licence. For Harrison it represented the ideal place to investigate working-class sexuality in the raw. This was the town where Boltonians went to escape the routines of factory life:

> Blackpool, an overwhelmingly working-class resort whose reputation for sexual licence was renowned throughout the country, was in many respects a material and psychological domain where bourgeois sexual fantasies could be played out.
>
> It was a place, so many observers thought, where working people, their lives cramped and confined by industrialism, could be nearer nature and thus express their 'animal' instincts. The unpublished book employed crude pseudo-Freudian concepts to organize and explain the material. The keystone was the theory of repression. Boltonians were repressed back home by the discipline of the factory and close-knit nature of family relationships, so the argument ran, but while at the seaside their desires could be given freer rein. The Blackpool study was consequently prefaced by chapters that described the ways in which the lives of workers in Bolton were structured and confined by work and particular orderings of time. (Gurney, 1997: 268)

In 1997 seven men in Bolton were arrested and prosecuted for consensual same-sexual activity under a law dating back to the last century. The case of the 'Bolton 7' attracted widespread media attention in both the gay and mainstream press. The seven were found guilty. Two received suspended prison sentences, the others were put on probation. The trial cost around £500,000. Three men lost their jobs; one had his house attacked. The case of the Bolton 7 was featured in a Channel Four documentary *Sex, Lies and Videotapes*. This documentary is a remarkable document of the events that took place in Bolton. The narrator and major actor in the documentary is Ray Gosling, a broadcaster who has a reputation, whether accurate or fair, as a professional Northerner, as a media icon, a documentor of Northern working-class culture. His documentaries in the past have been criticized for their romanticization of working-class culture and life in the North of England.

The film marks a rare inclusion of working-class lesbian and gay lives and perspectives in the media. The trial itself was notable for the ways in which the working-class background of the defendants was mobilized against them. Judge Lever sentenced the five 'younger' men first. In sentencing them he made a scathing attack on the 'unsophisticated' nature of their backgrounds:

> You are virtually all from deprived backgrounds, from broken families, ill-educated, not very intelligent, unemployed and – the words appear time and time again in the pre-sentence reports on each of you – immature, unsophisticated and woefully lacking in guidance to make something of your lives. It is highly significant that at least three out of the five of you are now apparently involved in serious relationships with the opposite sex at least to the extent that you are actually living with young ladies. Moreover, it seems that few if any of you had more than the haziest idea of the laws that you had broken. In terms of physical age you may have reached adulthood, but your behaviour in my judgement is little more than that of a group of smutty-minded schoolboys tipsily experimenting at sex. (Quoted in *Sex, Lies and Videotapes*)

These words demonstrate that these men were as much on trial for being inarticulate, for being working class, 'unsophisticated', as for being queer. They were also clearly being infantilized. In her essay 'The Theory of Infantile Citizenship', Lauren Berlant argues that it is the figure of the innocent girl who needs the protection of the state. She notes from the writing of Tocqueville that the American state tends to infantilize its citizens: 'democracies can ... produce a special form of tyranny that makes citizens like children, infantilized, passive, and overdependent on the "immense and tutelary power" of the state' (Berlant, 1997: 27).

The Mass Observation studies of working-class sexuality in Bolton and Blackpool were notable for the way in which sexuality was mobilized to articulate class difference. Judge Lever's comments in the Bolton 7 trial suggest that these representations of working-class sexuality as exotic and deviant remain central to distinction between classes today, just as the 'dirty girls' of 'Worktown' were seen as being lewd, crude and unsophisticated. In both cases these representations were contested and resisted. What was spectacular in the Bolton 7 trial was one defendant's determination that he had done nothing wrong and should not be punished.

In her famed essay 'The City of Desire' Pat Califia argues for a city where sexual experimentation – play – is possible. The challenges to the (re)construction of that city are manifold. One major challenge is that academic discourse on the sexualized city still fails to get to grips with the messy materiality of the city, fails to challenge the squeamishness, but rather favours a virtually normal form of distanced academic critique that reproduces the city as straight, vanilla, even asexual. The vision of the city imagined by Califia still requires much work, particularly in the queer sites outside metropolitan spaces. Richard Dyer argues that 'The project of making normality strange and thus ultimately decentring it must not seem to say that this has already taken place, that now masculinity, whiteness, heterosexuality and able-bodiedness are just images of identity alongside all others' (Dyer, 1993: 4). In middle-America, middle-England and the beyond the metropolises this work is particularly important, but also most difficult. We need to do more than articulate a metropolitan gaze on 'provincial' or non-metropolitan sexualities, but rather investigate the mutual interdependence of categories and identities such as provincial and metropolitan.

References

Altman, Dennis (1996) Rupture or continuity? The internationalization of gay identities. *Social Text* **48** (3), 77–94.

Bech, Henning (1997) *When Men Meet: Homosexuality and Modernity*, Polity Press, Cambridge.

Bell, David (forthcoming) Fragments for a queer city. In Bell, David, Binnie, Jon, Holliday, Ruth, Longhurst, Robyn and Peace, Robin *Pleasure Zones: Sexualities, Cities, Spaces,* Syracuse University Press, Syracuse, NY.

Berlant, Lauren (1997) *The Queen of America Goes to Washington City: Essays on Sex and Citizenship*, Duke University Press, Durham, NC.

Berlant, Lauren and Warner, Michael (1998) Sex in public. *Critical Inquiry* **24** 547–66.

Berube, Allan (1996a) The history of gay bathhouses. In Dangerous Bedfellows (eds) *Policing Public Sex: Queer Politics and the Future of AIDS Activism*, South End Press, Boston, MA., 187–220.

Berube, Allan (1996b) Intellectual desire. *GLQ*, **3** (1), 139–57.

Califia, Pat (1994) The city of desire. In *Public Sex: The Culture of Radical Sex*, Cleis Press, Pittsburgh, Pa. and San Francisco, Ca., 205–13.

Chauncey, George (1994) *Gay New York: Gender, Urban Culture, and the Making of the Gay Male World, 1890–1940*, Basic Books, New York.

Cohen, Peter F. (1997) 'All they needed': AIDS, consumption, and the politics of class. *Journal of the History of Sexuality*, **8** (1), 86–115.

Dangerous Bedfellows (eds) (1996) *Policing Public Sex: Queer Politics and the Future of AIDS Activism.* South End Press, Boston, Mass.

D'Emilio, John (1993) Capitalism and gay identity. In Abelove, Henry, Barale, Michele A. and Halperin, David M. (eds) *The Lesbian and Gay Studies Reader*. Routledge, London, 467–76.

D'Emilio, John and Freedman, Estelle B. (1988) *Intimate Matters: A History of Sexuality in America*, Harper and Row, New York.

Dyer, Richard (1993) *The Matter of Images: Essays on Representations*, Routledge, London.

Edelman, Lee (1992) Tearooms and sympathy, or, the epistemology of the water closet. In Parker, Andrew, Russo, Mary, Sommer, Doris and Yaeger, Patricia (eds) *Nationalisms and Sexualities*. Routledge, London, 263–84.

Elder, Glen S. (1995) Of moffies, kaffirs, and perverts: the control, and regulation of sexual discourse in apartheid South Africa. In Bell, David and Valentine, Gill (eds) *Mapping Desire: Geographies of Sexualities*. Routledge, London, 56–65.

Elder, Glen S. (1998) The South African body politic: space, race and heterosexuality. In Nast, Heidi J. and Pile, Steve (eds) *Places Through The Body*, Routledge, London, 153–64.

Gurney, Peter (1997) 'Intersex' and 'dirty girls': Mass-Observation and working-class sexuality in England in the 1930s. *Journal of the History of Sexuality*, **8** (2), 256–90.

Hannerz, Ulf (1996) Cosmopolitans and locals in world culture. In *Transnational Connections: Culture, People, Places*. Routledge, London.

Harper, Philip Brian (1997) Gay male identities, personal privacy, and relations of public exchange: notes on directions for queer critique. *Social Text*, **52/53**, (3/4), 5–29.

Herdt, Gilbert (1997) *Same Sex, Different Cultures: Exploring Gay and Lesbian Lives Across Cultures*, Westview Press, Boulder, CO.

Holliday, Ruth (1998) Review of Gilbert Herdt: *Same Sex, Different Cultures. Gender, Work and Organization*, **2** (3), 212–3.

Holston, James and Appadurai, Arjun (1996) Cities and citizenship. *Public Culture*, **8** 187–204.

Manalansan IV, Martin F. (1995) In the shadows of Stonewall: examining gay transnational politics and the diasporic dilemma. *GLQ*, **2**, 425–38.

Quilley, Steve (1995) Manchester's 'village in the city': the gay vernacular in a post-industrial landscape of power. *Transgressions*, **1** 36–50.

Robbins, Bruce (1998) Introduction to Part I: Actually existing cosmopolitanism. In Cheah, Pheng, and Robbins, Bruce (eds) *Cosmopolitics: Thinking and Feeling Beyond The Nation*, University of Minnesota Press, Minneapolis, MN, 1–19.

Rubin, Gayle S. (1993) Thinking sex: notes for a radical theory of the politics of sexuality. In Abelove, Henry, Barale, Michele A. and Halperin, David M. (eds) *The Lesbian and Gay Studies Reader*, Routledge, London, 3–44.

Rubin, Gayle S. (1998) The miracle mile: South of Market and gay male leather, 1962–1997. In Brook, James, Carlsson, Chris and Peters, Nancy J. (eds) *Reclaiming San Francisco: History, Politics, Culture*. City Lights Books, San Francisco.

Ryan, Jenny and Fitzpatrick, Hilary (1996) The space that difference makes: negotiation and urban identities through consumption practices. In O'Connor, J. and Wynne, D. (eds) *From the Margins to the Centre: Cultural Production and Consumption in the Post-industrial City*. Arena, Aldershot, 169–201.

Serlin, David (1996) The twilight (zone) of commercial sex. In Dangerous Bedfellows (eds) *Policing Public Sex: Queer Politics and the Future of AIDS Activism*, South End Press, Boston, MA., 45–52.

Smith, Anna Marie (1994) *New Right Discourse on Race and Sexuality: Britain, 1968–1990*, Cambridge University Press, Cambridge.

Sullivan, Andrew (1995) *Virtually Normal: An Argument About Homosexuality*. Picador, London.

Taylor, Ian, Evans, Karen and Fraser, Penny (1996) *A Tale of Two Cities: Global Change, Local Feeling and Everyday Life in the North of England. A Study in Manchester and Sheffield*, Routledge, London.

Weston, Kath (1995) Get thee to a big city: sexual imaginary and the great gay migration. *GLQ* **2**, 253–77.

Whittle, Stephen (ed.) (1994) *The Margins of the City: Gay Men's Urban Lives*, Ashgate, Aldershot.

Chapter 12

The new segregation*

David Theo Goldberg

The problem of the twentieth century, Du Bois warned in his memorable and all too prophetic words, would be the problem of the colour line. At more or less the same time, religious reformer Josiah Strong predicted with similar pre-science that the century's problem would be that of cities. In the United States, one might say, the problem of cities this century has been the indelible problem of the colour line, while the problem of race has been inextricably tied to the making and remaking of the contemporary – the modern – city. In the South under slave conditions, the prevailing articulation of racial division was social (Hirsch, 1993: 69); spatial separation was a secondary and more muted marker. Segregation is an urban condition, instituted and institutional-ized first in the South, as in South Africa, only with the urbanization of blacks moving off the plantations after abolition and the Civil War. White politicians in the southern Democratic Party machine secured their political power by shielding the white urban working class from competition by newly emanci-pated blacks, just as the National Party came to power in South Africa in 1948 by securing the wellbeing of the white working and poor classes on the promise of apartheid.

Black arrival in America was inserted through enslavement into Southern agriculture. The urban ghetto in America developed in the post-Reconstruction era as a spreading but contained, spatially and socially isolated place of sepa-rated black (or now black and brown) residence and institutions (Hirsch, 1993: 65). Massey and Hajnal (1995) demonstrate that what we have come to characterize as the contemporary urban ghetto developed thus in three stages. From 1880 to roughly the New Deal there was rapid black urbanization first into Southern cities and towns, and then via migration into Northern cities, coupled with large-scale European immigration largely to Northern cities. In these migratory shifts blacks began to find themselves increasingly segregated in urban settings. Nevertheless, ghetto areas, especially in Northern cities, still included a mix of blacks and European immigrants.

From 1933 to the late 1960s the African-Americanization of urban ghettos expanded dramatically, with the ghetto then distinguishable from that of the first period in both quantitative and qualitative terms. The second stream of black migration northwards during this period was mobilized initially by the depression and later the pull of work in war industries while blacks were pushed out of the South by the mechanization of agricultural production

(Hirsch, 1993: 83). It was over this period that blacks came to occupy urban neighbourhoods completely separated from white ones.

The third stage in the production of black ghettoization followed the Civil Rights Movement and has lasted to the present, producing what I will argue is a 'New Segregation' encouraged and rationalized by an ideology of a 'New Segregationism'. At Reconstruction's sunset, racial separation was set largely at the macro state-and-county levels; in the second stage, it became intra-urban and occurred mostly at the neighbourhood level. Today, while state- and county-level segregation has largely dissipated, neighbourhood segregation has solidified but it has been bolstered by a new form as a consequence of suburbanization: where blacks are located in large numbers, whites and blacks tend not only to live, work, school, and die in different neighbourhoods but in different cities.

Historically, cities are the magnets and media of migration, points of attraction and crossroads of movement both in terms of local and more distant (or 'alien') geography. It is what cities have to offer – in excitement, goods, work, services, and exchange – that renders them so attractive. In turn, the migratory matrix of cities makes them heterogenous and objects of appeal. At the same time, though, the perceived threats to homogeneity, to sameness and identity, to the expected and the usual, to order and control that cities suggest historically if not promote, have prompted the institution of controls over urban space, restrictions on entry and movement, limits on access and acceptability. Thus it is cities that gave rise to the perceived 'need' for, as they became the principal sites of, racial segregation. Late in the nineteenth century, accordingly, urban space became at once the summary motivation and purpose of the drive by whites to segregate – namely, to restrict heterogeneity and hybridity, to delimit intercourse, to control interaction and relation.

Segregation is now 'municipal' (cf. Massey and Hajnal 1995: 527). The past thirty years or so have witnessed the evisceration of the American working class, so much so that it is almost not too extreme to think of it as the outcome of a collusive and well co-ordinated attack, especially since the beginning of the 1980s. In the first half of the twentieth century deep struggles were played out between capital and labour, the clash of commercial interests with strong unions. Labour's power, such as it was, began to erode at the apex of success of the Civil Rights Movement, and for reasons at least partly related to local racial dynamics but also to wider and deeper economic shifts globally. What emerged was the erosion not just of the power of the working class in the US, but the erosion of the working class itself.

The traditional class formation of capital and labour has transformed into an expanding but still small cabal of the very (and increasingly) wealthy; an associated professional service class (doctors, lawyers, accountants, mutual fund executives); a middle class of technicians, managers, middle-level professionals, and academics, most of whom find it increasingly difficult to maintain their class position in very competitive markets; a fading working class and the (desperately) poor that include those characterized as the underclass, the working poor, minimum wage service workers, the retired working class, and the underemployed. The major manufacturing centres of urban America watched their manufacturing bases shrivel: from the mid-1950s to the mid-1980s, New York, Philadelphia, Chicago, and Los Angeles, among other cities – those with significant if not majority non-WASP working- and middle-class populations – each lost approximately half a million manufacturing jobs. Flexible production (Harvey, 1989) has meant that the US working class – and along with it the associated labour 'difficulties' – has been moved offshore, elsewhere in the world where labour costs are cheap, where the costs of labour reproduction are minimal for US capital and, when facing labour 'problems', capital can shift production swiftly to new (national) sites. Nike, the iconic contemporary American corporation, leads the field in the global labour safari.

One indication of the demise of the traditional working class is the fact that an increasing share of non-executive manufacturing positions requiring some technological skill is being cornered by those with college degrees in engineering and the like. And a sign of the tenuous condition of the middle class in America is the fact that middle-level professional jobs requiring considerable skill (e.g. computer programming) likewise are beginning to be shipped offshore where both direct and reproductive costs are significantly smaller. From 1968 to 1994 the top 20 per cent of American households increased their share of aggregate income from 40.5 per cent to 46.9 per cent, from roughly $74,000 to roughly $106,000, an increase of 44 per cent after adjustments for inflation. The bottom 20 per cent saw income increase by just $500 or 7 per cent (roughly $7,200 to $7,700). White men, while roughly 43 per cent of the workforce, continue to hold 95 per cent of senior management positions. At the same time, women generally, and white women in particular, have made significant economic headway, entering the professional classes especially in record numbers. Nevertheless, as poverty remains racially distinct, so too is it heavily marked in gender terms, women and children overwhelmingly filling its space. Thus, 48 per cent of those in poverty in the US are children, and 46 per cent of African American children live below the poverty line, almost overwhelmingly in cities. Poverty, always racialized in this country, has become increasingly feminized and infantilized.

By contrast, income soared for those fronting corporate organizations. There was little resistance to massive increases in chief executive compensation packages, now totalling more than $20 million per annum and 200 per cent more than the annual compensation of the corporation's workforce in the most notorious instances. From 1974 to 1994, chief executives in US corporations on average increased their earnings from $35 to $224 for every $1 earned by workers (the 1994 German ratio was 14:1). Chief executives take credit for corporate wellbeing, happy to see corporate profits grow by an average of 13 per cent per annum and to watch their stock value soar, but they are quick to request worker give-backs to achieve their profitable ends and to blame worker inadequacy or intransigence when profits drop and stock values plummet. This alienation effect, prompted by the shift from investment in housing stock to the faceless stock market, and coupled with recent programmatic dismantling of welfare assistance stretching back to the New Deal, has exacerbated the growing income gap in the US and the increased vulnerability and erosion of the working class. It has had dramatic effects on the prospects for black citizens, aiding the rise of at least part of the black bourgeoisie while braking dramatically the prospects of the black poor and working class.

Race has become a touchstone here, more so than ever. Sewn over the centuries into the seams of the social fabric, the idea of race (or, really, the ideas, for they are multiple) furnishes the terms around and through which a complex of social hopes, fears, anxieties, resentments, aspirations, self-elevations, and identities get to be articulated. It is not that race, for instance, is isomorphic with class, nor simply that class is articulated as race. Rather, classes themselves are now racially fractured, racial configurations cutting across class, rendering even intra-class alignments ambiguous, ambivalent, and anxious. This social fracturing makes any policy consideration concerning race, especially those that are racially explicit, fraught with the difficulties of fervent commitment and equally fervent denial. They become as a consequence difficult to negotiate, compromising to defend, apparently easy to denounce but not so easy to pass by or renounce. The racial considerations attendant on a social issue, the racial dimensions and determinations of it, when explicit and overt, tend to obscure virtually all other considerations; and the racial formulation or effects of a policy tend to drown out any other articulation or implication. These difficulties are further magnified by the variability of ethnoracial significance, the slipperiness of racial formation and reference, and (yet) by the overarching reduction of racial complexity to white and black, if not by whites tending to equate race with 'the black condition'.[1] Contemporary segregation has to be understood against the backdrop of these shifts in political economy and discursive formation.

Segregation, old and new

In 1900 the prevailing social setting and experience of blacks in America was 'Southern and rural'; for whites it was 'Northern and urban' (Massey and Hajnal, 1995: 531). In 1880 less than 13 per cent of blacks lived in towns and cities, while roughly 28 per cent of whites did. The typical black urban resident at the end of the nineteenth century lived in a ward 90 per cent white, and was more likely to share a neighbourhood with a white person than a black neighbour. Ninety per cent of blacks lived in Southern states while 75 per cent of whites lived in Northern states, mainly in urban counties. White *exposure*[2] to blacks outside the South was limited, and Jim Crow laws orchestrated contacts in the South. In 1900, then, the physical distance between blacks and whites was a function of the fact that they tended to live in different states and counties. Most blacks tended to live in relatively few states, the average black American living in a state the population of which was 36 per cent black (Massey and Hajnal, 1995: 531; Hirsch, 1993: 65–6). If *evenness*[3] of racial distribution across state lines is a value, nearly two-thirds of blacks would have had to change their state of residence to achieve it. In Southern states a majority of blacks lived on the land in counties significantly black. As Massey and Hajnal reveal, if cross-county racial evenness were to have been sought, it would have required nearly 70 per cent of all blacks to change their county of residence.

As black migration led to greater racial interaction across state and county lines, blacks were becoming progressively segregated within cities. The more black urbanization expanded, the more their racial segregation and restriction within cities was extended. Thus by 1930 the spatial location of segregation already had transformed perceptibly from region to neighbourhood. Black urban residents tended to live in wards 40 per cent black. From 1890 to 1930 black residence in New York surged nearly tenfold from 36,000 to 328,000; in Chicago nearly twentyfold from 14,000 to 234,000. Chicago neighbourhoods just 10 per cent black in 1900 were swept by the cold wind of segregation into neighbourhoods 70 per cent black just 30 years later (Massey and Hajnal, 1995: 533–4; Hirsch, 1993: 66).[4]

Already in 1940 ethnic white neighbourhoods were far from uniform in their ethnic composition. Neighbourhoods in which blacks lived tended much more to be overwhelmingly black (Denton, 1994: 21). Identifiably 'Irish' areas of cities included just 3 per cent of the total Irish population, and most of New York's Italians did not live in Little Italy, for instance. By contrast, 93 per cent of blacks lived in neighbourhoods that in the categorical formation of race in the United States can be characterized as majority black. The historical (re)production of contained Chinatowns reinforces the ethnoracial logic at work here (cf. Goldberg, 1993: 198, 201). Thus the conditions for the reproduction of European immigrant ghettos have never existed in the way they have for black ghettos. European immigrant segregation ebbed as their migration flow waned, while black segregation within the boundaries of confined black space increased, not primarily as a result of black housing preferences but of conscious white avoidance (cf. Denton, 1994: 22), as manifested in the use of physical violence, intimidation, and the creation of a dual housing market by way of racial covenants and the like. So white exposure to blacks was still self-determinedly minimized through ensuring black isolation in urban ghettos (Massey and Hajnal, 1995: 533–4). As Denton concludes (1994: 22), cities became instruments for European immigrant group advancement, but blocks for blacks not only residentially but educationally and economically also.

The post-war period saw the emergence of desegregating efforts, in the Army, the courts, on the streets, in buses and schools, prompted not only by moral and internal political imperatives but also by geopolitical Cold War competition. The 'national interest' and foreign policy demands necessitated public commitment to race-neutral governmentality (cf. Dudziak, 1995). On the face of it, the federal government also made a huge commitment to producing much-needed urban public housing, in 1949 authorizing 810,000 new units over six years. In fact what happened, as Denton points out, is that the federal government became principally engaged in reproducing segregation (Denton, 1994: 24). In the post-war modernization boom fuelling the economy, federal policy initiatives regarding mortgages and taxes fuelled the suburban housing explosion for middle- and working-class whites, while federal property appraisal policies rendered possible bank and mortgage company redlining of inner-city property purchase and development (Mohl, 1993: 15). Government, national and local, massively promoted private (re)development and gentrification of central business districts by underwriting loans; incentives to redevelop inner cities by building housing for the urban poor were almost wholly absent (Mohl, 1993: 26). By 1962 only 320,000 of the public housing units promised in 1949 (roughly 40 per cent) had been constructed. Much more inner-city housing in fact was bulldozed away in terms of the 1949 federal housing law than was actually built (Mohl, 1993: 15): shades of apartheid South Africa. From the end of the the Second World War until 1960, less than two per cent of new housing financed by mortgages guaranteed by federal insurance went to black home-owners (Hirsch, 1993: 91). From about 1950 on, then, segregation across and not just within cities began to increase. By the 1980s this trend had become evident: blacks (and one may also add Latinos/Chicanos and in a more limited sense recent Asian immigrants) tend now not just to live in different neighbourhoods from whites but in different cities (Massey and Hajnal, 1995: 527).

So it is that from 1950 on, starting imperceptibly and gathering speed, blacks and whites were becoming more segregated across municipal boundaries, living (to use Massey and Hajnal's apt language) not only in different neighbourhoods but also in different municipalities. At the very time there was growing expression of desegregation in the public sphere, one could say there was publicly subsidized resegregation in the private. Desegregation never stood a chance. By 1980, blacks living in cities found themselves in municipalities on average 35 per cent black; if black and white residents were to be evenly distributed across municipalities 50 per cent of blacks in cities would have had to switch places of residence with whites (Massey and Hajnal, 1995: 536–7). The suburban explosion that pulled whites out of the cities, transformed the countryside into sprawling suburbs. These suburbs eventually became small self-governing cities, the effect as much of the desire to be politically and fiscally autonomous from deteriorating old cities as of some purely administrative rationality.

In 1950 there were no central cities that were overwhelmingly or even largely black. No city with a population larger than 100,000 had a majority black population. Forty years later there were 14 such cities.[5] Eleven more cities had black populations of between 40 and 50 per cent.[6] Among cities larger than 25,000 in 1950 just two had majority black populations, a number that had exploded to 40 by 1990 (Massey and Hajnal, 1995: 537). Interestingly, the increase in segregation after mid-century is characteristic only of larger cities with large black populations. There was a noticeable decline in segregation in small cities with small black populations (Hirsch, 1993: 79). In the latter cases African Americans found themselves assimilated into dominant white space with little if any noticeable effect on prevailing urban arrangements or culture. By the end of the Civil Rights era, in contrast, geographic isolation of blacks in larger urban settings – the overwhelming majority of black folk – was nearly complete.

Prompted by a mix of fear, trembling, limiting competition, and cementing power, whites could embrace enthusiastically the ideological shift at mid-century from assimilation to pluralism. Pluralism – experienced as the commitment not only to different histories, cultural values and practices but also to the ideological cliché of 'live and let live' – made conceivable and legitimated differentiated urban spatial conditions, and, by the same token, their abandonment.

Old segregation, from post-Reconstruction to *Brown* (1954), was thus an *activist* segregation produced for the most part by an active intervention in politics, economics, law, and culture self-consciously designed to produce seg-regated city, town, and neighbourhood spaces. To combat this activism, the Civil Rights Movement likewise found itself called to action in every dimension. The period from the end of the Second World War to the 1970s, by contrast, was one of tension and contradiction, promise and projection, expectation and elevation, denial and dashed hope. It was a period of desegregating commit-ment and the seeds of a resegregating mobilization. The old segregation supposedly was swept aside, only to be replaced by the whisper of the new, the subtle and silent, the informal and insidious. From the mid-1970s on, then, we begin to witness the materialization – the increasingly bald expression – of this newly expressed (what I am calling the new) segregation, one no longer activist but conservative, a segregation in the literal sense conservationist.

This conservationist segregation proceeds by *undoing* the laws, rules, and norms of expectation the Civil Rights Movement was able to effect, attacking them as unconstitutional, moving to race-neutral standing, refusing to institute them as it embraces the letter of the only partially successful Civil Rights cul-ture. In the absence of the Civil Rights spirit, accordingly, the present period *conserves* (and deepens) the hold of segregation historically produced *as if it were the nature of things*. So the New Segregation is produced by doing noth-ing special, nothing beyond being guided by the presumptive laws of the market, the determinations of the majority's personal preferences. Even were contemporary discrimination to be wiped out, the volatile mix of historic insti-tutionalization and the cultural legacy of racially exclusionary preferences mandates segregated space far into the future (Hirsch, 1993: 92). Old Segregation was monolithic; New Segregation, while still manifest for blacks generally, nevertheless is fuelled by a differentiated and differential discrimi-nation. For it is experienced in a manner intensified by the dynamic of race–class–gender–age intersection: the racially defined poor (women and children particularly, black, Latino/a, and newly arrived poor Asian immi-grants) suffer the condition much more intensely than the black, Latino, and Asian-American middle classes, and especially the black, Latino, and Asian-American bourgeoisies.

In a provocative thought experiment in 1971, Thomas Schelling exempli-fied the segregating implications of personal preference.[7] Take a chess board: fill 10 per cent of its spaces with black pawns; fill 70 per cent of its spaces with white pawns. Assume each black pawn wants at least one neighbour to be a black pawn, and each white pawn wants at least one neighbour to be a white pawn. Segregation sets in within a couple of moves (try it). If each wants both neighbours to be like itself, segregation is produced all the more quickly (Schelling, 1971). So, rational choice theorists to the contrary, preferences are *not* naive, discursively unstructured, simply given, or unchanging. Preferences are *ordered* by the dominant discursive culture and terms.

In the case of the preference for segregated space, segregation is discursively (re)produced and ideologically massaged. Whites of all class stripes reportedly prefer to live in neighbourhoods which are at least 80 per cent white; blacks prefer to live in neighbourhoods that are 50 per cent white (which is to say 50 per cent not white). Given that those classified white make up over 70 per cent of the population, that leaves vast areas of possibility for all white areas. The age of 1980s deregulation has produced informally a more widespread segregation, a segregation less formally imposed than the more formal attempts of the old activist segregation but produced via the informalities of private preference schemes.

Nancy Denton demonstrates the debilitating effects of segregation thus. Take a city, not unrealistically, that is 25 per cent black with a white poverty rate of 10 per cent and a black poverty rate of 20 per cent. Where segregation is absent, the neighbourhood poverty rate is 12.5 per cent (whites make up three-quarters of the population). Where segregation is complete, the neighbourhood poverty rate for blacks becomes 20 per cent. Where class segregation intersects, the multiplier effect on the neighbourhood poverty rate of poor blacks doubles it to 40 per cent. In the face of economic downturns, the black poverty rate jumps from 20 to 30 per cent. The black neighbourhood poverty rate then would double to 60 per cent. That means almost two-thirds of the people in the black neighbourhood would live in poverty (Denton, 1994: 28).

The complex account of the conditions facing the racially marginalized today consists of three interactive conditions that call for immediate consideration: employment, housing, and education. Were I to add a fourth, thus making complicated matters significantly more complex still, this would include dominant media representations of blackness and whiteness, whether bald or subtle, explicit or implicit, brazen or nuanced. Resource availability determines educational opportunity, hence jobs, and so by extension quality of housing. Where one lives largely determines where one goes to school, the quality of education one receives, and so the quality of housing one can afford. The colour of ideological (mis)characterization tilts understanding one way or another, with dramatic material effect. Segregation accordingly is a totalizing condition: for segregation to be sustained in any one dimension there has to be segregation in every dimension. If we are to take seriously equalizing opportunity in America, we have to be serious about tackling this (un)holy alliance of factors.

Different theorists tend to take one or another of residential, educational, or employment segregation as the most basic form, in terms of which the other two are to be explained. So Massey and Denton (1993: 9) assume that 'Residential segregation is the principal organizational feature of American society that is responsible for the creation of the urban underclass.'[8] Orfield and Kozol both argue as though educational segregation is basic, and William Julius Wilson's recent work (Wilson, 1996) underscores the debilitating effects of employment segregation. But segregation, new as old, is over-determined. These three conditions and manifestations (for they are both) interact with each other to produce complex, intransigent structures of segregation. It is hard to think that there would be school segregation without residential or employment segregation; employment segregation without school or residential segregation; or residential segregation without school or employment segregation. The desire for segregation in one domain is consonant with the desire in the others, a combined outcome of the sentiments of dominant racist commitment.

From 1950 to 1980 industrial decline and global restructuring, tied to transformation in political power within cities, massive suburbanization, and white flight, caused industrial cities like St Louis, Baltimore, Washington DC, Detroit, Cleveland, Chicago, Pittsburgh and Boston to lose residents ranging from 20 to 50 per cent (cf. Mohl, 1993: 22). Once whites lost political control, they abandoned the political district, moving outside municipal limits, effectively leaving behind all black and brown municipalities, replicating conditions of the inner city with high poverty rates, excessive concentrations of public housing, elevated tax rates, revenue shortfalls and low levels of employment (Massey and Hajnal, 1995: 537–8).[9] It is now the North-East where segregation is highest, i.e., the region where no desegregation plan was thought necessary in the mid-1950s.

From 1970 to 1980, the period during which desegregation efforts were supposed to be at their height, there was virtually no improvement in residential integration in the old large Northern urban ghettos. In Boston, Chicago, Cleveland, Detroit, Gary, Philadelphia, Pittsburgh, and St Louis the decline in the segregation was minimal, and in New York and Newark segregation actually increased (Massey and Denton, 1993). The overwhelming majority of black Americans still live in deteriorating central cities. Where there has been black suburbanization, it has largely been into areas that effectively are extensions of central city black areas or into black suburbs slightly further afield like Prince William County outside Washington DC.[10] Residential segregation is almost as severe for those blacks earning more than $50,000 as for those earning less (Glazer, 1995: 70). Like urban segregation, suburban segregation is the outcome of personal comfort and structurally induced choice in the face of ongoing discrimination. A 1991 Urban Institute study, for instance, showed that African Americans face discrimination in purchasing a home 59 per cent of the time, and 56 per cent when renting (Hirsch, 1993: 84).[11]

Black suburbs, like black cities, tend to be areas of low socio-economic status and high population density, frequently older manufacturing suburbs. They have a weak tax base, poor municipal services, high level of debt, lower property values, higher crime rates, higher taxes, and spend a larger share of their revenue on social services (Hirsch 1993: 82). By contrast, in the 'new South' there has emerged a new type of black ghetto: all-black towns and cities in areas of black numerical strength, the problems of which are similar to black neighbourhoods and suburbs in the North. Also, as city politics began to be controlled by black politicians and as black mayors took over in the mid-1970s, they tended not only to hire black employees in existing positions but to add blacks significantly to the city payroll. This had a considerable impact (along with affirmative action) in expanding the black middle class, helping too in re-election campaigns. Washington DC under Marion Barry, for example, increased its employment payrolls in the years 1982 to 1990 from 39,000 to 48,700. With economic downturn in the early 1990s, cities with dominantly black administrations found themselves in dire financial straits. The political movement to scale back government entailed job retrenchments and lay-offs, exacerbating black unemployment, shrinking the tax base of black-run cities, and encouraging middle-class black flight from inner cities to black suburbs.

The socio-spatial make-up of the contemporary city renders invisible especially the racialized poor. Similarly, policy priorities and bureaucratic rationality of federal, state, and local administrations restrict the possibilities in addressing the problems faced by the (racially) marginalized (Goldfield, 1993: 170). Expanding opportunities and social services for the racially marginalized in periods of fiscal conservativism, externally imposed structural constraints, and downsizing imperatives, become at best a zero sum game, if not the city's political and economic suicide. Economic globalization has prompted demographic dislocation and massive migrations not only from country to city but from South and especially East to the West, geographic sites that assume geopolitical significance. Migrations of poor Latino and Asian populations to the United States (cf. Omatsu, 1994) in particular have underpinned dramatic transformations in the ethnoracial urban composition in this country, and increasing tensions among the marginalized in racially ordered cityspace. A growing share of a fixed, if not shrinking, racial pie means diminishing returns to those racially defined groups similarly structurally situated.

At the same time, segregation within cities concentrates crime – violent crimes, especially homicide – within the spatial confines of the city or neighbourhood. So those blacks who kill, kill other black people, for the most part, as too those Latinos who kill, and so on. In the wealthiest neighbourhood of Washington DC, for instance, which is 88 per cent white, there were no murders in the first half of 1996; just a couple of miles apart in the poorest part of the city, which is 91 per cent black, there were 30 murders. The overwhelming majority of homicides take place within ethnic groups. In Texas, for example, 86 per cent of murdered Latinos were killed by Latinos.[12] In the final analysis, it is not raw poverty that accounts for the high homicide rates among Latinos and blacks but income inequality, the fact of economic differentials – and they have been rising rapidly (Martinez, 1996). The wealthy get protected not only as the poorer do not, but at the expense, discomfort, and threat to the safety of the racialized impoverished.

Suburban expansion addressed the housing demands of whites (Denton, 1994: 24). By 1970 the white birth-rate was plunging and white flight to the suburbs was flourishing, prompted by the lure of larger houses, easy access to shopping malls, cheaper services, the abandonment of the cities for suburbs by white-collar employment, the lure of white-controlled school districts and municipal governments. So even if Northern urban school districts (largely confined to city limits anyway) had the will to desegregate, they lacked sufficient white students for the task by the time school desegregation and busing orders had taken hold (Kunen, 1996: 40). School segregation exacerbated residential segregation, as whites opted out of living in neighbourhoods with largely black and Latino schools. Detroit's public schools now are just 6 per cent white.

There is a good deal to say in favour of restricting state intervention in people's choice of residential location. In a deeply racialized society with a living racist history, however, we should recognize that the extent of current segregation is the inevitable outcome of the absence of even a modicum of state regulation of discriminatory preferences and privatized real estate practices (not to mention private schools). Where cities and suburbs co-exist in a single school district, by far the exception in the US, there is at least one avenue available for addressing this complex of issues concerning residential and educational segregation. In the early 1990s, the city of Detroit, the residents of which are largely black and relatively poor, went to court to try to force its surrounding and self-governing suburbs, largely white and relatively wealthy, into a single school district. The city lost. It seems that school desegregation is a value less important in the land of the free than racialized self-determination. And racialized self-determination usually entails a commitment to racial homogenization with its attendant exclusions and segregation, both spatial and in schooling.

Growing up in the inner city, indeed in cities (or parts) deprived of the social resources readily available to their neighbouring competitors, must be seen as a severe social handicap in a society that places such a high premium upon the markers of material opportunity and advancement. Inner-city adolescence 'statistically increases the likelihood of dropping out of high school, reduces the probability of attending college, lowers the likelihood of employment, reduces income earned as an adult, and increases the risk of teenager childbearing and unwed pregnancy' (Massey and Denton, 1993). Segregation intensifies racial marginalization even as it is made possible by it: it confines blacks and Latinos especially to identifiable areas so that 'divestment in the place is disinvestment in the people' (Massey and Hajnal, 1995: 528). The spatial distinction between blacks and whites weakens opposition to such divestment by dividing interests, racializing them, setting them apart, thus to set them off against each other.

Even where blacks, through the sheer force of numbers, control or have a major influence over urban electoral politics, whites control the urban economy (Goldfield, 1993: 173). Political success is often about who controls resources, and blacks as a consequence tend to have few (or far fewer) resources – media, campaign financing, consultants, economic leadership (Goldfield, 1993: 174) – than whites. The very fact that these racially dividing terms dominate political discourse in the US is tribute to the terrible success of the segregating logic. In all too many cities blacks and whites have acted in concert to reproduce partial segregation; ironically, indeed tragically, blacks have found themselves reduced sometimes to accepting social and spatial segregation as a means to greater self-determination, social respect, and fairer distribution of education resources.

The lack of available resources in inner-city space effectively means the lack of employment possibilities in the formal economy, which in turn exacerbates the devaluation of housing stock and the unavailability of educational resources. Besides criminalization, drug addiction and associated diseases as well as violent death are run-of-the-mill occupational hazards for young black men. In the late 1960s and early 1970s close to a generation of young poor black men were wiped out in Vietnam, followed in the 1970s and 1980s by heroin and then crack addiction and HIV infection, the latter afflicting poor black women also.

It seems generally agreed that welfare arrangements need to be reformed to promote greater self-respect through viable employment prospects. Nevertheless, the current delimitation of welfare benefits, and especially the workfare requirements that are so central to current reform, seem to regulate a reserve army of surplus labour that keeps the minimum wage rate in check, and thus the wage profile of the low end service sector. The prison industry likewise has become a form of colour-coded segregation, especially of young black men.[13] As prisons are being privatized, prisoners increasingly are being employed at sub-minimum wage on behalf of profit-producing private businesses. Welfare and prisons, then, have become cornerstones of the New Segregation, locking poor people of colour spatially as much as economically into lives of severe limitation. If the Old Segregation was largely about locking people *out*, the New Segregation is mostly about locking them *in* – into inner cities and prisons. The recent welfare 'debate' prises open the revised ideological terms, the New Segregationism, attendant to the (re-)production of the New Segregation.

Segregationism, old and new

It will help to distinguish between discourse as the prevailing terms of articulation and expression in an episteme, archaeologically described, and ideology as the self-conscious undertaking to define, design, and determine the terms of social conception. If discourse furnishes the conceptual schema for thought about objects in a domain, ideology is (as Zizek puts it) 'the circle within which we are determined by reasons, but only those which, retroactively, we recognize as such', what Hegel calls 'the positing of presuppositions'. This circle of presupposition and reason is embodied in a narrative that gives meaning to experience and pursuits, ordering past events and present or future undertakings (Zizek, 1994: 41). Here racial articulation is deemed a discourse spanning and serving in part to define modernity; segregationism is an ideology narrating the degrees of racial separation.

Old Segregation, the classic variety emerging in the US in the 1880s, not only was rationalized but was made possible and promoted by the scientific racism of the late nineteenth century and the eugenicism that gripped the scientific and socio-political establishment in the first third of the twentieth. These transformations both in racialized discourse and the experience of racism attendant on them are evidenced in the commensurate shift in the formal racial categories in US census taking. The multiplicity of racial categories in mid- to late-nineteenth-century census forms gave way to the 'one drop rule' in 1900 in terms of which a person was defined as black with any degree of African ('negroid') ancestry. The 'one drop rule' almost immediately was recognized as enumerationally problematic and slipped out of sight from census taking by 1910, but it continued to grip American social identity throughout the first half of the twentieth century, and its informal hold on the national imaginary resonates to this day (Goldberg, 1997: 36–7, 40–1). The 'one drop rule' more or less formally operationalized the definition of race for segregationist social policies until 1954. And the administrative reason of state, by instituting and institutionalizing the 'one drop rule' in and through its social programmes and technologies, ensured and secured solidification of segregation throughout American social life.[14]

Ideologically, then, classic segregation from the late nineteenth until the mid-twentieth century was rationalized through the intentionally misrepresentational doctrine of 'separate but equal'. 'Separate but equal' was offered brazenly as (constitutional and more broadly social) legitimation of the old segregation, the fact of separate because deemed inherently unequal barely hidden from self-consciousness by the bad faith of the Old Segregationism. By contrast, the banner of the New Segregationism might be said to consist in the balder and bolder, unapologetic but (because?) barely hidden insistence that 'unequal and therefore (to be) separated'.

The New Segregationism, then, is an emergent, if not yet fully emerged, worldview that at once reflects and represents – or narrates – the given and the presumed, the prevailing and the dominant. Yet it is a view that must trade necessarily on twisting truths, on stereotyping and gross generalization, on fabrication and misrepresentation. It mystifies by imagining a demonizing and demonized Other that needs accordingly to be set apart, isolated, circumscribed and constantly surveilled. In the renewed clarion call to assimilationist commitments, there is the articulation of a New Segregationism, one no longer purely and unambiguously racial but the complex and shifting product of the intersecting dynamics of race, class, and (even more silently) gender. Unlike the unambiguous and more or less formalized segregationism of the post-*Plessy* moment, this new mode of chequered segregationism reflects and represents the anxious, ambiguous, and privatizing legacy of the post-Fordist and post-industrial period. Where the former was prompted by urbanization, the latter has been brought on by suburbanization, subdivisional self-enclosures, and gated communities.

It is curious that where there is contemporary reference explicitly to a 'new segregation', it is by representatives of the 'new right' like Dinesh D'Souza, Philip Perlmutter, Barbara Lerner, Arch Puddington, Linda Chavez, Clint Bolick, and Shelby Steele. Their wrath seems directed singularly not at the structural segregation described above but at what they consider the ethno-racial balkanization of everyday life in America at the hands of cultural élites, especially on campuses (D'Souza, 1995; Puddington, 1995; Perlmutter, 1994; Lerner, 1996; Steele, 1993; Leo, 1995).[15] The New Segregationists pose as the arbiters of reasonableness in the face of demonic and irrational urban trends: those who have examined all the evidence in an unbiased and neutral way to arrive at the most reasonable conclusions. Their views – many insidious, warped, and quite literally incredible – accordingly become projected as representative of reasonableness and moderation. Behind this facade, history is rewritten, stereotypical generalizations established as inescapable truth, partial representation imparted as impartial knowledge.

The ideological and rhetorical components of what I suggest is the New Segregationism include a shrunken understanding of racism and its effects, a fixed understanding of biologically given races, the claim that European and Euro-American morality and culture are superior to all others, as well as a diminished representation of the effects of American slavery and a commitment to a viciously anti-Civil Rights and anti-affirmative action politics of anxiety and resentment.

Implicit in the New Segregationism thus is a set of barely stated assumptions. Firstly, segregation is taken as the norm, the assumed, the natural, the given. Integration, or at least desegregation, comes over as unnatural, literally absurd and irrational in the prevailing order of things, requiring intervention by the state at the cost of liberty (the freedom to choose where one lives or is educated, who one hires or works with, where one hangs out, worships or may be laid to rest). Secondly, standards are represented mainly as white (or those associated with whiteness) which are assumed as the norm, as the criteria of judgement, as representing excellence – as what everyone else should aspire to. Thirdly, whites are projected as the real victims of antiracist excess (of leftist antiracist racism, of political correctness, of the ideology of egalitarianism). And finally, those committed to affirmative action, those against the undoing of antiracist protections, are chided by the agents of rational choice and unfettered individual preference – by the New Segregationists, in other words – as the cultural élites (the very terms Bob Dole used to knock Democrats in his acceptance speech for the Republican nomination), out of touch with the 'real' concerns and interests of 'real', everyday, working – that is, street wary and weary – white people.

If these views make up the circle of presupposition, then the central tenets of the New Segregationist narrative are expressed thus: racism was neither the most egregious wrong suffered by people nor what now most requires social redress (cf. Levin, 1981). The Civil Rights Movement supposedly founded itself upon the 'misguided' liberal principle that all cultures are equal. This 'serious problem' disabled Civil Rights leaders like King from seeing that what was at issue regarding black inequality was not only 'racial discrimination' but also a poverty of culture which 'inhibited black competitiveness'. Black cultural pathology consists in that set of culturally acquired antisocial habits and behaviours supposed to promote and produce violent crime, drug addiction, teenage pregnancies, joblessness, in short, the 'underclass'.[16] The Civil Rights Movement accordingly spawned a professional class of self-serving bureaucrats and activists elevating only themselves on (largely false) charges of racism against any who stood in the path of satisfying their self-interest. Where blacks today are subjected to racism, it is thought to follow, they only have themselves (or their culturally élitist leaders) to blame. Moreover, 'black racism' against those not black has become a problem more serious than white racism. New Segregationists may admit that 'rational discrimination' against blacks – discrimination based on group-defined statistical aggregates – still exists. But because it is purportedly 'rational' they consider such discrimination justifiable. Black racism, by contrast, is 'explained' as 'a rationalization for black failure'. So the basic problem currently facing blacks is considered to be not white racism but black cultural pathology. Blacks are exhorted to cease relying on government and to acquire the values and habits – the culture – capable of making them 'civilized' (D'Souza, 1995: 22–4). Welcome to the New Segregationist neighbourhood.

Thus New Segregationists dismiss the Civil Rights legacy as relativistic liberal promiscuity. The Civil Rights Movement is dismissed for encouraging the emergence of race-conscious rather than race-neutral social policies, and blamed for the contemporary demise of black America (cf. D'Souza 1995: 167). The (il)logic of this position entails that the Civil Rights Movement was an *over*-reaction: if racism and its effects were not all that bad, if segregation was a form of protection, not disadvantage (D'Souza, 1995: 179), if blacks largely have themselves rather than racism to blame for their non-competitiveness, then the Civil Rights Movement went too far. The policy and legal changes it effected exacerbated rather than addressed or resolved the plight of black Americans, dividing the nation still further by elevating some over others on racial grounds rather than protecting and extending the rights of all. Thus, to a person, New Segregationists bemoan the 'fact, that there are two standards of civil rights today: one for white males, another for women, blacks and other minorities' (D'Souza, 1995: 223). The Civil Rights establishment has apparently mobilized a whole range of 'affirmative action groups … to act as pressure groups within the private and public spheres. They are formed under the aegis of racial preferences' (D'Souza, 1995: 231).

In this spirit, the desegregation decision in *Brown* is deemed not to be about integrating schools but simply about removing the legal dictation of segregation. Indeed, this distinction between 'acceptably' removing the vestiges of segregation and 'illegitimately' acting more affirmatively to promote integrating pursuits is at the heart of the New Segregationist project. The post-*Brown* insistence on integration by the Civil Rights establishment is dismissed as tantamount to a new form of racial discrimination – against whites (D'Souza, 1995: 224). 'Desegregation permits racial separation as long as it is *not compelled* by government. Integration, by contrast, is a state-mandated result' (D'Souza, 1995: 225, my emphasis). Quotas supposedly enacted in the name of affirmative action turn whites into victims.

I have discussed critically this circle of presupposition that makes up the New Segregationism in a lengthy critique of D'Souza (Goldberg, 1997: 175–226). I wish here to focus, with the view to sociological elaboration, only on those features 'foundational' to rationalizing the New Segregation. In this spirit, the decision in *Brown* may be read racially: whites generally seem to have interpreted *Brown* as D'Souza does, namely, as the narrow commitment no longer to mandate segregation by formal legal means. Blacks, by contrast, interpret *Brown* as holding out the reasonable hope of a more formalized commitment to equal treatment and access to equal resources, at least educationally. To say the least, shortsighted 'white' interests have prevailed, 'black' hopes dashed. Public schools, segregation-wise, today face *de facto* something like the *status quo ante*. Schools are about as segregated as they were in 1896 or 1954 though formalized segregation has been replaced by the informalized, privatized variety extended recently by the charter school, school voucher, and home-schooling lobbies.

In this context, the call by some blacks and black groups for separatism – separate black-controlled programmes, facilities, institutions for blacks (whether school classes, courses, residences, or student associations) – is not particularly new. Black separatism has been promoted during the past century in the name of black nationalism whenever white segregation and racism have been especially evident. What *is* new is that white liberal institutions and their administrations have given in to the demand, at least more readily. It is as though, in the name of self-empowerment, the whole political and representational establishment has seen as the sole (and unacceptable) alternative to white-dominated, defined, and ordered integration the call to segregate blacks 'for their own good'.

The heart of the New Segregationist transformation in racial rationalization concerns the relation of race and rationality. The standard liberal view during this past half-century has insisted that racial reference is irrational by conception, hence too racism. New Segregationists agree that racism is an ideology of inferiority based on biology. Racial *discrimination* by contrast is considered a practice of race-based differentiation. Discrimination can be rational, accordingly, when based on group conduct determined by culture and not on biology (D'Souza, 1995: 286). '[W]hites view racial discrimination today as a rational response to black group traits' (D'Souza, 1995: 246).

Expressly committed to colour-blindness, that is, to a standard of justice protective of individual rights and not group results, New Segregationists are ready to trash the rights of black individuals for the sake of the 'rational discrimination' of whites, who simultaneously rationalize their avoidance of black folk by appealing to statistical generalizations about racial (and gendered) groups. By 'rational discrimination' New Segregationists intend that the discrimination is instrumentally valuable: 'It is efficient, it makes economic sense' (D'Souza, 1995: 277). If discrimination in hiring, mortgage leasing, and criminal justice is mostly rational, then the dominant tradition in 'Western' moral doctrine suggests it is not racist (Goldberg, 1993: 14–40). The values of efficiency and economy – the fundamental(ist) foundations of rational choice reason – assume the status of empirically established truth, the force of which is unquestionable and, so, incontestable. The disvalue of racist discrimination is discounted in the calculus of maximizing personal preference schemes.

Banks are deemed rational to discriminate on the basis of statistical aggregates between loans to individual whites and blacks, charging the latter higher rates because of higher group default rates. Similarly, employers are considered rational – efficient and economical as a rule – not to hire black employees because of statistical 'evidence' that they supposedly make less reliable and more contentious employees than whites. Such a recruiting rule supposedly saves employers recruitment and hiring costs because of the higher risk of failure among black candidates. Indeed, fearing litigation in the wake of *Hopwood* v. *Texas*, universities throughout the United States now are carefully considering whether their hiring procedures discriminate against white men rather than against those groups explicitly enjoying legal protection as a result of the history of discrimination against them. And this shift no doubt will continue no matter the outcome of constitutional challenges to Proposition 209 in California. One could actually say that this shift in 'culture', rather than any specific legal outcome, was precisely the principal aim of lobby groups such as the California Civil Rights Initiative.

By the same token, New Segregationists consider the increased criminalization of African Americans over the past quarter-century to be a systemically rational response to heightened black criminality. They deny that the criminal justice system is more racist now than fifty years ago, thus predicting fewer blacks in prison rather than more. So the increased incarceration of African Americans is considered to lack any semblance of racist causation. Why, it must be asked, did the rates of incarceration explode after the 1960s? The largest difference between blacks and whites in respect of criminal justice is concentrated in the arrest rates. New Segregationists put this down to 'rational discrimination' once again: policemen especially in city settings 'know' that young black men are more likely than whites to commit crimes, so they pursue them more vigorously (D'Souza, 1995: 283–4). There is another explanation, however. Arrest is the most discretionary phase in the criminal justice system, the most open to abuse, and therefore potentially the most discriminatory. This is especially exemplified by the invocation by police of the racially over-determined category of 'symbolic assailant', a character that cues fear as much for the police as for the cab driver.[17] In light of the likes of Laurence Powell, Stacy Koons, and Mark Fuhrman, of deaths by choke hold or police trigger happiness, why is it 'rational' to think otherwise?

Black people thus are cut loose socially, set apart, and left to their own devices in increasingly abandoned cities. White America is freed from responsibility save for not discriminating, other than rationally, in their own self-interests. And at basis rational discrimination means living, schooling, and working in different cities. If a small percentage of black people succeed despite these overwhelming odds, rational discrimination will license exceptions. In their generosity New Segregationists seem not to mind an occasional black neighbour so long as they can afford it as a result of an honest day's labour and '*they* act white'. While white gangsters, corporate or organized, will not be excluded from the New Segregationists' gated ghetto, no gangstas – especially if they swagger like Tupac Shakur or bear a name like Snoop Doggy Dog – need apply. Too many black neighbours, I sense, will drive down the price of property. The accusation of black cultural pathology implies that order, security, and stability are established in the city through the white middle-class production of conformity in culture, values, habits, dress, look, and practice.

'Black culture' is represented repeatedly as exhibiting a 'vicious, self-defeating and repellant underside' (D'Souza, 1995: 486); 'underclass blacks [are characterized by] their gold chains, limping walk, obscene language, and arsenal of weapons' (D'Souza, 1995: 504). This is the image of the city as dangerous and threatening, a danger and threat created as much as invoked to partition the city along racial lines (see Goldberg, 1993: 185–205). The concept of pathological *culture* at work here is especially peculiar, sufficiently determining of what is considered pathological behaviour but unaffected or under-determined by the political, economic, social, legal, and psychological conditions on which it is founded and in which it is embedded. The slippage around blackness in New Segregationist articulation is readily apparent. New Segregationists chide the (black) *underclass* for pathological behaviour, then generalize to blacks ('many blacks') as such, thus reducing 'the black underclass' to pathology, blacks to the underclass, and the underclass to blacks, at once reifying the underclass.

In the land of Lockean libertarianism, one is and gets what one can pay for; those unable to pay, overwhelmingly black and Latino, tend to get locked away. Cultural capital and capital's culture are deeply intertwined. Three simple 'pathological' strikes (if one is not white) and you're out of sight, out of mind. For life. The powerful materiality of these (re)new(ed) ideological effects leaves blacks as segregated, and other racialized, marginalized persons virtually as excluded as they were a century ago.

Whistling in the wind(y city)

In terms of the institutionalization and reproduction of segregated cityspace, the New Segregation has managed to *informalize* what used to be formally produced, both to *realize* and *virtualize* segregation. Race still defines where one can go, what one can do, how one is seen and treated, one's social, economic, political, legal and cultural, in short, one's daily experience. Cityspace in the US is increasingly contradictory: as there is greater heterogeneity and multiplicity, so segregation is refined; as visible openness and accessibility are enlarged, exclusionary totalization is extended; as interaction is increased, access is monitored, traversal policed, intercourse surveilled. As boundaries and borders become more permeable, they are re-fixed in the imaginary, shifting from the visible to the virtual, from the formalized to the experiential, from the legal to the cultural at a time when the cultural, economically and socially, has become dominant.

New Segregationists refuse to consider an alternative reading of the social terrain: that the troubled and troubling behaviour of some among the most disaffected, hopeless, nihilistic and impoverished of young black and brown people, especially men, that we have watched emerge in the last 25 years or so developed out of a radically intensified impoverishment at the very moment of heightened expectation. As we have seen, this is a complex product of dramatic shifts in socio-economic structure and its attendant prospects for young black and brown people over this period. William Julius Wilson points out that employed black men in their late twenties experience similar rates of crime as employed white men of this age (Wilson, 1996). Not surprisingly, it is the jobless who tend to commit crimes, and (white anxiety notwithstanding) young black and brown inner-city men are much more likely to be out of work than their white peers. So when New Segregationists reduce the range of possible determinants of contemporary black social difficulties to genes or culture (or some combination) (D'Souza, 1995: 529), they reach once more for the superficial account. There are structural considerations over which young black and brown men and women have (had) no control: the history of racism has made blacks, in particular, more prone to the difficulties structurally produced, thus exacerbating the problems they face. One could say that the intersection of race, class, gender, and age with political economy gives precise location to the debilities at issue.

In response to New Segregationists, one might ask what social responsibilities there are facing individuals with or without institutional power to alter the social, political, and economic conditions underlying the reproduction of self- and other-destructive behaviour. New Segregationist representations are predicated on totalizing and vicious stereotypes, their only policy response being to end all government assistance and civil rights protections while insisting that individual black people have a special responsibility to stand up to the peculiar problems they or family or group members face. This response reifies existing commitments not just to already established forms of segregation but to the ongoing project of segregating those who do not assume and exhibit the culture of middle-class whiteness, a culture marked also in deep economic, political, and psychological ways by socio-spatial dimensions, one that tends to be suburban rather than urban, supposedly self-directed and expansive rather than curtailed and confined, optimistic rather than nihilistic.

The New Segregationist line of attack on the civil rights tradition in the US only makes sense in virtue of the now popular effort to efface history, by excising from historical memory the conditions that black and Latino people have suffered, with devastating contemporary effects for the racially marginalized. It is only by way of this aphasia that New Segregationists can mobilize the claim that civil rights policies – affirmative action, voting rights reapportionment to establish black-represented districts, integrated schools, and so forth – discriminate against white people.[18] The claim implicitly denies that it is black and brown people – those not white in general – who have suffered racial terror at the hands of whites; who have been discriminated against, excluded, denied equal access, violated persistently, devalued, dumped upon environmentally, ignored *qua* group identity and membership. This denial is why they spend so much ink trying to belittle – literally to diminish – the degradation black people continue to experience under the effects of racism (slavery wasn't so bad after all, segregation was designed to protect blacks from the likes of Ku Klux Klan radicals, in any case a small fringe minority, and then the Civil Rights establishment placed blacks in positions to discriminate against whites). What New Segregationists deny at every turn is the history in virtue of which racial power – in politics, education, employment, wealth creation, and cultural production – has guaranteed improportional representation, for whites.[19] To riff on Air Jordan's sage retort in *Space Jam*, the New Segregationists' commitments are not so much white – they're transparent!

The account I have offered suggests reasons also for current attacks on affirmative action. For one, these attacks are tokens of the increased anxiety among the white working and middle classes over job loss and increased competition for available educational and employment positions. In tightening labour and educational markets, whites tend to perceive that opportunities available to blacks have evened out. At the same time, in the globalization of the labour force not only is there pressure on the welfare state to rationalize costs. There is also pressure to produce a labour force that is 'Third World-like': cheap, unskilled, abundant, desperately poor – and black/brown by default. This places fiscal and political pressure on the availability of affirmative action programmes. Add to this the fact that in an increasingly heterogeneous world, at home as abroad, whites in general, and white men particularly, perceive their power to be dramatically diminished (partly because they can no longer harass racially and sexually with impunity). The standards of whiteness, as I mentioned above, no longer prevail (at least not unchallenged), and white terror – both experienced and expressed – has increased with this sensed loss of grip. Finally, the segregationist will or desire among whites (the white will to segregation, historically reproduced) is very much alive, if somewhat tempered in tone, in extremity of expression. It has become more informal, more discrete, a product more of preference formation than bare institutional mandate. Whites have seen what they regard as 'their' residential space, 'their' schools and playgrounds, 'their' work environments, and 'their' voting districts to have 'integrated' *enough*. So, for example, public universities, formerly exclusively white, tend now on average to have a student body that is roughly twenty per cent not white (consistent with expressed white preference for eighty per cent white residential space). Affirmative action thus is considered to have 'gone far enough' in the desegregative direction.

If integration is taken to be the contrast with segregation, it is given the gloss of whiteness. So integration supposedly is to be achieved by 'browning' white space, thus abandoning (and giving in to the assumption of) a black and brown space to be avoided or abandoned. The implicit acknowledgment is that black and brown space is unappealing, dangerous, to be avoided – and abandoned. This is an assumption that silently implies that white space tainted by (one drop) blackness will likely follow suit.

Integration, such as it is, hardly offers a way out, for its terms and standards are set by those already holding power, and historically the powerful in the US are overwhelmingly white (and male). Desegregation must mean, in its incorporative sense and if it is to have significance at all, the mutual making of decisions, the sharing of resources, the co-definition and co-production of space and place, norms and rules, the common experience of an interactive state not in the blindness to racial definition but in the sensitivity to mutual racial recognition and respect, where racial definition is self-chosen as an important ingredient of one's identity without imposing that choice on others.

The prevailing racial divide at city limits (material and imagined) is no longer, as it once was, between town and country but between one municipality and another: one wealthier, the other likely poor; one resource-heavy, the other resource-lacking; one relatively safe and secure, at least for its residents, even as they are haunted by the image of rampant criminality, the other far more dangerous; one overwhelmingly white, the other exclusively black and brown.

The modernist city emerged (if it was not conceived) as containable, a container, a space with limits and so a limiting space. That image must now be challenged by considering the city uncontainable, unrestrictable, as the refusal to be divided: not simply as desegregable but as indivisible, inseparable, *un*segregable – both within and across established divides. Urban heterogeneity mobilizes an urbanity in and through transnational and inter-urban movement. By generating new energies and (insisting on new) insights of established place, new views of the city and urban practices, the city assumes an identity unlike any other, giving rise to the possibilities of dialogical transformation, an urban uniqueness in the face of the now localized and globalizing logic of sameness and marketized uniformity. Indeed, the urban uniformity promoted by globalizing commodification further promotes segregating pressures on those spaces presumed antithetical to the conditions of globalized commodification. The perceived unruliness, the otherness, of black and brown urban space, marked by definitional difference, is placed under pressure to conform or be rendered invisible via spatial containment and confinement.

The inhabitants of black/brown space for the most part are deemed unemployable in the dynamic, expanding, information-based formal economy; they are employed overwhelmingly (where employed at all) in the derivative, castaway, informal, low-end service sector, the economy of minimum wage work. Space, place, and the experience of daily life mirror the increasing divide of income and wealth. Renewed and renewable racializing discourse orders the urban imaginary into segregable divisibility, delimiting the possibilities of conceiving desegregated urban space. If segregation is about the container city, desegregated urbanity must promote city flow, heterogeneous movement and attendant multiplicity. Segregation is extended through restricted electronic access, by the unavailability of the material media of technological exchange. Lack of access to the Internet may secure kids from its excessively commercialized framing but it is the equivalent of denying them the possibilities enabled by the Industrial Revolution. In cities linked with other cities through informational flows, those out of the loop lose out by becoming once again *invisible*. And while Fanon shows that invisibility may sometimes fashion a site to counter racist and segregationist projects, Ralph Ellison makes abundantly clear the costs of that social state, especially in the absence of any possible alternative.

The city is no longer, if it ever was, a vigorous trans- or cross-racial meeting place. At best, and at worst, its image has been turned into a space of avoidance and denial or the crucible of racial collision. As I was ambling through downtown San Francisco recently, it struck me how urban spatial conditions and culture are sewn together to enable or disable heterogeneous flourishing. The people of many US cities are now increasingly diversely constituted (one might see this as the appeal of cities *per se*). Two principal conditions of urban life accordingly promote their exhilarating intercourse or (in their absence) the separateness of their policed apartness. The first feature is that residential and commercial space should not be divided, that heterogeneous people interact in their daily and nightly lives on the streets and in their social and commercial exchanges. The second, relatedly, is that the city should promote pedestrian contact, thrusting people together unmediated by the insularity of automobiles. Very few cities in the United States meet both conditions: New York, Chicago, San Francisco, cities marked by renewed migration and immigration. Rich and poor, black, brown, and white, Chinatown and the cultural avant garde live cheek by jowl. It is not without reason, then, that these cities have long, progressive, if sometimes troubled and always contested, political traditions. Cities of the South and West, by contrast – Los Angeles, Phoenix, Houston, Dallas, Atlanta, San Diego, and the like – are marked by weak centres inhabited, if at all, by the racially marginalized at the centre's faded and forgotten edge, and traversed by car culture (suburbs only became viable with the modernization of automobiles and highways and of centralized heating and cooling systems in the 1940s and 1950s) – that is, by the very conditions promoting the possibility of insularity.

So, to conclude by opening up a set of questions: where black, brown, white or whatever cross paths, can the clash of cultures and experience be redirected from 'crash' (to use J. G. Ballard's title) to (self-)critical civility? From collision to mutual sensitivity? From competed-for to shared resources? From ignorance to intercourse, power to interactive possibility, dismissiveness to mutual engagement? From the production of policed homogeneity and purity to transracial heterogeneity and proliferation? From boundedness, confinement, and fixity to boundlessness, openness, and transformability? From the pessimism of Spike Lee's *Do the Right Thing* to the sensitive humour and humility of Wayne Wang and Paul Auster's gentle ode to Brooklyn, *Blue in the Face*. In short, from the narrative of Hobbes to the anti-narratives, the fragments, of Heraclitus?

References

Black, Donald and Reiss, Albert (1970) Police control of juveniles. *American Sociological Review*, **35**, 63–77.

Bolick, Clint (1996) Discriminating Liberals. *New York Times*, 6 May, A15.

Denton, Nancy (1994) Residential segregation: challenge to white America. *Journal of Intergroup Relations*, **XXI**, (2) (Summer), 19–35.

D'Souza, Dinesh (1995) *The End of Racism*, The Free Press, New York.

Dudziak, Mary (1995) Foreign policy and desegregation. In Delagado, Richard (ed.). *Critical Legal Studies*, Temple University Press, Philadelphia.

Ford, Richard (1995) *Independence Day*, Knopf, New York.

Glazer, Nathan (1995) Black and white after thirty years. *The Public Interest*, **121** (Fall) 61–79

Goldberg, David Theo (1993) *Racist Culture: Philosophy and the Politics of Meaning*, Blackwell, Oxford.

Goldberg, David Theo (1997) *Racial Subjects: Writing on Race in America*, Routledge, New York.

Goldfield, David R. (1993) Black political power and public policy in the urban south. In Hirsch, Arnold and Mohl (eds) *Urban Policy in Twentieth Century America*, Rutgers University Press, New Brunswick, 159–82.

Harvey, David (1989) *The Condition of Postmodernity*, Blackwell, Oxford.

Hirsch, Arnold (1993) With or without Jim Crow: black residential segregation in the United States. In Hirsch, Arnold and Mohl (eds). *Urban Policy in Twentieth Century America*, Rutgers University Press, New Brunswick, 65–99.

Jordan, Bill (1996) *A Theory of Poverty and Social Exclusion,* Polity, Oxford.

Kunen, James (1996) The end of integration. *Time Magazine*, **147**, (18), 29 April, 39–46.

Leo, John (1995) Just say No to the New Segregation. *US News and World Report* **119**. (23), 11 December, 39.

Lerner, Barbara (1996) Black future. *The National Review* **48** (1), 29 January, 51–5.

Levin, Michael (1981) Is racial discrimination special? *Journal of Value Inquiry*, **15** (3), 225–34.

Martinez Jr., Ramiro (1996) Latinos and lethal policy: the impact of poverty and inequality. *Social Problems*, **43**, (2) (May), 131–45.

Massey, D. and Denton, N. (1993) *American Apartheid: Segregation and the Making of the Underclass*, Harvard University Press, Cambridge Mass.

Massey, D. and Hajnal, R. (1995) The changing geographic structure of black–white segregation in the United States. *Social Science Quarterly*, **76** (3) (September), 527–42.

Mohl, Raymond (1993) Shifting patterns of American urban policy since 1900. In Hirsch, Arnold and Mohl (eds) *Urban Policy in Twentieth Century America*, Rutgers University Press, New Brunswick, 1–45.

Omatsu, Glenn (1994) The 'Four Prisons' and the movements of liberation: Asian American activism from the 1960s to the 1990s. In Aguilar–San Juan, Karen (ed.) *The State of Asian America*, South End Press, Boston, 19–70.

Perlmutter, Philip (1994) Segregation: coming back in style? *Journal of Intergroup Relations*, **XXI**, (2) (Summer) 41–2.

Pilivian, Irving and Briar, Scott (1964) Police encounters with juveniles. *American Journal of Sociology*, **70** 206–14.

Puddington, Arch (1995) Speaking of race. *Commentary*, **100**, (6) 21–5

Roediger, David (1996) White looks: hairy apes, true stories, and Limbaugh's laughs. *Minnesota Review*, **47** (Fall).

Schelling, Thomas (1971) On the ecology of micromotoves. *The Public Interest*, **25** (Fall), 1971.

Skolnick, Jerome (1973) Justice Without Trial

Steele, Shelby (1993) Rise of 'The New Segregation'. *USA Today*, **121** (2574) (March) 53–5.

Warr, Mark (1990) Dangerous situations: social contexts and fear of victimization. *Social Forces*, **69** (3) 891–908.

Wilson, William Julius (1996) *When Work Disappears: The World of the New Urban Poor*, Knopf, New York.

Zizek, Slavoj (1994) Identity and its vicissitudes: Hegel's 'Logic of Essence' as a theory of ideology. In Lacalu, Ernesto (ed.) *The Making of Political Identities*, Verso, London, 40–75.

Notes

* This chapter was first prepared as a plenary address for the City Limits Conference, University of Staffordshire, England, September 1996, organized by Azzedine Haddour and Laura Peters. I am grateful for their helpful comments and discussions, also to Mary Romero, Gray Cavender, Paul Gilroy, to the African American Studies colloquium at the University of California, San Diego, especially to Grace Hong for provoking some thoughts about Asian migration, and to my research assistants Melissa McDonald, Sam Michalowski, and Ruth Butler.

1   The New Segregation is rendered additionally complex by considering the partly differentiated experience of other marginalized groups, especially Latino/as. Latino experience obviously is historically differentiated from that of African Americans, but it is internally also much more varied. The experience of Puerto Ricans, Cuban Americans, Mexican Americans, South and Central American immigrants, while sharing some commonalities in the US, involves significant distinctions. Moreover, the prevailing governmental presumption about 'Hispanic' racial formation for the past fifty years has been that they are racially white unless 'self-evidently' not. Thus while barrios have become significant spatial domains of American cities, especially in the South and West, their history of emergence and production is not quite isomorphic with that of black urban space upon which I concentrate in this chapter.

2   Massey and Hajnal (1995: 529) define 'exposure' as 'the degree of potential contact between blacks and whites within geographic units'. Exposure 'measures the extent to which group members are exposed to one another by virtue of sharing a common area of residence'.

3   Massey and Hajnal (1995: 529) define 'evenness' as 'the degree to which blacks and whites are distributed uniformly across geographic units'. With evenness, 'the black percentage of each geographic unit equals the black percentage as a whole'.

4   By 1920 a majority of whites had been urbanized, compared to only a third of blacks; by 1950 a majority of blacks were city folk, and by 1960 a greater percentage of blacks than whites lived in cities. Between 1920 and 1980 the blacks living on the land and working in agriculture declined by 96 per cent, and by 1981 their numbers had dropped to one per cent of their original level (Hirsch, 1993: 66–7).

5 As Massey and Hajnal note, these included Atlanta, Baltimore, Detroit, Gary, New Orleans, and Washington.

6 These included Cleveland, St Louis, and Oakland.

7 Schelling uses a checkerboard with dimes and nickels though I think the colouring of the chess pieces more provocative.

8 'Residential segregation, as embodied in the black ghetto at its extreme form, is the key institutional arrangement that ensures the continued subordination of blacks in American society' (Denton, 1994: 21).

9 By 1970 Washington DC and Atlanta had black majorities; ten years later so did Detroit, Baltimore, and New Orleans. In 1960 the New York suburbs were 4.8 per cent black; by 1980 despite all the desegregation noise they were just 7.6 per cent so; in Chicago the increase went from 2.9 to 5.6 per cent. Desegregation clearly was making waves! Between 1976 and 1996, the Washington DC population dropped from 706,000 to 554,000 (over 20 per cent) and is now at its lowest since the early 1930s. Residents of the wealthiest ward of that city are 88 per cent white, only 6 per cent black, for the poorest ward 91 per cent black and 8 per cent white (for the city as a whole it is 66:8 per cent black to white). Median income is nearly $50,000 in the wealthiest but just $21,000 in the poorest ward, with massive education, violent crime, single-parent homes, and out-of-wedlock differentials. In 1970 roughly 58 per cent of all blacks lived in inner cities while by 1980, for all the fine talk about desegregation and the elevation of the black middle class, this figure had dropped off less than 3 per cent to 55.7 per cent (Hirsch, 1993: 66–7). At the same time federal support for cities dropped from 11.5 per cent of city budgets in 1980 to 3.8 per cent in 1990.

10 Between 1980 and 1990 the black population of Prince George's County grew 63 per cent, 60 per cent of whom were leaving DC (in contrast to 20 per cent of white residents of County). As a result of this black in-migration and the attendant white flight, Prince George's is now the first majority black suburban county in US.

11 Richard Ford's *Independence Day* (1995, the novel, not the movie) provides a rich and fascinating characterization of the segregated and segregating effects of the real estate industry. It is not that integration with whites is an appealing value in virtue of some inherent goodness about whiteness (quite the contrary, if the recent literature on whiteness is even close to the mark). Integration (in contrast to 'incorporation', see Goldberg, 1993: 219–21) in a society like the US is a value only in so far as it makes more readily accessible resources and power for those to whom they are not otherwise available.

12 Comparable figures in Miami were 85 per cent, New York 78 per cent, and Chicago 82 per cent.

13 More than a third of young black men between the ages of 17 and 35 are in some subordinate disciplinary relation to the criminal justice system: arrested and awaiting arraignment or trial, in prison, on bail or parole. Of the one million federal prisoners, 33 per cent are black; of the half-million additional state prisoners 46 per cent are black. The data for Latinos is comparable.

14 In 1978 Louisiana became the last state in the Union to strike out the 'one drop rule' as the basis of state racial classification.

15 Unsurprisingly, many of these interventions were published in the *National Review* and in *Commentary*. See also the symposium in *Commentary* on 'The National Prospect', vol. 100, No. 5, 1 November, 1995, 23–94. Rush Limbaugh offers a more populist expression of this (see Roediger, 1996). While representative expression of the 'New Segregationism' can be found in publications by all

those mentioned here, Dinesh D'Souza offers the fullest elaboration and I will concentrate on his articulation.

16 The corollary of this is to discuss poverty, social inequality, and the underclass in Anglo-American societies with no mention whatsoever of their racialized production or configuration (cf. Jordan, 1996).

17 On the notion of 'symbolic assailant', see Skolnick (1973); on 'fear cues' and race, see Warr (1990); on police discretion in arresting juveniles, see Black and Reiss (1970) and Pilivian and Briar (1964).

18 For a representative sample, see Bolick (1996).

19 If any popular doubt remains about the depths of continuing racist sentiment and segregating rationalization, Ted Koppel's *Nightline* series on 'America in Black and White' (ABC, 20–24 May 1996) surely must have dispelled it. Rachel Ward, a black single mother who moved into Bridesburg, a white working-class neighbourhood of Philadelphia, was run out within a week not only by the clamour of racial slurs and threats she immediately faced but as much by the 'polite racism' of those 'ordinary people' quietly committed to keeping their neighbourhood white.

Chapter 13

A critique of integration as the remedy for segregation

Iris Marion Young

In an opinion piece entitled 'An Integrationist Manifesto' which appeared in the *Wall Street Journal* in 1994, Tamar Jacoby tells a story of conversing at a dinner party about the state of race relations in the United States.

> We were talking about what we thought was wrong, and I found myself describing my nostalgia for the ideals of the 1960s. 'What ever happened to integration', I asked, 'to the idea that we are one community?' No-one answered the rhetorical question, but I noticed the black woman at the other end of the table staring intently. 'What happened,' I went on, 'to the idea that underneath the skin we are all the same?' The woman, whom I'd never met before that evening, jumped up from the table. 'We woke up,' she answered icily and walked out of the room.
>
> The evening ended soon after, but I could not shake my astonishment. Did the woman's racial resentment really loom that large – so large that she had to leave the room when I mentioned our common humanity?[1]

For Jacoby an ideal of integration is so obviously the solution to a segregated society riven with conflict and injustice that she fumes with righteous indignation when a black woman questions it. The woman's response only confirms what Jacoby suspected: that today people of colour, not white liberals like herself, are responsible for continued racial divisions. I find Jacoby's attitudes typical of many white Americans who think of themselves as liberal. The response of the black woman Jacoby reports is, I believe, also common though not universal. Many people of colour today reject or at least distrust the ideal of integration that propelled the Civil Rights Movement of the 1950s and 1960s. While I am less familiar with the tensions and rhetoric of race and culture outside the United States, I have seen what seems to be a similar divergence in attitudes elsewhere. While official German policy toward migrants calls for their 'integration' into the workforce and civic life of the cities to which they come, for example, I have heard migrants of colour in Germany question this goal of integration.

In this essay I aim to sort out what's going on here. I will side with the black woman in Jacoby's story, and argue that despite a noble history, an ideal of integration today is not the best alternative to the harms of segregation. To make that argument, I will define segregation and distinguish it from separation. Even in the absence of legal enforcement, segregation, most especially residential racial segregation, is an ongoing reality in many places, bringing significant harms. I define the norms of integration frequently taken as the proper remedy for these harms, and argue that this ideal of integration does not well address these harms and can produce others. I argue instead for an ideal of social and spatial relations which I call 'together-in-difference'. Unlike the ideal of integration, the ideal of together-in-difference does not imagine that social equality calls for dissolving people's group affinities. Unlike segregation, however, the ideal of together-in-difference affirms the undecidable overlap, fluidity, and interdependence of differentiated but socially equal groups.

What is segregation?

The ideal of integration aims to remedy the wrongs of segregation. What, then is segregation, and what are these wrongs? Most starkly and obviously, segregation consists in an enforced separation of groups that confines members of some groups to specific areas, or excludes members of a group from specific spaces, institutions, or activities, or regulates the movements of members of segregated groups. Anti-Semitic policies in Eastern Europe in the early twentieth century, enforced segregation of transportation, school attendance, employment, recreational and other institutions in the American South before 1960, the policies of apartheid in South Africa all exemplify segregation in this sense. Segregation in this sense of legally enforced separation is wrong because it brings the coercive apparatus of the state to enforce separation and restrict freedom for the sake of preserving privileges of some while seriously depriving others.

Segregation in this sense is still alive in the world (in the policies of Israel, for example). Nevertheless, legally enforced group segregation is rare in the world today and is generally considered a violation of human rights. Thus today what legal writers have come to call *de facto* segregation should command the attention of social and moral theory. In many societies without official policies regarding who should live where or participate in what kind of institutions, whose stated norms promote formal equality among persons and groups, there are nevertheless strong patterns of racial or ethnic concentration in residence, employment, schooling and other settings, and these patterns are associated with privilege and deprivation. Many people in societies which exhibit such patterns but which lack a history of enforced segregation either fail to notice these patterns or deny that they should be identified with the wrongs of segregation. I suspect that many in England, for example, would deny that the patterns of racial concentration in some major British cities should be called segregation. Despite what seems to myself and some students of German cities as patterns of segregation in Frankfurt, Hamburg, and Berlin, to take another example, many Germans are innocently puzzled at the suggestion that they might have this 'American' problem. In the United States it is difficult to deny the existence of *de facto* segregation, but many interpret it as an effect of past *de jure* segregation.[2] Since the patterns of racial concentration are generally more pronounced today in Northern cities which did not have explicit policies of racial exclusion than in the Southern cities that did, however, it does not seem plausible to interpret most current patterns of segregation in the United States today as a legacy from the days of enforced segregation.[3]

For the purposes of this essay I will assert without much argument that group-based residential segregation is common in modern democracies with self-consciously differentiated groups. Where it exists, it is the product of class and income differential combined with a variety of discriminatory actions and policies of individuals, private institutions, and governments. In her study of residential segregation in Britain, for example, Susan Smith documents how the combination of limited income and the discriminatory housing allocation policies of public housing authorities from the 1950s to the 1970s forced many South Asian and Afro-Caribbeans into private rental housing markets in the inner cities. Continuing private and public discrimination forces blacks to pay relatively more for older, deteriorating housing in crowded neighbourhoods with relatively poor services and few amenities.[4] The intersection of market forces, group affinity, public and private discriminatory behaviour, reproduce and exacerbate patterns of racial concentration in cities as diverse as Auckland, New Zealand, Birmingham, England, and Pittsburgh, Pennsylvania. Whatever the country, neighbourhoods with high concentrations of those groups thought of as 'black', 'coloured', 'strange', or Other, by the economically and socially dominant whites contain poorer quality housing, fewer amenities and services, more crowding, worse transportation access to other areas, than do other neighbourhoods, often even other neighbourhoods of equal median income.

But what exactly is wrong with these typical patterns of racial or group concentration? Many appear to identify the wrong of segregation as group clustering itself. Under this conception, integration, understood as the residential mixing of group members roughly in proportion to their numbers in the general population, becomes the only and necessary means of righting the wrong. I will argue shortly, however, that group clustering is not wrong in itself and can be a positive good. The primary wrong of segregation, I argue, is not clustering itself, but rather that it gives many white people privileged access to many resources and benefits at the same time that it denies these to people of colour. Segregation is also wrong because the process that produces this relative privilege and disadvantage limits choice for both groups, but especially for people of colour. It is wrong, finally, because the very process that produces privilege also obscures the fact of privilege to those who have it.

Studies like those of Massey and Denton for the United States and Smith for Britain detail ways that segregation produces and magnifies the effects of a seriously unequal distribution of resources, benefits and opportunities between whites and people of colour. Processes of the exclusion of people of colour from certain communities and neighbourhoods to some extent restrict the competition for housing in the neighbourhoods and communities with the highest quality housing. People of colour have restricted access to the best housing opportunities. Even though demand is thus restricted, segregation helps keep housing prices high by adding to them a desirable amenity over and above location and quality of structures: the whiteness of the neighbourhood. High property values in white neighbourhoods encourage investment in the neighbourhood, thus keeping the property values high. Thus segregation gives to the better neighbourhood conveniences and amenities such as shopping and entertainment, transportation action and green spaces.

By contrast, people who live in neighbourhoods with a high concentration of people of colour often have a worse quality of life than others. Their neighbourhoods are usually more densely populated and the quality of housing poorer. Yet the restricted housing market produced by processes of segregation often means that they must pay more for housing relative to its size and quality. The combination of poorer quality structures and segregation means, furthermore, that owners are not inclined to invest in their property and new investment is difficult to attract. Thus buildings are not maintained, decline in quality and value, and businesses tend to leave, with few coming in. Thus residents of neighbourhoods with relatively high concentrations of people of colour tend to have fewer conveniences and amenities than other neighbourhoods, in the way of stores, banks, restaurants, cinemas, etc. As a consequence, people of colour often must travel further than white people to conduct their daily business. The businesses that do exist in the segregated neighbourhoods, moreover, can charge higher prices for comparable goods and services than those in white neighbourhoods. Segregation often produces disadvantage in transportation access to jobs, because people of colour are sometimes forced to live further from major commercial and industrial centres than whites. Because knowledge of job openings depends significantly on being part of a network of people who know about such openings, segregation also tends to limit knowledge of job opportunities.

In all these ways, then, segregation tends to give privileged access to resources and benefits to many whites, while severely restricting the access of many people of colour. One might hope that public officials would notice this unjust result of private action and try to compensate for it by providing better parks, playgrounds, police and fire service, transportation, lighting, schools and community centres to the relatively disadvantaged neighbourhoods. The sad truth is, however, that the public sector is usually at least as reluctant as the private to invest in worse-off neighbourhoods or communities. To the extent that white privileged communities are separate municipalities with separate public services and tax base, it seems to be wholly appropriate to put public investment only in their own community. In larger cities a similar dynamic occurs with politicians preferring better-off neighbourhoods and often ignoring the worse-off, because they look for votes or tax base.

The exclusion from benefits which segregation enacts occurs at least partly through processes that restrict choice. Residential racial segregation is reproduced through legal and illegal discrimination by landlords, home-owners, real estate agents, and government policies. Whether technically illegal or not, many property owners believe that they are entirely within their rights to

decide who will or will not live in their property, according to whatever criteria they choose. Real estate agents often lie, falsely or selectively advertise,
and 'steer' white clients to some neighbourhoods and people of colour to others. People of colour are denied mortgage loans far more frequently than
whites of comparable income. Public housing administrations in Britain as
well as the United States, and I suspect other countries as well, often allocate
units according to their ideas of what sort of people 'belong' in them, and
often have allocation rules that are discriminatory in effect even when apparently neutral.

The very same process that produces relations of privilege, moreover,
obscures that privilege from the awareness of those who have it. In order to
see themselves as privileged, the white people who live in pleasant white
neighbourhoods must be able to compare their environment with others. But
this comparison is rarely forced upon them because those excluded from
access to the resources and benefits which they themselves have, are spatially
separated, out of sight. Another place defines their lives. Whites usually avoid
experiencing those other places, but usually we do not even need to think
about such avoidance, because our daily lives and social spaces are so constructed that we have no reasons to go where the others live. Segregation
produces very different life experiences for different people, but it also
impedes their ability to communicate about their lives and their perspectives
on the operations of society.

Distinction between segregation and separation

I define segregation, then, as the exclusion of members of a group from
access to the benefits and opportunities others have, enacted partly through
spatial concentration of the excluded group. Even in the absence of enforced
policies of exclusion, I have suggested, segregated patterns that produce privilege and disadvantage are reproduced by means that restrict choice. But in
the absence of policies of enforced separation, how do we know that patterns
of racial residential concentration do not simply reflect a preference that people have for living among those with whom they most identify? Many people
today, especially in societies without a history of legally enforced segregation,
argue that patterns of racial concentration are not wrong because they arise
from such preference.

Cities in modern mobile societies do in fact experience people clustering
voluntarily in terms of affinity groups defined by ethnicity, religion, language,
or lifestyle. People settle in a new city near family or friends, or near those
with whom they believe they will be most comfortable with their particular
tastes, language, religious practices, and so on. They seek to enter friendly
networks for locating housing or jobs, and these are often particularized by
affinity groupings. Residential and civic clustering by social groups is not
wrong when it occurs through processes of attraction rather than exclusion,
and when it does not produce or reproduce systematic privileges and disad-
vantages. To understand the specific wrongs of segregation, and especially to

understand why the ideal of integration implicit in Jacoby's dinner party questions is problematic, we need to distinguish segregation from separation.

Maria Lugones makes such a distinction through images of eggs and mayonnaise. Making mayonnaise requires separating egg whites from yolks. The cook aims to split parts of the egg by cutting between the yolk and the white with the edge of the shell, and achieve pure white, unmixed with any yolk. She often fails, however, and slices through the yolk, which then runs through the white making the separation impossible. The making of mayonnaise, on the other hand, requires mixing the yolk with oil and water, beating them hard so they become one. Lugones aims with these images to distinguish two meanings of group separation: splitting and curdling. Splitting is an effort to purify by forcefully segregating one group from another. It requires special effort to exclude some people from their neighbouring attachment to others, and it often fails, requiring even stricter measures of exclusion. Curdling, on the other hand, is a more passive process in which people are attracted to one another because of their similar constitution, so they cluster. But this process of separation never produces a pure homogeneous group; instead people from a group may noticeably dominate a space which others inhabit as well.

For the purposes of this essay I interpret segregation as the group differentiated process Lugones analogizes with splitting, and separation as the process she evokes with the image of curdling. Separation consists in a process whereby people tend to associate with people with whom they have cultural affinity or affinity of social position and life circumstance. This affinity grouping produces group-specific residential clusters, as well as group-specific associations and institutions.

The important conceptual difference between separation and segregation is the direction of agency. The acts that produce and reproduce segregation are directed outward, towards persons to be excluded in order to produce or maintain a 'pure' space. In acts and institutions of separation, on the other hand, people look inward to affirm an affinity, culture, or way of life. The other-directed processes of segregation often involve aversion or repulsion, and position certain people as despised – Others not worthy of benefits or association with oneself. Separation, on the other hand, is motivated by an attraction that people have for one another. As the eggy images suggest, segregation is a process of exclusion that requires active force, whereas separation, like curdling, is a result of attractions present in the absence of force, and such loosely separated groups always remain intermingled, especially at the edges.

In itself associative and residential clustering is not wrong, and may be a positive good.[5] Several recent theorists of multiculturalism argue that affinity grouping according to language, culture, religion, history – and I would add sexuality, culture, physical ability or gender – are important 'sources of the self'. People do not develop a sense of value and meaning except in specific social contexts with others, and thus their sense of what is important to them is conditioned by group affinities that they have a moral right to affirm.[6]

Separation in this sense of affinity grouping can be especially valuable to the lives and sense of self of those in oppressed or disadvantaged – that is, often segregated – groups. People discriminated against and excluded from benefits by dominant groups often find resources for survival, self-help and resistance in clustering with and affirming their affinity with one another, even where that affinity is defined first by the fact of their exclusion rather than by cultural practices they share. Segregation thus sometimes motivates separation.

The problem, however, is that, like purifying the white of the yolk, it is difficult to make a pure distinction between separation and segregation. On the one hand, segregation usually relies on desires for separation. If affinity groupings of people in living space and association did not exist, then it would be difficult for a dominant group to recognize itself as a group and to pick out those to exclude from benefits. A benign and affirmative situation of association or residential clustering, moreover, can slide into segregation. Separation is or becomes wrong whenever it aims to draw borders that determine who is in and who is out, wherever it aims explicitly or implicitly to exclude or purify. Separation becomes segregation when the positivity of group affinity turns into an essentialist effort to say what attributes or dispositions a person should have to be a 'real' member of the group. Separation can become segregation when its motivation is to avoid those with whom one feels less affinity, more than to find those with whom one shares affinity. Most especially, separation becomes segregation when affinity groupings aim to create or preserve their access to resources, benefits or opportunities by excluding or removing those whom they identify as different.

Integration is not the solution

Lugones interprets her image of separating the white from the yolk and of curdling in the failed mayonnaise, but what of the effort to make mayonnaise itself? At least some interpretations of an ideal of integration, I suggest, can be thought of as like making mayonnaise. They aim by a process of active mixing to produce a substance in which the starting constituents are indistinguishable. As with mayonnaise, integration 'fails' if any noticeable curdling occurs.

The 'integrationist manifesto' I quoted earlier invokes an ideal of 'one community' and that 'underneath the skin we are all the same'. This integrationism identifies the primary problem as differentiation and separation, for which the solution is mixing. In this ideal of integration the desirable configuration of residential location, occupation, institutional membership, and so on, is a statistical random distribution. In a well integrated city, according to this ideal, no neighbourhood would be dominated by a minority group, nor would any neighbourhood be inhabited exclusively by a majority group. Instead, every neighbourhood would contain people of different groups in rough proportion to their incidence in the general population. This ideal of integration aims to bring about a social condition in which everyone treats skin colour, or religion, or heritage, or language, as accidental attributes of individuals who are essentially the same inasmuch as each is a unique individual.

In arguing against the appropriateness of this ideal of integration as a remedy for segregation, I assume a situation in which the segregated, relatively disadvantaged groups are also numerical minorities. This situation describes many of the societies in which segregation is a historical or growing contemporary problem; for example, societies of European majorities or majorities of European descent that have privilege in relation to people of colour in their societies. Under these circumstances, the ideal of integration I have articulated above translates into an image of dispersion. In order for people of colour to gain social equality and access to resources and opportunities that whites have, the segregated enclaves in which they live should be dispersed, and whites must relinquish the homogeneity of their surroundings to allow entrance of people of colour – to neighbourhoods, clubs, workplaces.

There are several problems with this ideal of integration. Firstly, especially when the socially privileged group is also a numerical majority, attempts to bring about integration tend to leave the dominant group relatively undisturbed while requiring great changes from the excluded groups. Pro-integration housing policies, for example, usually involve the movement of blacks to predominantly white neighbourhoods, rather than the reverse. Integrating the corporate workplace usually means that formerly excluded people of colour obtain opportunities in white-dominated settings to whose cultural styles they must learn to adapt. More generally, practical efforts at integration too often mean that the socially dominant groups set the terms of integration by which the formerly segregated groups must conform to the expectations of the dominant group.

Secondly, an ideal of integration rejects the validity of people's desire to live and associate with others for whom they feel particular affinity. Separation in the sense of curdling or clustering without borders or exclusion, that is, is inimical to the ideal of integration. This ideal seems to imply the break-up of Caribbean or Pakistani neighbourhoods, the rejection of Afro-centric schools, and encouraging everyone to interact in a culturally neutral mainstream. The segregated groups whose affinity grouping is thus threatened by such an ideal, however, do not perceive that mainstream as neutral, but as dominated by norms more associated with the dominant group.

Consider the following survey results. Several studies in the United States ask both blacks and whites to describe the racial composition of a neighbourhood in which they would most like to live. Liberal-minded 'integrationist' whites tend to depict a neighbourhood that is between 10 and 20 per cent black. Blacks, on the other hand, most often want a neighbourhood that is roughly 50:50.[7] The dissonance between these two images of a desirable neighbourhood has many implications, but one is this: blacks regard the white image of integration as continued dominance, and whites regard the black image as an illegitimate effort to remain separate. In this picture, the practical effect of the white image of integration is to allow affinity grouping only for whites.

The third problem with the ideal of integration usually implicit in discussions of race and residence, therefore, is that it is likely to fail, and when it fails, the fault seems to lie with the formerly segregated group. To the extent that integration requires members of the formerly segregated group to change their lives and conform to the expectations of the dominant group, it puts the onus for success on the relatively more disadvantaged groups. The project of integration may fail, largely because serious commitment and resources have not been devoted to it. Resistance of the formerly segregated group to dispersing its affinity groups, or their failure to measure up to the dominant norms and expectations seem to be more noticeable obstacles to successful integration. Members of the dominant group committed to the project of integration then throw up their hands and blame the subordinate group members who cannot or will not integrate on these terms. Clearly there is the main message of Jacoby's 'integrationist manifesto'. For thirty years liberal whites have kept the integrationist faith; if segregation and conflict about race continue to divide Americans, that is because people of colour have given up the faith in antiwhite resentment.[8]

Finally, and most importantly, the ideal of integration tends to focus on the wrong issue. According to this ideal, the problem of segregation is also the problem of separation, and the remedy is mixing in proper proportions. The primary issues become diversity and mixing as such, rather than the production and maintenance of privilege and deprivation which I have argued are the main harm of segregation. Some pro-integration housing policies exhibit this problem. Public housing authorities have sometimes instituted quotas on the numbers of members of particular groups who can be accepted for leases in particular buildings, with the aim of producing a more salutary mix of diverse groups. Not infrequently such policies have the perverse effect of reserving places for whites for which there is little demand while creating longer waiting lists for the people of colour whose housing options are already more limited.

Diversity or multicultural programmes in workplaces and universities sometimes also seem aimed more at racial and ethnic mixture in courses, on panels, in workspaces, etc., than at justice in the relations of power and the distributions of benefits and burdens. Policies that aim to correct the wrongs of segregation should respond to acts and structures which exclude some people from benefits and opportunities to which others have access. This may or may not entail directly promoting a mixture of members of different groups in the same spaces. Sometimes policy should transfer resources, rather than people. An important redress for disinvestment in neighbourhoods with high concentrations of people of colour, for example, is investment in the housing stock, commercial property, and public spaces of those neighbourhoods. Serious investment might well encourage more whites to move to those neighbourhoods, but this is not the main objective. The main objective would be to redistribute resources and enlarge opportunities for those who have been unjustly deprived because of discriminatory acts and institutions.

Another ideal: together-in-difference

Segregation is the exclusion of members of a group from access to benefits and opportunities others have, enacted partly through spatial concentration of the excluded group. Segregation is wrong because it limits choice, and produces privilege and oppression while obscuring the fact of that privilege from those who have it. Separation, on the other hand, is the tendency for people to live among and associate with those with whom they share language, taste, religious practice, or other everyday ways of living. When not enforced, such separation does not entail exclusion of others. The ideal of integration usually articulated as an alternative to segregation tends to identify the wrongs of segregation with separation. By focusing on a process of mixing members of formerly segregated groups, attempts to enact an ideal of integration may wrongly inhibit a legitimate desire to associate through affinity grouping. Such efforts also may be treating symptoms – racial composition – rather than causes – the desire to maintain differential access to benefits and opportunities. My argument against an ideal of integration, however, assumes that segregation perpetuates deep wrongs that should be remedied. I propose a different ideal for such remedy, which I designate with a phrase rather than a word: together-in-difference (*Differenzzusamenheit*). The image of a mayonnaise that has curdled corresponds to my ideal of together-in-difference. Social space is noticeably differentiated, with people often clustering according to shared practices or ways of living. But the differentiations are not pure. While the clusters are noticeable, there are many mestiza spaces and many mestiza people. The space contains no clear borders, and the differentiations are fluid and changing. As Lugones says,

> When seen as split, the impure/multiplicitous are seen from the logic of unity, and thus their multiplicity can neither be seen nor understood. But splitting can itself be understood from the logic of resistance and countered through curdling separation, a power of the impure. When seen from the logic of curdling, the alteration of the impure into unity is seen as fictitious and as an exercise in domination: the impure are rendered uncreative, ascetic, static, realizers of the modem subject's imagination. Curdling, in contrast, realizes their against-the-grain creativity, articulates their within-structure-inarticulate powers. As we come to understand curdling as resisting domination, so we also need to recognize its potential to germinate a nonoppressive pattern, a mestiza consciousness, *una consciencia mestiza*'.[9]

I imagine an urban region with many neighbourhoods and communities generally recognized as group-differentiated – Jewish, or Turkish, or African, or gay, or straight White European neighbourhoods. None of them is homogeneous or nearly homogeneous, however; while some may have a dominant differentiated character, statistically speaking the neighbourhoods are hybrid. This urban space also has some hybrid neighbourhoods with little noticeable group clustering. A traveller in this urban region of spatial separations without exclusions finds no clear borders between neighbourhoods; they flow into one another without the abrupt border between fancy facades and boarded windows that appear in many American cities. In my imaginary urban region, such borders are not noticeable partly because public policy tries to equalize the places more than to move the people. Instead of pretending that the market is indifferent to place, thus allowing resources to accumulate more where they already are and drain from where they are already lacking, public policy is conscious of class and group differences of place. Public–private partnerships co-operate to promote a decent quality of life in all the region's locales. It takes measures to invest in deteriorating places while not displacing people and being sensitive to the local use of space. It also promotes income diversity in every locale. In this imaginary metropolitan region, everyone has their home place, the place of their immediate residence and where they do their local shopping and playing, but no one feels that another part of the city is closed to them because of the behaviour and attitude of its residents. Public policy encourages people to travel to different parts of the urban region by its decisions to locate offices, museums, and recreational facilities in diverse neighbourhoods. In this city of togetherness-in-difference, everyone has options. To a large extent, people live in an area because they want to be near a workplace or family or particular civic institutions or they like the architecture, and not because they are avoiding certain kinds of people, or because their neighbourhood has the best services, or because their level of income and discriminatory attitudes of others confine their choice to their neighbourhood.

I will conclude by sketching an account of the normative principles of this ideal of together-in-difference as these might guide policies that aim to promote desegregation, equal opportunity, and wellbeing. I find five principles.

1 *Political togetherness* – Because social and economic processes in modern mass society are such that people's actions and situations are causally related to one another over large spatial areas, political consciousness and institutions ought to acknowledge the fact that the fates of diverse groups and locales are bound to one another. Racial and economic justice cannot be decentralist, but rather should promote strong regional governance structures. These should be combined, however, with local participatory structures.

2 *Recognition of difference* – Political consciousness and institutions should explicitly recognize the cultural and social differentiation peopling a region. To recognize the other culture, practices or forms of life with which one dwells together in a region is to do more than merely tolerate them. When people take a stance of tolerance toward gay life or Muslims, then they promise to treat them with basic respect and not to interfere with their pursuit of the good as they define it. Tolerant public policy does not repress or curtail the liberty of laws. Such respect for liberty and the pursuit of happiness is quite compatible, however, with utter indifference. Tolerance is a stance of mutual non-interference: you leave me alone and I will leave you alone.

Recognition, on the other hand, is a more affirmative stance towards others with differing practices, values, or ways of living. Recognition keeps a respectful distance, not claiming, as Jacoby does, that 'underneath we are all the same'. Yet people and groups that recognize one another also want to communicate, to understand something about the specificities of one another's lives, and to seek bases for political coalitions and problem solving. To recognize others is to affirm the validity of their way of living even when it is different from one's own and one would not wish to adopt theirs. I take such a stance of the recognition of difference to be in tension with an ideal of integration that wishes to say that moral respect for others should ignore all social and cultural differences and simply treat each person as an individual.

3 *Prohibition of exclusion and discrimination* – The primary wrong of segregation is that it excludes some people or groups of people from benefits that others have. As an alternative to segregation, then, an ideal of together-in-difference presumes a prohibition on activities of exclusion. Both formal rules that say certain persons or kinds of persons are barred entry to spaces, institutions, occupations, and informal acts and attitudes that put people at risk, humiliate them, or make them uncomfortable are generally wrong. This means, for example, a presumption against walled and gated communities, as well as a presumption against residing in a certain place to use public facilities.[10]

More generally and controversially, principles of together-in-difference question the imposition of clear borders separating people and activities, or preventing free movement. An anti-border principle has obvious implications for international relations and the movement of people globally, which I cannot detail or defend here. For now, I will say that there is no reason for the ideal of together-in-difference to apply only within what are now nation-states, and good reasons why the ideal should apply globally.

4 *Individual liberty in a regulated institutional framework* – Segregation restricts individual choice through discriminatory action that limits the ability of some people to pursue goals or benefits. Social policies aimed at desegregation and at promoting norms of together-in-difference should open opportunities for individuals and allow them liberty to pursue their own chosen goals for their personal lives, especially their place of residence. People should not be forcibly moved or forced to move, for example, to achieve desegregation objectives, any more than they should be removed for the sake of segregation. The allocation of housing should rely primarily on price-regulated or subsidized markets; where housing is assigned it should be as much as possible according to preferences. While promoting liberty of individual movement in this way, policy that promotes an ideal of together-in-difference must also regulate markets, and the actions of landlords, banks, developers and other private agents with the power to affect the social meaning of urban space.

5 *Equalization of resources and opportunities across space* – The primary wrong of segregation is the exclusion of some people from benefits that others have. The most important remedy for segregation, then, is the reduction of the inequality of resources, amenities, services, and opportu-

nities across space. One of the main strategies of redress for the past wrongs of apartheid, for example, is the reallocation of resources from privileged to less privileged areas. Rather than attempting to break up the townships and integrate their members into areas formerly exclusively white, for example, local government is merging township service provision with the resources of rich white suburbs. Equalization of resources and opportunities across space does not amount to creating income equality, and it leaves many other inequalities in place. But the ideal of making every place roughly as good as any other removes one of the central motives for the exclusionary and discriminatory actions and policies that produce and reproduce segregation. Most pro-integration policies envision encouraging individuals to move to places from which they have either been excluded, or which they have found unattractive. Such people-oriented integration policies are necessarily slow in their effect, and most often involve only the well-off of the formerly segregated group and the relatively well-intentioned and/or less well-off of the formerly privileged group. Policies aimed at equalizing place work more quickly, and address issues of class as well as race difference. They are likely to bring with them a desire on the part of some members of formerly segregated groups to move from place to place.

Some might object to my account that I have not really offered an alternative to the ideal of integration, but instead a specific interpretation of that ideal. The ideal of together-in-difference is, like the ideal of integration, an ideal of desegregated urban space. If all desegregation is integration, then I will accept the claim that together-in-difference is a specific ideal of integration. Because I believe with Tamar Jacoby that the idea of integration entails the belief that we are all one community and underneath the skin we are all the same. However, I think it is less confusing to think of together-in-difference as a distinct ideal.

Notes

1   Tamar Jacoby (1994) 'An integrationist manifesto', *Wall Street Journal*, 4 October, editorial page.

2   See, for example, Randall Kennedy (1996) 'On racial integration', *Dissent*, Summer 1996, 47–52.

3   In their massive study of these residential patterns of cities in the US Northeast and Midwest, *American Apartheid* (1993, Harvard University Press, Cambridge, MA.), Douglas Massey and Nancy Denton document the evolution of current patterns of racial residential segregation as the result of some policies not explicitly directed at segregating, discriminatory behaviour by individuals and institutions, and the magnification of the economic disadvantage initially associated with racial concentration.

4   Susan J. Smith (1989), *The Politics of 'Race' and Residence: Citizenship, Segregation and White Supremacy in Britain*, Polity Press, Cambridge.

5   Maria Lugones (1994), 'Purity, impurity, and separation', *Signs: A Journal of Women in Culture and Society*, **19**, (2), Winter 1994, 458–79.

6 See Charles Taylor (1992), *Multiculturalism and the Politics of Recognition,* Princeton University Press, Princeton; Will Kymlicka (1995) *Multicultural Citizenship,* Oxford University Press, Oxford; Yael Tamir (1993) *Liberal Nationalism,* Princeton University Press, Princeton.

7 Reported in Massey and Denton: see note 3.

8 I first heard this particular 'blaming the victim' argument against the ideal of integration made by an Indian immigrant woman in Frankfurt at a conference on women and migration in the summer of 1995.

9 Lugones, op. cit. 468–9.

10 I have phrased this principle as a *presumption* against exclusion, to allow for the possibility that certain institutions of disadvantaged or culturally endangered groups might be exclusive for the sake of the survival of the group or promoting the empowerment of its members under circumstances where they suffer disadvantage. Thus all-female schools or Native American preservation organizations might be justified in policies of exclusion.

Chapter 14

Otherness and citizenship: towards a politics of the plural community

Anthony Gorman

> JIMMY: Do you know the Greek word *endogamein*? It means to marry within the
> tribe. And the word *exogamein* means to marry outside the tribe. And you
> don't cross these borders casually – both sides get very angry. Now, the
> problem is this: Is Athene sufficiently mortal or am I sufficiently godlike
> for the marriage to be acceptable to her people and to my people? You
> think about that.[1]

In a recent work, Murray Bookchin attempts to reinstate the classical concept
of the city as a political community.[2] He contends that the city is first and
foremost an *ethical* union of citizens and only secondarily a discrete system of
space.[3] However, as Iris Marion Young has pointed out, Bookchin's attempt to
rehabilitate the civic ideal is open to two fatal criticisms.[4] Firstly, it is hope-
lessly utopian. Any attempt to restore the classical city-state in a modern
context would require such an extensive reorganization of social and political
relations as to render it practically impossible. Secondly, given this is the case,
the revival of the civic ideal in the face of its manifest impossibility can only
serve to promote a conservative political agenda that seeks to exclude difference and otherness in the name of an artificial civic consensus. Although
these criticisms are well founded, we must not, in my view, simply dismiss the
civic ideal altogether, as being at best an irrelevant nostalgia and at worst a
pernicious ideology. To do so would be to reduce the concept of the city to a
mere association of conflicting interests, devoid of ethical coherence and transcendent unity; and Bookchin is right, in my view, in contending that such a
conceptual reduction would represent a profound contraction of human possibility and meaning.

What is required, then, is a concept of the city as a political community that
is able to accommodate the fact of otherness and diversity. That is to say, the
generation of a form of civic consensus capable of fully acknowledging the
prevailing moral dissensus of the modern city. In this chapter, I shall suggest
that the necessary desideratum may be found, not through the repudiation of
the civic ideal, but via its reconceptualization on the basis of a different concept of alterity to that which has been employed hitherto in the application of
critical theory to the study of urban life. To this end, I shall contrast two
concepts of the Other; a neo-dialectical concept, ultimately Sartrean in prove-
nance,[5] and a phenomenological account derived from the work of

Emmanuel Levinas. To place my analysis in a concrete context, I have taken the Irish community in Britain as the focus of my study of the relationship between otherness, community and citizenship. The chapter is divided into three parts. In Part 1, I critically examine Mary Hickman's studies of the Irish community in Britain.[6] In Part 2, I proceed to briefly expound Levinas' concept of alterity and to suggest how it might be radically modified so as to provide a basis on which to rethink the civic ideal in a way which does not imply the exclusion of difference. Finally, I return to the example of the Irish community to illustrate how this ethical ideal may be made to 'traverse' the reality of the modern city.

Part 1

Mary Hickman, in her pioneering work on the socio-historical formation of the Irish community in Britain, draws heavily upon contemporary cultural theory. The key critical tool of her analysis is the concept of the 'Other'. Hickman argues that the identity of the Irish in Britain has been defined in and through the process of opposing and resisting the 'incorporativist' strategies of the British state.[7] From the beginning of Irish mass-immigration into Britain, the British state consistently posited the Irish as 'Other' to the host nation. The education system was the most important means employed to further this end, with the active assistance of a Catholic Church keen to render its Irish working-class congregation 'respectable' in the eyes of the British establishment. Together they sought to 'de-nationalize' Irish children by means of the systematic elision of all reference to the Irish nation in the life of the school. This was done not despite but because of the fact that Catholic schools were overwhelmingly Irish in their composition. The desired result was that the pupils would internalize the implied inferiority of their Irish identity, and so become 'other' to themselves. More recently, representations of Irish people as an innately inferior race have given way to their portrayal as a 'suspect community', altogether alien to the British way of life.[8]

The Irish in Britain responded to their representation as a subversive and alien people by proceeding to construct, in Hickman's words, 'a community life based on the very features that encapsulated the threat they represented, religion, national politics and class organization'.[9] Indeed, there can be very few Irish in Britain whose identity has not been formed out of the nexus of these three factors. However, Hickman further points out that the Irish community, in common with all communities in modernity, is 'imagined' rather than real;[10] in other words, its unity is not founded on ideal–typical *Gemeinschaft* relationships but, rather, upon a set of common allegiances to a range of contested, symbolic representations. It follows, therefore, that the symbols around which the Irish community coheres are mutable and subject to an ongoing process of renegotiation. In recent years, under the impact of wider economic, social and cultural changes, this process has intensified, resulting in an ever-increasing pluralization of the Irish experience in Britain.[11]

Hickman summarizes these developments in terms of a two-stage process in which the Irish community in Britain has undergone the transition from being passively defined as 'Other' to actively defining *itself* as Other.[12] That is to say, it has advanced from identifying itself in *opposition* to its construction by the hegemonic culture to defining itself in a way that is not simply determined by this opposition but allows for the indeterminacies of its own internal differences.

As is clear even from this brief summary of her analysis, Hickman's study implies a social ontology in which the community is prior to the individual. This has definite implications at the level of public policy. Firstly, the identification of the Irish in Britain as a separate community evidently lends support to the claim that the Irish in Britain are entitled to recognition as a distinct ethnic group on a par with other ethnic minorities. Secondly, to the extent that there is evidence to show that the Irish as a community are socially and economically disadvantaged simply as a result of their minority status, then this provides clear grounds for the introduction of remedial measures to redress this social injustice, again on a par with that granted to other ethnic groups. In short, it legitimizes the inclusion of the British-Irish in what I shall call, following Charles Taylor, a *politics of recognition*, which demands that dignity and respect be afforded equally between different social groups and ethnic communities *and* that each social group and ethnic minority be recognized in its own uniqueness and specificity.[13] Now, although I believe that the social ontology implicit in Hickman's analysis is valid at the level of sociological analysis, and although I broadly endorse the policy implications that flow from it, I believe it still needs to be further refined. For, as it stands, it results in a concept of the Irish community in Britain that is, paradoxically, at once too inclusive and too exclusive, such that it places unnecessary limits on the process of pluralization taking place within and beyond its borders.

On the one hand, the concept of community employed is too *inclusive* because it does not sufficiently take into account the capacity of modern subjects to reflectively reconstitute the conditions of their initial self-formation. It must be acknowledged that this capacity is at least as *constitutive* of the self as the process of induction into the norms of a given communal life.[14] For example, a necessary part of being Irish in Britain is to have absorbed many 'British values' and affiliations. The internalization of these values and affiliations is, for the most part, compatible with an Irish identification and need not be interpreted as a disguised form of 'self-hatred'. What is required, then, is a definition of community that encompasses such forms of 'impurity'.

On the other hand, Hickman's concept of the community is too *exclusive* because, as said, it leads to a politics of recognition focused on the demands for ethnic self-determination and social compensation. This form of political action necessarily runs the risk of creating a political culture of *ressentiment* and is, moreover, open to corruption. But, equally, to reject the demand for redress on these grounds is to simply acquiesce in the face of manifest injustice. However, my main concern here is not with the empirical outcomes of the implementation of such policies but with the question of the wider implications of a 'politics of recognition' for the *concept* of the political community as such.[15] If this conceptual dimension is not taken into account, the pursuit of a 'politics of recognition' becomes a self-defeating exercise in so far as the concept of a political community is the condition of possibility for the 'politics of recognition'.[16] The problem is this: reciprocity is a necessary condition of social and political recognition at both a group and an individual level. Self-respect requires recognition by others; but, in order for this to be sincere and meaningful, it must be granted universally to all persons. This implies a collective socialization into the impartial, universal norms and procedures of justice. However, the 'politics of recognition' rests precisely on the denial of the genuine universality of these norms and procedures, regarding them merely as the disguised expression of concealed, partial interests, serving to suppress difference.[17] Thus, the achievement of communal recognition in such circumstances is purely formal, i.e. not *real* recognition at all. The 'recognition' afforded under such conditions is that of power rather than right; it is merely the power to compel the re-cognition of what is already *de facto* the case. Of course, the empowerment of the previously powerless often represents a moral advance, but it must also be acknowledged that it may equally bring new forms of injustice and powerlessness into being.

We appear to have reached an aporia: the universal excludes difference and otherness, giving rise to a politics of recognition; but the success of the politics of recognition rests on the claims of the excluded being recognized in terms of the very universal norms that exclude them. The upshot is that marginalized groups enter into compromises with the discredited universal, resulting in their internal schism, and the production of new forms of exclusion, and so on to infinity. May a way be found to break this bad infinite by developing a concept of the civic ideal that does not rest on the denial of alterity?

Part 2

I shall now explore the possibility that a way out of this impasse may be found on the basis of extending Levinas' understanding of alterity. For Levinas, the primary relationship to otherness is not mediated through the dialectics of intersubjectivity. That is to say, the 'I', in the first instance, is not identified in terms of its relations to other egos but through its immersion in the 'elements' – Levinas' term for the world prior to its 'construction' through consciousness.[18] Therefore, the subject in itself does not coincide with its consciousness. Conversely, the Other is not defined through its relation to the 'I'; it is rather, *absolutely* Other, deriving its alterity solely from itself.[19] The relation between the 'I' and the Other is therefore one of absolute *immediacy* (for it is wholly unmediated by the interposition of concepts), and of absolute *transcendence* (for the absence of a common medium necessarily entails that the 'relation' is radically asymmetrical). Levinas' term for the primary I–Other relation is the *face to face*.[20]

The total alterity of the other is expressed in the fact that he reveals to me the limits of my knowledge and power. The face calls my spontaneity into

question, paralysing my enjoyment and possession and summoning me to justify myself. Nevertheless, Levinas insists that the ethical demand addressed to
me by the Other is *not* a violence; for the relation to be violent the Other and
I would have to share a common boundary; but it is precisely Levinas' point
that the face of the Other transcends the ontological binary of the 'same' and
the 'other'. Hence, the Other exerts over me not a physical but a purely *moral
force*.[21]

The exteriority of the Other is therefore not constituted by the I; it is rather
revealed in the ethical approach. The ethical approach does not disclose a
transcendent being, as it were, 'behind' appearance; it reveals or expresses
absolute alterity in appearance. Infinite alterity is, thus, inseparable from the
Other who, over and above all the ontological descriptions that may serve to
qualify his unique strangeness, nakedly faces me. The essence of the ethical
relation, therefore, resides in my substitution for the other, a going towards
the *other in me* which is at once a going beyond myself towards alterity, without quitting the locus of my own incarnation in creation.[22] It is, so to speak, a
turning of myself inside out for-the-other. In this way, my response to
absolute alterity turns into *responsibility* for my neighbour. I must substitute
myself for the other; but no one can substitute himself or herself for me.[23] My
responsibility is always greater; I must suffer the whole weight of creation,
and no one can relieve me of this infinite responsibility.

Evidently, Levinas has a radically different concept of the Other to that
found in contemporary cultural and critical theory, of which Hickman's work
is a representative example. To show the full depth of this contrast we must
consider the question of Levinas' understanding of the relationship between
the primary ethical dyad and the third party. Levinas conceptualizes this relation in two different ways. In his later work, he maintains that the face to face
is from the outset, so to speak, over-determined by the presence of others
standing in 'synchronic' relations with one another.[24] Although Levinas con-
cedes that the existence of social structures and systems is necessary to the
accomplishment of social justice, he nonetheless maintains that justice is oriented by the perpetual 'interruption' of the 'trace' of ethical asymmetry within
the 'symmetrical' space of the city.[25] In his earlier work, however, Levinas presents a markedly different account of the relation between the face to face
and the third party. In *Totality and Infinity*, Levinas makes a clear distinction
between two dimensions of the relation of the face to face to pluralism or 'fraternity'.[26] On the one hand, Levinas speaks of the face to face as inaugurating
what he variously calls 'language', 'discourse' or 'conversation'.[27] The calling
into question of the 'I' announced in the face, produces the primary dispos-
session of the self that allows for a life lived separately yet in common with
others.[28] However, although it is impossible for the 'I' not to hear the ethical
summons, it nonetheless retains the freedom to ignore or 'forget' it (as well it
must if immorality is to be possible).[29] Levinas singles out the State as the
principal structure which systematically induces the 'forgetting' of the ethical
relation.[30]

On the other hand, over and above the 'discourse' that calls in question my freedom and establishes a common world, the face to face is also 'sermon, exhortation, the prophetic word'; in short, a 'teaching', that *commands* a supererogatory form of response in which the revelation and the welcome of the face coincide and the temptation to 'forget' the irrecusable ethical obligation is incessantly overcome.[31] In Levinas' formulation, I receive the command in the face as a command to command the Other – not to serve me – but to 'join me' in ethical service.[32] Together, we constitute an ethical rather than an 'imagined' community; united not by a common commitment to a symbolic identity but by our shared ethical vocation. Obedience to ethical 'teaching', therefore, signifies an 'election' into a society of equals dedicated to ethical service.[33] This election confirms the 'I' in its absolute unicity and judges it in its interiority outside of the totality and history.[34]

Moreover, for Levinas, ethical election is grounded in the *ethical* institution of the family.[35] Levinas holds paternity, and specifically the father–son relationship, to be the prototype of ethical love.[36] Paternity is an ethical relationship that transcends its biological foundation. The father and son stand in a non-causal and non-possessive relation of radical dependence and independence. The paternal relationship is also the prototype for the ethical community as a whole.[37] Fraternity refers to the fact that each elected being is individuated in their response to the summons of ethical responsibility (and not by their participation in a genus or a concept) and that they all share a common (transcendent) 'father'.[38] The generational perpetuity of the family, and the family of families, the ethical nation or religion, in an infinite time outside of history, ensures the ultimate convergence of morality and reality.[39]

There are two salient aspects of the above outline to which I want to draw the reader's attention. First, although Levinas is perhaps right to say that the face of the Other places me under an unchosen moral obligation, this cannot *of itself* directly determine the supererogatory response required for ethical election. Obedience to the summons of infinite responsibility must retain an ineliminable moment of decision, of crisis and turning, even if it is only the second-order decision of resisting the temptation to be tempted to remain closed up in one's egoism or in reiterating one's absolute passivity in the face of the Other.[40] In this respect, the ethical community is also a political community, or at least contains an irreducible political dimension, and, as such, it is both exposed to history and must itself have a history.

Secondly, we note that on Levinas' own account, the ethical relation is not *absolutely* non-reciprocal, insofar as the common orientation towards a transcendent paternal love encompasses, without thereby totalizing, the primary asymmetry between members of the ethical elect. Of course, the form of 'reciprocity' involved here is of a special kind; it involves no element of *quid pro quo*, indeed, quite the reverse. Nevertheless, ethical service frees the 'I' from its primary egoism and, as observed above, this liberation has an invisible, communal dimension. The ethical elect constitutes, so to speak, a reciprocal non-reciprocity or, in Levinas' words, a 'society of equals'.[41]

These two modifications to Levinas' account introduced above (viz. that the ethical community involves a moment of free decision and is therefore in history not outside it, and that it is further grounded in a special form of *recognition*) together entail that Levinas' concept of the ethical is open to be reinterpreted such that it is no longer conceived as the 'interruption' of the political community, as it were, from 'above' but as being *immanent* within it from below. When seen from this latter point of view, although it remains the case that no one can relieve me of my ethical responsibility, I nonetheless possess, as it were, a supraconscious awareness that I am not alone in bearing witness. My self-transcendence towards absolute alterity through ethical response and responsibility is indirectly reciprocated by the ethical self-transcendence of the Other towards me and all the others. I say 'indirectly reciprocated', because the primary asymmetry of the ethical relation is not abrogated, since as Levinas' account of paternity shows, the transcendent mutuality of ethical responsibility is accomplished through the asymmetrical relation itself. Thus, in the measure that I bear witness I enter into communion with the Other as witness who faces me, that is, as my friend, who confirms me in my being, beyond self-presence.[42] And, in the measure that we bear witness to all the others that face us, that is, to our neighbours or fellow citizens, then they are recognized in their absolute alterity, i.e. not socially, but *absolutely*, as created selves. We may therefore define the political community or city as a freely constituted ethical association founded on the indirect absolute recognition of all by each. In sum, *the city is the site of reciprocal alterity*. However, in saying this, it must be emphasized that the concept of the political community as defined above presupposes rather than supplants the liberal idea of the state.

The structure of absolute recognition, therefore, entails that ethical transcendence is a necessary moment in the actualization of the concept of the political. The citizen exists not only for the Other, but also for the Other's *sake*, taking an interest in his or her aspirations, projects and possibilities. This ought to apply not merely on an individual level but also on a communal basis. In this way, it is possible that what I shall call a *politics of conciliation* may supplement and orient the *politics of recognition*.

Part 3

How can the ideal city adumbrated above be said to be 'immanent' in modern urban reality? Have we not simply exchanged one wild utopia for another? Of course, I cannot hope to answer this question here but only to provide an indicative example of how such an answer might, as it were, materialize. I return once again to the Irish community, and specifically, the Birmingham Irish community. Over the past five years the Irish community has organized a political forum and has campaigned for the right to be granted ethnic minority status. In addition, it has also lobbied for the Digbeth area of the city, which is the part of the city where the first Irish migrants settled, to be officially designated as the 'Irish Quarter'. This is partly a demand for parity with other ethnic groups in the city – there is, for example, a 'Chinese Quarter' in existence – but it is essentially a claim for the public recognition of the contribution of the Irish community to the history and life of the city. Although Digbeth has long since ceased to be a residential district, it contains a high concentration of Irish pubs, social clubs and businesses, and a parish church is still based there. For all practical purposes it is an 'Irish Quarter' already.

This campaign provides an example of how the traversal of the real by the ideal may be said to manifest itself. Firstly, it is a campaign involving religious leaders, political activists, community workers and local businessmen. The fact that it is Catholic religious leaders involved in this instance is merely circumstantial and does not detract from the wider point that a commitment to a supererogatory ethical stance is in principle shared by all the major faiths. Secondly, the mobilization around this issue constitutes an active participation by the community in the political process, an example, that is, of 'active citizenship'. However, significantly, the goal of this activity, the claim to the name, is not predicated on exclusion. For the centrepiece of the Quarter is the pub life, which has been augmented in recent years by the addition of a number of so-called Irish 'theme' pubs. Their combination of the traditional, the ersatz and the modern gives them an appeal that transcends their core Irish constituency. By the same token, it provides a focus for the multiple identities that constitute the Irish community itself. In this instance, therefore, the market is giving expression to that element of free subjectivity constitutive of the modern personhood which is yet predicated on the wider attachment to community. The official designation of Digbeth as an 'Irish Quarter', therefore, need not alienate non-Irish citizens, for it would be a source of pride for all of Birmingham's citizenry. Of course, such initiatives will come to nothing if structural inequalities of class, race, and gender continue to exclude a significant minority from effective participation in the city's cultural life. Poverty remains the ultimate form of non-recognition. But the struggle for social justice must nonetheless be *oriented* by a civic sense, a spirit of conciliation and forgiveness, that transcends the plurality of communities – which are themselves plural – that constitute the city, if we are not to suffer that contraction of human meaning and possibility which Bookchin so prophetically warns against.

Notes

1   Brian Friel (1981) *Translations*, Faber & Faber, London, 68.
2   Murray Bookchin (1992) *From Urbanization to Cities: Towards a New Politics of Citzenship*, Cassell, New York.
3   Ibid., 8.
4   Iris Marion Young (1990) *Justice and the Politics of Difference*, Princeton University Press, Princeton, 233–6.
5   Jean-Paul, Sartre (1948) *Anti-Semite and Jew: An Explanation of the Etiology of Hatred*, trans. G. J. Becker, Schocken Books, New York, 143: 'it is not the Jewish character that provokes anti-Semitism but, rather, it is the anti-Semite that creates the Jew. The primary phenomenon, therefore, is anti-Semitism, a regressive social force and a conception deriving from the pre-logical world.'
6   Mary Hickman (1995a) *Religion, Class and Identity: The State, the Church and Education of the Irish in Britain*, Avebury, Aldershot; and (1995b) 'Difference,

boundaries, community: the Irish in Britain, in Ziff, Trista M. (ed.) *Distant Relations*, Santa Monica Museum of Art, Santa Monica, 44–5.

7 Hickman (1995a: 13): 'The concept of incorporation is being used to denote the active attempts by the state to regulate the expression and development of separate and distinctive identities by potentially oppositional groups in order to create a single nation-state ... Incorporation is distinct from assimilation in that it assumes state and institutional intervention in the regulation of the experience and identity of significant ethnic groups.'

8 Hickman (1995b: 48) is here drawing attention to the continuing impact of the colonial legacy on the perception of the Irish in Britain. See Patrick Hillyard (1993) *Suspect Community: People's Experience of the Prevention of Terrorism Acts in Britain*, Pluto Press, London.

9 Ibid.: 48–9.

10 Ibid. The allusion here is, of course, to Benedict Anderson's (1983) *Imagined Communities: Reflections on the Origins and Spread of Nationalism*, Verso, London.

11 Ibid. Hickman cites, *inter alia*, the decline of the industrial proletariat; changes in the social basis of emigration – now more middle-class educated than working-class unskilled; mass Irish Protestant emigration for the first time; the disaffection of Irish youth with the Catholic Church; a sharp trend towards partisan dealignment amongst Irish voters; and the emergence of Irish women, gay and lesbian groups, as the most significant contributory factors towards a greater degree of communual pluralization and heterogeneity.

12 Ibid. Hickman acknowledges her debt to Stuart Hall's analysis of recent changes in the politics of identity. In his essay 'New ethnicities', in James MacDonald and Ali Rattansi (eds) (1991) *Race, Culture and Difference*, Sage, London, Hall observes a recent shift in Black cultural politics 'from a struggle over the relations of representation to a struggle over representation itself' (p. 253). He argues that this signals the end of the unitary black subject and the emergence of 'a new cultural politics which engages rather than suppresses difference and which depends, in part, on the cultural construction of new ethnic identities'.

13 Charles Taylor (1995) 'The politics of recognition', in *Philosophical Arguments*, Harvard University Press, Cambridge, MA. 225–56. Taylor concisely delineates the nature and genesis of this distinctly modern form of politics. He argues that the transition from a social differentiation based on the positional good of honour to modern societies that formally admit the equal dignity of all persons, has necessarily been accompanied by a general preoccupation with the need for recognition across all social classes and groups. Although this in itself is not unprecedented, what *is* new is that, in contrast to pre-modern societies wherein individuals are afforded recognition on the basis of their ascribed status, in modernity, recognition has become a task to be achieved. In Taylor's elegant formulation: 'What has come about with the modern age is not the need for recognition but the conditions in which the attempt to be recognized can fail' (p. 231). In the first phase of this new politics, which has continued into the present, the demand for recognition assumed a universal form. To cite Taylor again: 'With the move from honour to dignity has come a politics of universalism, emphasizing the equal dignity of all citizens and the content of this politics has been the equalization of rights and entitlements. What is to be avoided at all costs is the existence of first and second-class citizens. Naturally, the actual detailed measures justified by the principle have varied greatly, and have often been controversial. For some equalization has

only affected civil rights and voting rights; for others, it has extended into the socioeconomic sphere. People who are systematically handicapped by poverty from making the most of their citizenship rights are deemed on this view to have been relegated to second-class status, necessitating remedial action through equalisation' (p. 233). In more recent times, however, Taylor goes on to argue, the universal demand that every individual be recognized in her uniqueness has, ironically, generated an anti-universalist politics of difference: 'The politics of difference is full of denunciations of discrimination and refusals of second-class citizenship. This gives the principle of universal equality a point of entry within the politics of dignity. But, once inside, as it were, its demands are hard to assimilate to that politics. For it asks that we give acknowledgement and status to something that is not universally shared. or otherwise put, we give due acknowledgement only to what is universally present – everyone has an identity – through recognizing what is peculiar to each. The universal demand powers an acknowledgement of specificity' (p. 234).

14 Here, I am not endorsing a Kantian view of the self as existing prior to its ends; rather, I simply acknowledge that within modernity individuals possess the *formal* capacity to be recognized in abstraction from the natural determinants of their identity, and so to reconstitute themselves on the basis of new *formal* self-interpretations. This is entirely compatible with a communitarian social ontology. See A. Sandel (1982) *Liberalism and the Limits of Justice*, Cambridge University Press, Cambridge, 152.

15 In accordance with Aristotle's classical definition in the *Politics* (1252b): 'When we come to the final and perfect association, formed from a number of villages, we have already reached the polis – an association which may be said to have reached the height of full self-sufficiency; or rather [to speak more exactly] we may say that while it *grows* for the sake of mere life [and so far, and at that stage, still short of full self-sufficiency], it *exists* [when once fully grown] for the sake of a good life [and is therefore fully self-sufficient]' *The Politics of Aristotle*, trans. Ernest Barker, Clarendon Press, Oxford, 1946. For a full exposition of Aristotle's concept of the political community, see A. C. Bradley, 'Aristotle's conception of the State', in David Keyt and Fred D. Miller, Jr. (eds) (1991), *A Companion to Aristotle's Politics*, Blackwell, Oxford, 13–56.

16 Taylor (1995: 241) formulates this essentially Hegelian insight as follows: 'Each consciousness seeks recognition in another, and this is not a sign of lack of virtue. But the ordinary conception of honor as hierarchical is crucially flawed, because it doesn't answer the need that sends people after recognition in the first place. Those who fail to win out in the recognition stakes remain unrecognized. But even those who win are more frustrated, because they win recognition from the losers, whose acknowledgement is by hypothesis not really valuable, since they are no longer free, self-supporting subjects on the same level with the winners. The struggle for recognition can find only one satisfactory solution that is a regime of reciprocal recognition among others.' However, Liberalism, both as a political philosophy and a political movement, is incapable of resolving this dilemma, for it is only able to understand recognition in formal terms. Taylor accepts this point; see, for example, the conclusion of his monumental study *Sources of the Self*, Cambridge University Press, Cambridge, 1989, 521, where he argues for the need to retrieve the 'central promise' implicit within the Judaeo-Christian tradition of a 'divine affirmation of the human, more total than humans can ever attain unaided'.

17 Taylor (1995: 236–7) sums up the antinomy of the politics of universalism and

difference as follows: 'These two modes of politics, then, both based on the notion of equal respect, come into conflict. For one, the principle of equal respect requires that we treat people in a difference-blind fashion. The fundamental intuition that humans command this respect focuses on what is the same in all. For the other, we have to recognise and even foster particularity. The reproach that the first makes to the second is just that it violates the principle of non-discrimination. The reproach that the second makes to the first is that it negates identity by forcing people into a homogeneous mold that is untrue to them. This would be bad enough if the mold itself were neutral – nobody's mold in particular. But the complaint generally goes further. The complaint is that the supposedly neutral set of difference-blind principles is in fact a reflection of one hegemonic culture. As it turns out, then, only minority or suppressed cultures are being forced to take alien form. So the supposedly fair and difference-blind society is not only inhuman (because suppressing identities) but also, in a subtle and unconscious way, itself highly discriminatory.'

18 Emmanuel Levinas (1979) *Totality and Infinity*, trans. Alphonso Lingis, Martinus Nijhoff, The Hague, 131.

19 Ibid.: 38–9, 251.

20 Ibid.: 39.

21 Ibid.: 171.

22 Emmanuel Levinas (1981) *Otherwise than Being or Beyond Essence*, trans. Alphonso Lingis, Martinus Nijhoff, The Hague, 126.

23 Ibid.

24 Ibid.: 159.

25 Ibid.: 160–1.

26 Levinas (1979), Section III, B.6. 212–15. See also, Emmanuel Levinas, 'The ego and totality', in (1986) *Collected Philosophical Papers*, trans. Alphonso Lingis, Martinus Nijhoff, Dordrecht, 25–46.

27 Levinas, 1979: 173.

28 Ibid.

29 Ibid.: 172–3: 'But the separated being can close itself up in its egoism, that is, in the very accomplishment of its isolation. And this possibility of forgetting the transcendence of the Other – of banishing with impunity all hospitality (that is, all language) from one's home, banishing the transcendental relation that alone permits the I to shut itself up in itself – evinces the absolute truth, the radicalism of separation.'

30 Ibid.: 178, 298.

31 Ibid.: 213.

32 Ibid.

33 Ibid.: 279.

34 Levinas, 1986: 44.

35 Levinas, 1979: 280.

36 Ibid.: 287.

37 Ibid.: 300.

38 Ibid.: 214.

39 Ibid.: 306.

40 Levinas, 1981: 143: 'For subjectivity to signify unreservedly, it would then be necessary that the passivity of the exposure to the other not be immediately inverted into activity, but expose itself in its turn.' Levinas provides a more detailed account of his understanding of the nature of ethical obligation in his Talmudic commen-

tary, 'The temptation of temptation', *Nine Talmuduc Readings*, Indiana University Press, Bloomington 1994, 30–50. The essay is a commentary on the Tractate Shabbath, pp. 88a and 88b, which is in turn a reading of Exodus 19:17. Levinas begins the essay by contrasting two forms of self-transcendence: knowledge and ethics. Self-transcendence through knowledge involves a dialectical movement of engagement and disengagement. The self exposes itself to otherness in order to absorb difference into its own ego. Knowledge, and *a fortiori* philosophical knowledge, is accomplished via this circular movement of externalization and return into self. The knowing subject needs to experience evil in order to realize the good. As such, it is subject to the 'temptation of temptation' – the need to experience everything whilst keeping itself intact. According to Levinas, Christianity follows Western philosophy in repeating this basic trope. Ethical self-transcendence, on the other hand, constitutes an inner rather than an outer form of self-overcoming. In Levinas' words, it is a 'going within oneself further than oneself' (p. 34). Philosophical experience is oriented in advance by knowledge – knowing precedes doing – hence it is only able to engage the other as she appears under its own pre-determined categories; ethical experience, by contrast, is radically open to the other – doing precedes knowing – thus it is able to acknowledge the other as *wholly Other*. Levinas' essay consists of an extended meditation on the following Rabbinic commentary on a passage from Exodus 19:17 ('Then Moses led the people out of the camp to meet with God and they stood at the foot of the mountain'): 'Rev Abdimi bar Hama bar Hasa has said: this teaches us that the Holy One, blessed be he, inclined the mountain over them like a tilted tub and that He said: if you accept the Torah, all is well, if not here will be your grave.' Levinas maintains that we are mistaken if we interpret the Rabbi to have understood the biblical passage as presenting, so to speak, God's Hobbesian ultimatum to the Israelites; for the further Rabbinic commentaries in the Tractate establish that the choice here is not between the law and death but between meaningful and meaningless existence. However, since the law must be received in order to make possible the freedom to make this choice, the law itself cannot be chosen. And yet, the acceptance of the law is eminently rational in so far as it makes possible a meaningful existence. Levinas employs an analogy with artistic inspiration to illuminate the paradox of a non-chosen obligation: in the same way as an artist only discovers the archetype that inspired her work *in* the process of creation, so, the ethical self only understands and consents to the law *through* the act of obeying it. The deed, as it were, creates the law. Levinas contends, therefore, that the law is accepted before it is known – the doing precedes the knowing, the response comes before the question. Thus the law is adopted neither through fear or coercion nor as the result of a freely given choice; rather it issues from a 'non-freedom that is beyond freedom' (p. 40). Levinas further contends that 'The excellent choice that makes doing go before hearing does not prevent a fall. It arms not against temptation but the temptation of temptation.' This comment makes explicit the full implications of Levinas' implicit distinction between the ethical elect and the world. The elect have made a pact with the good prior to the distinction between good and evil, hence they can fall from goodness but they cannot disavow the good itself (p. 43). That is to say, they can be tempted to 'withdraw into egoism' but they cannot be tempted to enter into the world and become subject to the vicissitudes of its history. Indeed, it is only if the ethical community abstains from history that there can be history at all, since it is the ethical community that provides the world with its ethical orientation and meaning. Now, even if we were to accept Levinas' account of the covenant, it still remains the case that the ethical elect are continually faced with the choice whether to affirm or renounce their allegiance to the ethical covenant. In Levinas' own terms, and contrary to what he himself says, nobody is immune to the 'temptation of temptation'. The negative affirmation that one is ultimately not of this world, is the affirmation of a decision, even if it is only to reaffirm a prior obligation. Therefore, Levinas' absolute asymmetry of the elect and the world cannot be sustained. Indeed, it *must* not be sustained, for it leads to the dangerous conclusion that Israel as an ethical community is *a priori* innocent of history. See footnote 42 below for a further estimation of the deleterious implications of Levinas' account of the covenant for his politics.

41 Ibid.: 64: 'Politics tends to reciprocal recognition, that is toward equality; it ensures happiness and political law concludes and sanctions the struggle for recognition. Religion is Desire and not struggle for recognition. It is the surplus possible in a society of equals, that of glorious humility, responsibility and sacrifice, which are the condition of equality itself.' Levinas here draws a clear distinction between ethical and political equality, and makes the former the condition of possibility of the latter. See also his statement: 'Equality is produced where the other commands the same and reveals himself to the same in responsibility; otherwise it is but an abstract idea and a word' (p. 214).

42 Here I am attempting to show that the ethical as Levinas defines it is already political. In this respect, we may note a certain parallel with the thought of Carl Schmitt. On the reading I have developed here, Levinas *mutatis mutandis* identifies the ethical, no less than Schmitt identifies the concept of the political, with the 'friend'. We may remind ourselves in this respect of Levinas' comments in response to the massacres at Sabra and Chatilla in September 1982: 'My *self*, I repeat is never absolved from responsibility towards the Other. But I think we should also say that those who attack us with such venom have no right to do so, and that consequently, along with this feeling of unbounded responsibility, there is certainly a place for defence, for it is not always a question of "me", but those who are close to me, who are also my neighbours. I'd call such a defence a politics that's ethically necessary. Alongside *ethics*, there is a place for *politics*' ('Ethics and politics' in Sean Hand, (ed.) (1989) *The Levinas Reader*, Blackwell, Oxford, 293). In the same interview, Levinas also states: 'The other is the neighbour, who is not necessarily kin, but who can be. And in that sense, if you're for the other, you're for the neighbour. But if your neighbour attacks another neighbour or treats him unjustly, what can you do? Then alterity takes on another character, in alterity we can find an enemy, or at least then we are faced with the problem of knowing who is right and who is wrong, who is just and who is unjust?' (p. 294). Compare this with Schmitt's definition of the concept of the political as resting on the antagonism between friend and enemy: 'The specific political distinction to which political actions and motives can be reduced is that between friend and enemy' (Carl Schmitt (1996) *Concept of the Political*, trans. George Schwab, Chicago University Press, Chicago, 26). Compare it also with the following development of this criterion: 'The political enemy need not be morally evil or aesthetically ugly; he need not appear as an economic competitor, and it may even be advantageous to engage with him in business transactions. But he is, nevertheless, the other, the stranger; and it is sufficient for his nature that he is, in a specially intense way, existentially something different and alien, so that in the extreme case conflicts with him are possible. These can neither be decided by a

previously determined general norm nor by the judgement of a disinterested and therefore neutral third party. Only the actual participants can correctly recognize, understand, and judge the concrete situation and settle the extreme case of conflict. Each participant is in a position to judge whether the adversary intends to negate his opponent's way of life and therefore must be repulsed or fought in order to preserve one's own form of existence' (p. 27). If we substitute 'neighbour' for 'friend' it is clear that Levinas and Schmitt are essentially in agreement on the nature of the political.

Chapter 15

'Not a straight line but a curve', or, Cities are not mirrors of modernity*

Nigel Thrift

> If we think of the world's future, we always mean the place it will get to if it keeps going in the direction we can see it going now and it doesn't occur to us that it is not going in a straight line but a curve, and that its direction is constantly changing. (Wittgenstein, 1998: 5e)

> It is precisely this sense ... of the immanent transitoriness of the world that introduces myth into social theory. Despite the fact that we have no idea what our historical possibilities will be, every theory of social change must theorise not only the past but the present and future as well. We can do so only in a non-rational way, in relation not only to what we know but to what we believe, hope, and fear. (Alexander, 1995: 9–10)

Introduction: hidden cities

In this chapter I want to lay to rest some of the most prevalent myths about modern Western cities and, at the same time, I want to provide some of the elements of an alternative account. In so doing, I am aware that I may appear as a figure, rather like Coleridge's person from Porlock, who punctures the magic of the moment with mundane, even tiresome, observations. But my argument is that the everyday life of the modern city is already shot through with magic. There is no need to stir more in.

This chapter is therefore organized into three main parts. In the first and briefest part, I want to explain my theoretical position. This is crucial since it informs the rest of the chapter. In the second I want to outline some of the myths about the modern Western city which have come down to us from the nineteenth century and which we still cannot seem to shake off. Finally, I want to show how, once these myths are set aside, a number of aspects of the modern city start to come into view which had previously remained hidden, aspects which provide new resources for talking and writing 'city limits'.

Before I begin, it seems sensible to make it clear how I want to see the world, and therefore cities. This 'vision' can be distilled into four main propositions. First, I am a long-run historian. I believe that human practices change slowly and that many of the practices that we assume are 'new' are in fact of considerable antiquity. But I also believe that we do not want to believe this, because we only have a short time to live and we want to believe that our time is the most significant time in history, because novelty is a value we have been taught to cleave to, and because, as a consequence, we have found it difficult to find a language which can describe certain aspects of practice as 'new' and others as 'old': it is all or nothing (Serres, 1995; Serres and Latour, 1995). None of this is meant to imply that I believe in long cycles of history. Rather I assume that history is uneven, open-ended and non-redemptive: there are no foregone conclusions (Bernstein, 1994; Morson, 1994). So, secondly, I therefore believe that time is a multiple phenomenon; many times are working themselves out simultaneously in resonant interaction with each other. 'Absolute, calibrated time – like its negation, pure stochasticity – is an illusion, reinforced by our own species-specific dimensional scaling, and self-referentiality' (Davis, 1996: 65). Thirdly, I am a theorist of practices. That is, my main concern, unlike the bulk of those operating in the human sciences, is with the nonrepresentational aspects of human life, the embodied noncognitive activity which is the mainstay of how we go on. That means that my interest is caught up with the world of emotions, desires and imagination, with embodied or practical knowledge, and, more generally, with the infinitude of sensuous real-time encounters through which we make the world and are made in turn. In other words, I want to talk about 'thinking' as both doing and inhabiting. Fourthly, when I consider how practices are formed, maintained and changed, I think in terms of networks of associations between unlike actors which can exert power for change (even when that change is only intended to hold the network in its current state). These networks are more or less durable over space and through time according to their ability to translate actors to their cause. They are, therefore, always radically incomplete and require constant semiotic-cum-material work to maintain.

How do these four principles influence how I view cityspaces? I think that there are three main things I would want to note. Firstly, when I look at what is going on in contemporary cities I try always to think about how future historians of this era will see these happenings, once the natural enough emphasis upon the uniqueness of our own time is overtaken by our deaths, and once new urban fabric has been tempered by ageing, dereliction and decay. Secondly, I always try to think of cities as performative, as *in use*, and therefore I see urban landscapes as essentially incomplete and only rarely in the hands of just one network of association. Thirdly, I see cities as bubbling over with human creativity, which means that 'twenty minutes into the future, the radically new development is already being absorbed, being made to respond to the exigencies of specific cultural contexts' (Collins, 1995: 5).

In turn, these three rules have three consequences. Firstly, I am suspicious of concepts like 'modernity'. Though such concepts may be treated as simply a kind of historical shorthand, I believe that they are ultimately dangerous. Whether 'modernity' is interpreted as one or more periods of expansion in sense of self, a period of constant change and motion, the gradual triumph of ordered and secularized systems, or a new paradigm of spatio-visual experience, it always leaves out too much and, thereby, too often writes the West as 'the stuff of saga, a vast saga of radical rupture, fatal destiny, irreversible good or bad fortune' (Latour, 1993: 48). In arguing against modernity, and the simple divisions in time that the concept works to achieve, I am opting instead for:

> an end of history and the beginning of many histories. This end is radically different from the closure envisaged by Fukuyama: it is neither the climactic self-negation of Spirit in free-market capitalism nor the shattering of the global into a myriad of dispersed and local narratives. It is, precisely, a question of the linkage of unequal times in the contingent, shifting and relatively unstable orderings – political, economic, cultural – which make up our entangled world, and, which, while organised as goal-seeking structures, drive towards no predetermined end. (Frow, 1997: 10)

Secondly, I am also suspicious of concepts like everyday life and lifeworld. The danger of these concepts arises less from their existence than their implicit opposition to another abstract and systematic order which is responsible for oppression. This order is thereby reified, when it is just as anthropologically practised as any other order. As Latour (1993: 125–6) famously puts it:

> Take some small business owner hesitatingly going after a few modest shares, some conqueror trembling with fever, some poor scientist tinkering in his lab, a lowly engineer piecing together a few more or less favourable relationships or forces, some strutting and fearful politician: turn the critics loose on them, and what do you get? Capitalism, imperialism, science, technology, domination – all equally absolute, systematic, totalitarian. In the first scenario, the actors were trembling, in the second, they are not. The actors in the first scenario could be defeated; in the second they no longer can.

In other words, I want to look towards a sense of the world as always in process, dragged into existence by the work of many.

Thirdly, it follows that I am therefore clear that human creativity is not being 'damped down'. Life is fundamentally uncertain and people make their peace with that uncertainty in many 'rational' and 'irrational' kinds of ways. It is only a surprise to intellectuals who want to believe that the Enlightenment project has been carried through – or thwarted – that we live in an era which has seen, for example, a rebirth of evangelical Christianity and paganism, in an age when many people still take astrology seriously and many others at least take note of it, in an age in which many people believe in UFOs and others at least follow *Star Trek*, in an age when near-death experience has its own academic journals and death still has its very definite rituals, in an age in which older religious traditions like New Thought, Theosophy and Spiritualism, stirred and shaken by the psychological and eastern religious organizations of the 1950s and 1960s countercultures have seen a rebirth as New Age, and in an age in which the study of 'implicit religion', which recognizes the religiosity of everyday life, has rightly become an important area of research. The whole fabric of everyday life, in other words, is shot through with dreams, fantasies, superstitions, projections, religious yearnings and millenarian movements (Thrift, 1996).

Like Probyn (1996: 19), then, 'I want to reinvigorate the idea that living is bewildering, strange, and sometimes wonderful [and] I also want to emphasize

the magic of ordinary desires.' The magic has not gone away. Indeed quite
the reverse. Two hundred or more years after the Enlightenment, the majority
of ordinary people around the world remain as faithful as ever to the 'beliefs'
that many members of the Enlightenment were attempting to expunge:

> In the United States (the country to which these scholars of the Enlightenment
> looked before all for the revolution in ideology that would mark the escape from
> religious superstition) census data for the 1980s show that 95% of the population
> still believe in an active God, 88% in a human soul and 71% in the survival of the
> soul after bodily death. In Europe 75% believe in God, 61% in a soul and 43% in
> survival after death.
>
> If these beliefs in God and the soul were merely tools of a shallow faith in 'some-
> thing over and beyond', such levels of credibility might not be so remarkable.
> However, other surveys show that this theism and soulism is paralleled in modern
> society by the continuance of many more particular beliefs in what we may loosely
> call a 'counter-scientific reality'; that is to say, beliefs in specific supernatural entities
> and powers for which science has discovered no empirical basis and whose exis-
> tence would in many cases flatly contradict scientific theory.
>
> It is found, for example, that in the developed world, between a third and two
> thirds of the population still attest to the reality of such phenomena as telepathy,
> precognition, interaction with spirits of the dead, reincarnation and the paranormal
> effects of prayer.
>
> Typical of recent findings are those from a study of a representational cross-
> section of people in the town of Reading: ... Respondents were asked: 'Do you
> think it's true?' of a series of statements relating to paranormal phenomena.
> Excluding the small number of 'don't knows', the proportion of people who said
> definitely 'Yes' were as follows:
>
>> 'It is possible to know what someone else is thinking or feeling even if they are
>> out of touch by ordinary means' – 63%. 'Dreams can foretell the future' – 71%.
>> 'Prayers will sometimes be answered' – 71%. 'It is possible to make someone
>> turn around just by looking at them' – 66%. 'Some people can remember past
>> lives that they have lived in other bodies' – 54%. 'It is possible to get a message
>> from the dead' – 37%. 'Some houses are haunted by ghosts' – 69%.
>
> Despite the generally high level of belief, it could perhaps still have been the case
> that there was a substantial minority who were either unsure or who didn't believe
> in any of it. Not so: 88% of the population said definitely 'Yes' to at least one of the
> statements and 68% said definitely 'Yes' to at least three of them. (Humphrey,
> 1995: 4)

Having set out the background, I am now in a position to begin to identify
where I believe some current writings on cities have gone wrong. In this way,
I hope to be able to foreground what has often been regarded as background.
I want to begin this task by setting out four of the chief myths about current
urban societies.[1] I should add straightaway that these narratives of the contemporary metropolitan condition are hardly new: they all date back to the
nineteenth century or even earlier. But these myths are undoubtedly powerful: they are difficult to see around because they tell us that we already know
what urban society is about. This cannot be. It is not.

Four myths _____

Myth 1: All cities are becoming instants in a global space of flows

We live, so it is said, in a globalizing world and one of the key elements of this world is the existence of a space of flows brought into existence by modern information technologies. This space of flows is 'the space of information. This proliferating and multi-dimensional space is virtual, densely webbed and uniformly complex; a vast and sublime realm accessed through the mediations of our imaginative and technical representations' (Davis, 1994: 86). And, in turn, the existence of the space of flows is changing our apprehension of space, time, and subjectivity. Places move closer together in time, time becomes instantaneous, and the subject becomes decentred, strung out on the wire.

The two main contemporary tellers of this myth are David Harvey and Paul Virilio. For Harvey, the world is in the grip of another round of 'time–space compression':

> the processes that so revolutionise the objective qualities of space and time that we are forced to alter ... how we represent the world to ourselves ... space appears to shrink to a 'global village' of telecommunications and a 'spaceship earth' of economic and ecological interdependencies ... and as time horizons shorten to the point where the present is all there is ... so we have to learn how to cope with an overwhelming sense of *compression* of our spatial and temporal worlds. (Harvey, 1989: 240)

Thus time–space compression is a story of the simultaneous marked increase in the pace of life brought about by modern transport and telecommunications *and* the upheaval in our experience of space and time that this speed-up brings about as people, images, capital, and information all speed more and more rapidly around the world. The result is that 'time-space compression ... exacts its toll on our capacity to grapple with the realities unfolding around us' (Harvey, 1989: 306), most especially by challenging our sense of identity and our ability to preserve tradition.

Virilio is willing to go further still with this story. For him, Harvey only describes the first effects of speed-up in which physical displacement still presupposes a journey: the individual makes a departure, moves from one location to another, and so arrives. But now, with the 'instant' transmission made possible by electronic technology, a new 'generalised arrival' has occurred, in which the element of a journey across space is lost. The individual can be in two places at once, acting as both transmitter and receiver. Thus we have arrived in a historical period in which there is 'a crisis of the temporal dimension of the present':

> one by one, the perceptive faculties of an individual's body are transferred to machines, or instruments that record images, and sound; more recently the transfer is made to receivers, to sensors and to detectors that can replace absence of tactility over distance. A general use of telecommunications is on the verge of achieving

permanent telesurveillance. What is becoming critical here is no longer the concept of three spatial dimensions, but a fourth, temporal dimension – in other words that of the present itself. (Virilio, 1993: 4)

In this new order, cities become interruptions in the space of flows, transient moments in the circulation of capital. Their future is to act as waystations for dominant organizational forces making their wishes known 'through the powerful medium of information technologies' (Castells, 1989: 6).

Here is a myth, a story which is overdrawn and overdone, and for at least four reasons. Firstly, these revelations, which are presented as of the hour, are certainly of considerable antiquity. Through history and around the world, innovations in transport and communications have been heralded as proof positive that the world was speeding up, that places were moving closer together in time, that the world was shrinking. For example, in the eighteenth century a number of nervous disorders were thought to be the result of the faster pace of life (Porter, 1993). In the nineteenth century, as the stage-coach was replaced by the train, the telegraph and the telephone, so the theme of the 'annihilation of space and time' became a favourite meditation of many writers (Thrift, 1995). In the twentieth century, the idea surfaced yet again; one author even went so far as to calculate the rate (in minutes per year) at which places were converging on one another, a phenomenon which he christened 'time–space convergence' (Janelle, 1969). These different variants on the theme of a speeded-up world were nearly always associated with a generalized crisis of identity. From the nervous disorders of the eighteenth century onwards, the acceleration of everyday life has been thought likely to lead to volatile, fragmented, and spread-out people whose identity is in question because of the shallowness of their lives. Writing in the late nineteenth century the philosopher Nietzsche (cited in Prendergast, 1992) described the 'tropical tempo' of the modern world in a way which is typical:

> Sensibility immensely irritable ... the abundance of disparate impressions greater than ever; cosmopolitan in foods, literatures, newspapers, forms, tastes, even land-scapes. The tempo of this influx *prestissimo*, the impressions erase each other, one intuitively resists taking in anything ... A kind of adaptation to the flood of impressions takes place: men unlearn spontaneous actions, they merely react to stimuli from the outside.

Secondly, this myth is based in a technological determinism which unproblematically reads off the characteristics of the technologies involved onto society. For example, there is little or no sense of technologies as culturally mediated. Thus, whilst it cannot be denied that 'a dynamics of thought is not separable from a physics of traces ... the medium ... is but the ground floor. One cannot rest there' (Debray, 1996: 11). Therefore, 'when these technologies are mastered ... their functions and cultural resonances change fundamentally' (Collins, 1995: 16).[2] In the case of the 'new' information technologies, there is manifold evidence that this process is already taking place:

> the bloom of unlimited possibility is already passing from them. The Internet begins to shrink back from its ecstatic characterisation (a web for democracy, a veritable

sea of data, an instrument of expanded perception) and settles down as a very large database-cum-techie salon. The smart idolatry is moving on. (Spufford, 1996: 271)

Again, too often the medium is assumed to be one when it is multiple. Thus work on communications networks too often ignores the content of communication – from bank statements to love letters – which illustrate that the medium is in fact a set of different but intersecting elements of a whole range of actor-networks.

Thirdly, this myth places too much of a premium on technologies which conform to dominant cultural stereotypes of what is 'new'. But, as Mark Edwards (1996: 10) points out, 'in a survey last year, Americans were asked which piece of recent technology had most altered their lives. The most common reply was not the Internet, not even the personal computer. First – by some way – came the microwave, followed by the video recorder.'

Then, fourthly, cities cannot be seen as places which are leaking away into space of flows. This is to fundamentally misunderstand the way in which new information technologies have normally acted as a supplement to human communication rather than a replacement. Innovations like the telephone, the fax, and the computer are used to extend the range of human communication, rather than act as a substitute. It is not a case of either/or but both/and. For example, in my work on the City of London (Thrift, 1996), I have shown the way in which the growth of information from new communication technologies has presented fundamental problems of interpretation for workers in the City which have forced greater, rather than less, face-to-face communication: the City has become, even more than formerly, a key story-telling node for the world as a whole, and, as a result, its importance as a spatially fixed centre has, if anything, been boosted.

Myth 2: Cities are becoming homogenized

In this myth, cities are depicted as increasingly bleak and interchangeable places. Usually such a view is connected in some way with a rising tide of commodification which acts as a block on creative citizenship. The rise of shopping malls is often taken as the most visible sign of the landscape as simply a moment in the circulation of commodities. The landscape is increasingly constructed in the image of the commodities. Worst of all, the landscape itself becomes a commercial package through the growth of a 'heritage industry' which packages the past of places to sell them in the present. Thus, places increasingly become infected by the condition of 'placelessness' (Relph, 1981) which apparently characterizes sites like shopping malls: 'we are in the midst of a desert of shops, a wasteland of services, a chaos of commerce. If not nowhere, we are in an extremely shallow somewhere' (Casey, 1993: 268–9).

However, there are some serious problems with this viewpoint. First, the icons of placeless urbanity may not travel. For example, high-rise office buildings, which are often regarded as symbols of American invasion, are few and far between in Britain. Even in the United States, high-rise central business districts were rare in the late 1960s: only New York, Chicago, Detroit,

Philadelphia and Pittsburgh had more than two buildings over 25 stories, and two of these cities – New York and Chicago – had produced quite specific capitalist vernaculars.

Secondly, as has now been shown in numerous empirical studies, what may look like similar urban spaces can be used in quite dissimilar ways. Thus a growing literature on shopping malls in Britain shows that although they are growing in number and size, they are simply not used in the same way as in the United States. For example, fewer malls, lower car ownership, and limited opening hours mean that these malls are only rarely used in the same way as by American youth (Thornton, 1995). In any case, British traditions of shopping, which have a long history, are sometimes quite different (Mort, 1996).

Thirdly, judgements on placelessness are bedevilled by a lack of historical depth rooted in their narrative of an all-consuming newness. Take the example of the shopping mall (which is usually anchored by one or two department stores, outposts of an earlier mode of retailing). As Savage and Warde (1993: 143) note in their discussion of malls as icons of placelessness:

> If the shopping mall appears new and placeless today, this is because it has not yet been integrated back into its surrounding urban fabric, either by wear and tear, by feats of imagination or by reputation. Urban dwellers of the nineteenth century regarded innovations such as the subway as a bewildering and new, placeless realm. Today, these have been moulded into their contextual environments.

Fourthly, it might even be argued that, if anything, cities have, over time, become richer place experiences. This last point deserves expansion. Most cities have offered, over the course of history, an increasing range of experiences which it is possible to use as imaginative resources. For some time the city has had an underground landscape (Williams, 1991; Prendergast, 1992). It has a sky which is often filled with activity. It is made of all kinds of materials that let us sense in new ways; for example, glass (Armstrong, 1996). This is to ignore the new means of apprehension, from photography to film to video to CD-Rom, which have allowed us to touch the city in new ways, to memorize it, to rewrite it, to make it tactile, 'once more with feeling' (Benjamin, 1969, 1979). And these 'new magics' are chiefly examples chosen from the visual register. The same kind of richness is, *pace* writers like Corbin, being invested in other sensory registers too.

It is worth underlining this point: if we could measure the process, perhaps our experience of places has thickened, not thinned. One example of this process might be the advent of different forms of artificial light which, in turn, have extended our experience of place into the night. Of course, the night has never been just a blank space in which people slept. All societies have had some activity at night, usually aided by the light from fires, which itself has been a potent source for the imagination over the centuries (Bachelard, 1964). However, for all that, night-time activity was generally limited by a general lack of light. Thus when Dr Johnson walked around night-time London in the early eighteenth century, 'we can instantly imagine the scene: the cobbled streets, the stinking rubbish, the tavern signs, the shuttered house-fronts; the moonlight and the dark alleys; the slumbering beggars, the footpaths and the night watch' (Holmes, 1993: 35). This was the 'City of Dreadful Night', still physically present but transformed by danger and lack of light, except for the moon and guttering tapers, into an alternative landscape, peopled by young rakes, the desperate poor and criminals (for a time in the eighteenth century, it was a capital offence to be caught out at night with a blackened face). In turn, the landscape of night stood for many things but especially the unknowable, symbolized by the candle in the window which was knowledge struggling to be born, and the mysterious, symbolized by the shadows cast by what little artificial light existed.

But over the succeeding centuries all this changed. As the use of artificial light spread, so urban landscapes began to appear that before had only been figments of the imagination. At first it was the glow of gas-light that lit up the city. Then, from the late nineteenth century, it was the gradual spread of electric light. Finally, it was the spread of light attached to mobile objects – cars, aeroplanes, satellites, and so on. The night-time city becomes an active landscape which, furthermore, has become actively peopled:

> Now there is a whole after-hours community – everything from evening classes to supermarkets, night clubs, discos and massage parlours, as well as a great array of maintenance people who service and repair the daytime world while its inhabitants sleep. The defence establishment, the financial markets, broadcasting, transport, communications now work on a 24 hour day schedule. (Alvarez, 1995: 20)

The night-time urban landscape is one in which many people work (for example, 14 per cent of the UK workforce is involved in shifts) and in which, for all the real dangers, many people play. Thus, what we have also seen is the growth of a specific 'night life'. Night life is hardly a new invention: the Vauxhall pleasure gardens and the theatres of seventeenth-century London were an early manifestation. But night life has now grown to major proportions: theatres, pubs, night-clubs, shops, cafés, restaurants. Thus, Worpole (1995) can write of a '24 hour city' of culture and entertainment.

The point is that this extension of human activity into the night resulting from the profusion of artificial light, provides us with *new* imaginative resources. In particular, it has produced a whole set of alternative, technologically induced landscapes that were simply not there before; edited and highlighted landscapes which have now become 'second nature' to us (some, like Nye (1995), have even talked of a 'technological sublime') and which have radically altered our appreciation of many places. For example, for painters at the beginning of the twentieth century, 'the night cityscape was a new visual reality signifying a break in the continuity of lived experience that the painter confronted along with the rest of society' (Nye, 1990: 76). Further, as the century went on, so the appreciation of this cityscape at different times of the night began to suffuse into paintings – for example, places in the dusk when lights begin to appear in windows and street lights are switched on, or places in the deep of night when the glow of powerful electronic light can create a world of intense colours. In other words, painters were able to see *new* places and, in turn, they have helped to make these perceptions into a normal part of the currency of our vision (Zajonc, 1993).[3]

Myth 3: Cities (or parts of cities) are inauthentic

This is a myth that dates from at least the founding fathers of urban studies like Simmel and Wirth and no doubt before. In that work, these writers depicted the city as a mirror of modernity. The city, in opposition to the small-scale, tradition-bound community, was the main locale in which new impersonal social relationships, the calculative attitudes of the money economy, and social fragmentation would be observed. According to some commentators the same kind of division can be found in Benjamin's work: experience (*Erfahrung*) is replaced by instrumental reaction (*Erlebnis*):

> In the former state, found in preindustrial societies, experience is based in habit and repetition of actions, without conscious intention. These experiences are bound to traditions, the socially constructed and legitimated ways of acting, which gain their authority by their uniqueness and specificity. In the latter state, found in modern industrial societies, the mass reproduction of commodities and symbols disperses tradition, so that individuals simply react to the stimuli of the environment and develop instrumental ways of thinking in order to cope in such a changed environment. (Savage, 1995: 201)

Nowadays this kind of thinking seems to have been replaced by a different kind, one in which authentic experience can still be found, but now located in the city in a residual sphere called 'everyday life', which, pressed in on all sides by an alienating and monolithic capitalism, is still the last best hope for humankind. Here the authentic 'when' has been replaced by an authentic 'where':

> superior activities leave a 'technical vacuum' between one another which is filled up by everyday life. Everyday life is profoundly related to all activities, and emphasises that with all their differences and their conflicts, it is their meeting place, their bond, their common ground. And it is in everyday life that the sum total of relations which make the human – and every human being – a whole takes its shape and its form. In it are expressed and fulfilled those relations which bring into play the totality of the real, albeit in a certain manner which is always partial and incomplete: friendship, comradeship, love, the need to communicate, play, etc. (Lefebvre, 1995: 97)

Whilst it might well be necessary to retain some notion of the everyday as a means of promising 'the possibility of other sorts of non-exploitative solidarities' (Taussig, 1992: 141), it is hard not to avoid the conclusion that Lefebvre's notion of everyday life is ultimately romantic. Much the same has been said of de Certeau's momentary notion of 'tactics' in *The Practice of Everyday Life*. As Ahearne (1995) points out, for all of the understated subtleties of de Certeau's work, figures of indeterminacy and of a bewildering multiplicity are continually converted into visions of 'the night side of societies', 'an obscure sea' or 'the oceanic murmur of the ordinary'. These images work to conceal as much as they reveal and, at worst, can be counted as an uncritical hero worship, changing incomprehension into aesthetic pleasure, to no real political end.

As Lefebvre and de Certeau were well aware, there is no quotidian walled garden to be found in cities, in however liminal a register. But that such a notion can become current stems from the assumption that sometime or somewhere authentic spaces have existed. In fact, outside or inside the city, this assumption has deadly consequences. It distracts attention from the sheer work involved in constructing alternative spaces. It cedes too much to capitalist and other 'distinct specialised, structured activities' (Lefebvre, 1995) by depicting them as somehow asocial. It splits people's lives into acceptable and non-acceptable bits. It erases difference 'in much the same way as do modern European-derived notions of the public and the masses' (Taussig, 1992: 141). And it thereby provides a licence to ignore much of the outpouring of ethnographic and other research on cities which shows urban life as ambiguous, fragmented, dilemmatic, and thereby creative.

Let me briefly illustrate this latter point by turning to recent work by British and North American authors on kinship and friendship in cities. One of the things that is remarkable about this work is that it is largely ignored by many writers on cities, in part at least because it simply does not fit their preconceptions about modernity and the city. What this work shows is that, historically speaking, rather than living in a period when kinship and friendship relationships are being smothered by individualism and privatism, we may actually be living at a kind of high-water mark of interaction. Take the example of families: 'Families are now much more complex than at any time in the past. As long as we do not make the silly mistake of confusing household with family, it is arguable that the family and family relations are now, in the 1990s, stronger than at any time in the past' (Pahl, 1995: 179).

We can illustrate this statement through the seminal work of Janet Finch (1989; Finch and Mason, 1993). She shows that families are probably more complex than in the past. Though fewer children may be born, longer life, divorce and so on mean that family relations are often much more extensive. But she also shows that family relationships are still central to most people's lives. Of course, these relationships are not matters of automatic allegiance: they are a product of sustained and thoughtful interaction over long periods of time; often involving all manner of dilemmas, negotiations and compromises, and, even, extensive reading (Giddens, 1992). Further, the increasing geographical distance between family members seems to be of only minor consequence. Thus:

> Colin Bell's (1965) work ... found that middle-class families, although more geographically dispersed than their counterparts in traditional working-class communities, nonetheless used their greater financial resources to retain contact over larger distances, and gave both practical and financial support to each other at key points rather than on a day-to-day basis. Cars and telephones made this possible for middle-class people. The spread of these commodities more widely through the population 20 years later probably has reduced the class difference, but

by the same token it means that more people have the capacity to keep regular contact with geographically distant kin.

A rather different example of support being retained over very long distances can be found in Wallman's study of eight London households, where the key emotional support for one of her respondents continued to be provided by the woman's mother who lived in the West Indies, and whom she had not seen for 20 years ...

The example of migration over long distances shows very clearly the fallacy in the argument that kin support is necessarily reduced in proportion to the distance between relatives. Indeed in some circumstances this kind of migration can, if anything, strengthen structures of support. (Finch, 1989: 95)

Pahl (1995: 178) puts it more succinctly:

When family members migrate to America or Australia, they do not necessarily lose contact with those they leave behind. Regular letters can come in a few days; telephoning is becoming progressively cheaper and the cost of air tickets from London to New York is not much more expensive than an ordinary return ticket from London to Aberdeen by train.

Threading through kin networks are friendship networks – the two are not exclusive – which seem as or more concentrated than at any time in the past (see, for example, Wellman and Berkowitz, 1988). Indeed, it might be argued that for some communities (such as the lesbian and gay communities) the importance of friendships is so great that they have become the 'families we choose' (Weston, 1991).

To summarize, it seems that a story of greater privatism and individualism – and less and less authenticity – does not really hold, as Allan (1996: 129) makes clear:

Finally, there is the question of whether people lead socially isolated lives. The answer to this should be clear. It is, of course, that most people do not. Most of us have personal networks which contain large numbers of others, and within which a smaller number of relationships are particularly important to us. Some of these important relationships are with kin and some are with non-kin. The patterns here do vary depending on a wide range of circumstances. And equally, some people do lead quite restricted lives and are socially isolated. The point, though, is that this is relatively unusual rather than the normal state of affairs in contemporary society.

Yet in a sense Wellman is right to argue that the individual relationships to which people are party have themselves become 'privatized'. This, though, does not mean that lifestyles under contemporary social and economic conditions are privatized in a fuller sense. What it means is, firstly, that sociable relationships are frequently enacted away from public gaze and scrutiny. This certainly applies to most primary kin ties. However, many friendships are also organized in this fashion, with the private sphere being one of the key areas in which friends engage with one another. Secondly, it means that social relationships overall often have lower densities than would otherwise be the case. Although the subset of kin ties within the personal network will be highly connected, friendships are often more individual. An individual's different friends are quite likely to have met each other through that individual, but they are not themselves necessarily friends or involved with each other in any other way. Thus these points of the network will have relatively low densities.

Myth 4: One city tells all

There is a fourth myth, that in each era there is a paradigmatic 'celebrity' city which sums up that era, the place where it all comes together. In recent years, out of many such cities (Chicago, Vienna, New York, Zurich) two cities seem to have been particularly prominent examples. So far as the nineteenth century is concerned, Paris is taken to be the birthplace of all the most significant signs of modernity – department stores, mass leisure, *flânerie*, urban crowds, and so on – and these historical assets have become one of the means of ensuring the cultural ascendancy of Paris in the twentieth century (see Sheringham, 1996).

Taking up the baton, and pointing to the future, is the city of Los Angeles, 'the capital of the late 20th century' (Scott and Soja, 1986: 249) where 'it all comes together' (Soja, 1989: 8). By 1990, Davis was able to note that a so-called LA School had emerged, consciously founded on the Chicago School, who 'have made clear that they see themselves as excavating the outline of a paradigmatic post-fordism, an emergent twenty first century urbanism' (p. 84). Meanwhile, social theorists like Baudrillard, Eco and Jameson made pilgrimages to this new cradle of urban civilization, adding to its mystique and, as with the case of Paris, reproducing its story around the world.

But alighting on one city as a means of telling the tale of an era (and thereby producing that era) is dangerous. Four problems become immediately apparent. First of all, such a choice can nearly always be empirically undone. For example, the choice of Paris as the leading urban edge of the nineteenth century is increasingly being challenged. Thus, the expanding history of consumption and shopping in Britain makes it clear that many of the developments thought to be special to Paris in the mid-nineteenth century were common in Britain before that date (Glennie and Thrift, 1996). Again, the history of art suggests that many of the key elements of Impressionism were also able to be found elsewhere than in Paris. Secondly, as these examples make clear, the choice of a paradigmatic city deflects attention from other cities which may be just as significant but remain unexamined, as what may well be an exception becomes the norm through which all other cities are measured. For example, were it not for a remarkable series of articles by Richard Walker (1994, 1995, 1996), extraordinary events in San Francisco would have been wiped from the map by the new Angeleno hegemony. Thirdly, it seems unlikely that any city can bear the weight of this kind of interpretative load. Los Angeles is a case in point. The depiction of the city as the site of a utopian or, more normally dystopian, future stretches interpretation beyond what we can know into the realms of myth. At the same time, it can divert attention (although it does not in the case of authors like Soja) from many of the struggles that are constantly taking place in the city, over the environment, over homelessness and poverty, over ethnic identity, and so on.

'Rather than address these realities the media draws a portrait of Los Angeles entering a *Blade Runner* epoch' (Acuna, 1996: xviii). Fourthly, and finally, it can divert attention from the fact that events in cities are often linked to events in other cities. As Paul Gilroy and others have spent much time working through, even (or especially) in imperial systems cities are points of interconnection, not hermetically sealed objects. What is important, then, is often a city's connectedness to events in other cities which are not necessarily or even preponderantly hierarchical in nature.

Other cities

Why are these four myths so problematic in studying cities? For three reasons, I think. Firstly, they cleave to a particular form of theory which sees great forces everywhere; the task of theory is to delineate and delimit these forces. In other words, this kind of theory refuses the idea of the city as a partially connected multiplicity which we can only ever know partially and from multiple places (Strathern, 1991; Harraway, 1995): 'Here's a set undefined by elements or boundaries. Locally, it is not individuated; globally it is not summed up. So it's neither a flock, nor a school, nor a heap, nor a swarm, nor a pack. It is not an aggregate; it is discrete' (Serres, 1995: 4–5).

Secondly, and relatedly, these myths downgrade everyday human practices in cities to the everyday, thereby stripping their magic from them. Not only does this mean that the embodied and performative aspects of these practices are lost, but equally that the 'domestic' is downgraded by an enforced domestication: the 'local' is localized by being opposed to the 'global', and the linguistic (whether spoken or written) is privileged, when so much agency cannot be captured by linguistic re-presentation (Callon and Law, 1995). Thirdly, these myths therefore block our view of so much that is currently transpiring in contemporary cities. They obscure many of the most interesting things that are going on arising out of the sheer inventiveness of these cities' inhabitants, blocking off emergent effects by trapping them in old theoretical amber.

I cannot write everything so in the remainder of this chapter I want to note just three of these different kinds of city which can come to our attention and which together begin to provide a sense of a city that is constantly changing (even to stay the same), that does not necessarily hold together, that is both little and large. In making these three choices, I am acutely aware that it may appear as though I am negating other vibrant areas of work. Nothing could be further from my mind. For example, I would, if I had the space and time, have wanted to point to a whole series of other areas of work, including the relationship between the city and the visual register, as found especially in the work of feminist writers like Giordana Bruno (1993) and Griselda Pollock (1988), the relationship between the city and memory, as found in the work of Raphael Samuel (1994) and Christine Boyer (1995), and the relationship between the city and the rise of geographical information systems which sometimes seems to prefigure a mimetic impulse of Borgesian dimensions.

And this would still have been to ignore massively important work on ethnic identity, and on environment and 'urban nature'. But choices have to be made. Here I will concentrate on three of these different kinds of city: the embodied city (play), the learning city (knowledge, identity and soft capitalism) and the unjust city (unequal relations of power, new forms of injustice and alternative financial institutions), each of which is concerned with different means of disclosing cities (as embodied, as capitalist, as alternative).

The embodied city (play)

The first element of the undiscovered city must be embodiment. There is, of course, currently a very large amount of work on the body. But it is surprising how little of this work locates the body, even though it is difficult to think of the body except as located. There are, I think, two reasons for this. One can be found in Grosz's (1996) excellent set of papers on the body and the city. Too often she writes in an almost overwhelmingly abstract way about something which by definition cannot be counted as abstract. In part, this may be because Grosz regards the city as just one more element of explanation of the social constitution of the body.[4] Yet in the following passage, she gradually moves away from this view:

> The city in its particular geographical, architectural and municipal arrangements is one particular ingredient in the social construction of the body. It is by no means the most significant (the structure and particularity of, say, the family is more directly and visibly influential); nonetheless the form, structure and norms of the city seep into and alter all the other elements that go into the constitution of complexity. It affects the way the subject sees others (an effect of, for example, domestic architecture as much as smaller family size), the subject's understanding of and alignment with space, different kinds of limited spontaneity (the verticality of the city as applied to the horizontality of the landscape – at least our own) must have effects on the ways we live space and this on our corporeal alignments, compartment and orientations. It also affects the subject's focus of corporeal exertion – the kind of terrain it must negotiate day to day, the effect this has on its muscular stature, its nutritional context, providing the most elementary focus of material support and sustenance for the body. Moreover the city is also by now the site for the body's cultural saturation, its take-over and transformation by images, representational systems, the mass media, and the arts – the place where the body is representationally re-explored, transformed, contested, reinscribed. In turn, the body (as cultural product) transforms, reinscribes the urban landscape according to its changing (demographic) needs, extending the limits of the city even towards the countryside that borders it. As a hinge between the population and the individual, the body, its distribution, habits, alignments, pleasures, needs, and ideals are the ostensive subject of governmental regulation, and the unity is both a mode for the regulation and administration of subjects but also an urban space in turn reinscribed by the similarities of its occupation and use. (Grosz, 1996: 109)

The second reason is that the body in the city is too often thought of as the individual body, rather than an 'intersubjective' body occupying a space with other bodies which necessarily responds in a variety of sensuous and differently skilled ways according to prevailing cultural notions of body, space and time. And this is to ignore the presence of other objects (to which Merleau-Ponty was so keen to draw attention)[5] which are the missing terms of this tactile collection.

Are there any authors who have genuinely attempted to situate the body in the city? I can think of three. The first is Henri Lefebvre, whose 'rhythm-analysis' is an attempt to capture the interactive temporality of the city as, amongst other things, the to and fro of a crowd of crowded bodies. Thus when Lefebvre looks out of his window onto crowded streets he remarks:

> What this window which opens onto one of the most lively streets of Paris shows, what appears spectacular, would it be this feeling of spectacle? To attribute this rather derogatory character to this vision (as dominant feature) would be unjust and would bypass the real, that is, of meaning. The characteristic features are really temporal and rhythmical, not visual. To extricate and to listen to the rhythms requires attentiveness and a certain amount of time. Otherwise it only serves as a glance to enter into the murmurs, noises and cries. The classical term in philosophy, the 'object', is not appropriate to rhythm. 'Objective'? Yes, but a spilling over the narrow framework of objectivity by bringing into it the multiplicity of the senses (sensorial and meaningful).

The second author is Torsten Hägerstrand, who in his 'time geography' attempted to produce a musical score of bodies and things. His diagrammatic scores were attempts to describe something that could not be written, but could be seen and felt. Though they produce uncomfortable echoes of Laban's work on worker effort, these scores certainly have an effect. The third and final author is Richard Sennett, who has tried to consider bodies in the city as a series of differential mapping of bodies onto spaces and spaces onto bodies via cultural intermediaries like rituals. He sees one of the main problems of current cities as the passive relationship between body and environment and in *Flesh and Stone* (1995) he sets out how a more positive relationship might be installed.

In the work of each of these three authors there is, then, a sense that the space of the body (or, more properly, embodiment) is important. Most particularly, there is a sense that space is important because it produces what Probyn (1996) calls singularity from the range of specificities (race, class, sexuality, gender) that we inhabit. Cultural geographers have spent considerable time attempting to juggle with the singularity of specificity, most especially in work on sexuality (Bell and Valentine, 1995). Here I want to point to one of the areas that has so far been curiously neglected: the example of dance, an area of study which is now starting to become central to work on the body and the city and for obvious reasons (Thomas, 1995, 1996).

The importance of dance becomes self-evident in the light of the concerns of writers like Lefebvre, Hägerstrand and Sennett. This importance can be indexed in five main ways. Firstly, dance represents a conscious formulation of the sensuous inter-bodily movement that we usually take for granted, and so do not write down. Dance has itself been called knowledge because dance does not pass through the sign but is itself contextualised in movement. (Of course, some caution needs to be exercised here or dance can be represented as a pure phenomenology of immediate experience – indeed many forms of modern dance, prior to the 1980s, were intended to convey exactly this – but, even so, it cannot be denied that, in part, dance is a gratuitous outpouring of force, an expenditure without return (Lingis, 1994).)

Thus, secondly, it is important to approach dance as a socially constructed ritual, which, like other forms of artful and skilful moving together in time (such as drill, see McNeill, 1995) act to produce or strengthen connection. Indeed, in certain cases, dance has become a vital part of modern spectacle (and not always to the good, as Wollen (1995) has shown). Thirdly, dance is usually linked to other media, most especially music but also quite often visual display (from special costume to light) and artificial enhancers (such as alcohol and other drugs). Fourthly, dance is normally carried out at predetermined sites. These may be concert halls and other formal sites (thus underlining the links between dance and other artistic modes) but for most people, in most of their lives, these sites are more likely to be more mundane settings in which they are quite likely to be participants, 'from street dancing to dance halls, discos and raves, to parties, dinner dances, weddings and church socials' (Thomas, 1995: 3). Fifthly, as these settings make clear, dance is simply a very widespread part of the urban commercial sphere (Mort, 1996), ranging all the way from ballet to modern dance to ballroom dancing to visual theatre, to film and TV and video, to keep fit clubs and raves, and so on. As Thornton (1995) notes, for single younger people admissions to 'dances' rank far above admissions to sports, live arts or cinema. In other words, dance is a significant urban economic activity because for so many people it is a crucial part of 'leisure activity, entertainment and sexuality' (McRobbie, 1984: 130–61).

Why then, given its widespread nature (surely equivalent to the breadth of musical activity found by Finnegan (1989) in Milton Keynes), is there so little work on dance in relation to urban life?[6] Five main explanations come to mind. Firstly, there is the simple fact that issues of sensuous body movement in the city have only just begun to be considered, especially as 'non-representational theories' (Thrift, 1996) have begun to bite. In particular, dance is an instance of *expressive* embodiment: the body is not just inscribed, it is itself a source of inscription which can be used to conjure up virtual 'as if' worlds which can in turn become claims to 'something more' (Pini, 1996). Thus, in the case of dance:

> dance can be considered as the fabrication of a 'different world' of meaning made with the body. It is perhaps the most direct way in which the body-subject sketches out an imaginary sphere. The word 'imaginary' is used here in the sense 'as if', suggesting a field of potential space. The dance is not aimed at describing events (that is, it is not representational) but at evolving a semblance of a world within which specific questions take their meaning. (Radley 1995: 12)

Secondly, dance chiefly belongs to the sphere of *play* and therefore lives beyond the rational auspices of purposive means–end systems like work. In

Western societies, play is often regarded as peripheral to the real business of life because it is gratuitous, free (if one is forced into such a practice, it is no longer play), and non-cumulative (Huizinga, 1949; Geertz, 1972; Bateson, 1973; Winnicott, 1971; Bauman, 1993). Play is simply the irrational precursor to the real business of life. Thirdly, dance, unlike other, more durable aspects of human experience, is comparatively rarely recorded. Fourthly, dance is nearly always considered in relation to music and it is the music that is privileged (although, recently, much popular music has become subordinated to dance). Of course, musical soundscapes are a crucial element of dance but dance cannot be reduced to them. Then, finally, dance is often regarded as 'high culture' and therefore as of little interest to cultural studies. Yet, even a cursory inspection of contemporary dance practice shows the tenuousness of judgement: children's ballet classes, square dancing, ceroc, line dancing, morris dance, step dancing, ice dance ... the list goes on and on.

The learning city of knowledge, identity and soft capitalism

Nowadays, cities are often seen as repositories of knowledge, as pedagogic engines. Certainly pedagogic institutions, from schools to museums to libraries, are scattered through them. It is no surprise then, that there has been an outpouring of work on the educational institutions of the city, sometimes, but not only, stimulated by writers such as Bourdieu and Foucault. For example, we now know that there are cities of science. Writers in the sociology of scientific knowledge have revealed the sheer diversity of institutions and institutional contexts through which 'science' is produced and reproduced, from laboratories to lecture rooms to libraries to public houses.

Then, there have been studies of academic institutions. Not least, there has been a rapid growth of work on the economic, social and cultural effects of universities (which have turned out to be substantial) and equally, of late, the economic, social and cultural effects of students. Then there is work on what makes a city 'creative'. Writers like Ken Worpole and Franco Bianchini have pointed to the cultural creativity that some cities seem to enjoy which, at least in part, seems to be a function of their pedagogic power.

But, in all this, what has not been realized has been that capitalism itself has been receiving a pedagogic boost. The four myths of urban life tend to support the authorized version of a 'global' capitalism which is interpreted as a vast and monolithic presence, now able to dream global dreams through the advent of the space of flows: cyberspace becomes a metaphor for the flow of capital (Buck-Morss, 1995). This authorized version of global capitalism has been repeated many times since the cultural turn has swept across the social sciences and humanities, most probably because it allows capitalism to be portrayed as something always already accounted for. Its presence can therefore be acknowledged but then everyone can get on to the more interesting, more cultural things. The irony, of course, is that those who manage capitalism no longer see it in this way. Beginning in the 1960s a new discourse of how capitalism is has come into being. Spun out of the interaction between events, managers and business schools, this discussion interprets the world in which capitalist organizations must survive as profoundly *uncertain*: capitalists don't know what is going on either. The business organization's 'environment' is framed as multiple, complex and fast-moving, and therefore as ambiguous, fuzzy, plastic or even chaotic. To survive, the business organization must attempt to form an island of superior adaptability able to 'dance' its way out of trouble – in fact, dancing is a favourite metaphor of the new capitalism. This means three things in particular.

Firstly, in a business world that is increasingly constructed by information, knowledge becomes crucial. Increasingly, knowledge is seen as the key to competitive advantage. Thus, whereas managers

> used to think that the most precious resource was capital, and that the prime task of management was to allocate it in the most productive way, now they have become convinced that their most precious resource is knowledge and that the prime task of management is to ensure that their knowledge is generated as widely and used as efficiently as possible. (Wooldridge, 1995: 4)

In Drucker's (1982: 16) famous words, 'Knowledge has become the key economic resource and the dominant, if not the only source of comparative advantage.'

Secondly, it follows that organizations must foster and preserve their knowledge base. This will be achieved via a number of means. These include the introduction of an emergent 'evolutionary' or 'learning' strategy which is 'necessarily incremental and adaptive but that does not in any way imply that its evolution cannot be, or should not be analysed, managed, and controlled' (Kay, 1993: 359). Such a strategy will be based on what are seen as the particular capabilities of a business organization which are then amplified via informal methods of control which rely on a much greater grasp of the issues involved, and which also mean that whole layers of bureaucracy, most of whose time was taken up with oversight, can be shrunk or, in the jargon, 'delayered' (Clark and Newman, 1993). Another means of fostering knowledge is to pay much greater attention to the skills of the workforce. Thus the organization pays much closer attention to enhancing formal skills. It will also likely become involved in experiential learning which involves placing the workforce in situations which demand co-operative responses to the uncertain and unknown (Martin, 1994). Most importantly, the organization will also pay close attention to the resources of *tacit* (familiar but unarticulated) *knowledge* embodied in its workforce and to the generation of trust, both within its workforce and with other organizations. Work on tacit knowledge has been almost entirely generated from the writings of Michael Polanyi (Botwinick, 1986) (rather than, for example, Heidegger, Merleau-Ponty or Bourdieu) who, in turn, drew on the ideas of gestalt psychology. Polanyi's (1967:20) most famous saying, 'We can know more than we can tell,' has become a vital part of business discourse, as a way into the problem of mobilizing the full bodily resources of the workforce. In turn, Polanyi's work has underlined the need to generate *trust* or (as Polanyi often called it) confidence, since 'the overwhelming proportion of our factual beliefs continue to be held at second hand through trusting others' (Polanyi, 1958: 208). Thirdly, the business organization is mainly therefore seen as a culture which attempts to generate new representations of itself (new metaphors, new traditions) which will allow it to see itself and others in new, more profitable ways.

This, then, is a radically different economic narrative in which the capitalist organization is framed as always in *action*, 'on the move, if only stumbling or blundering along' (Boden, 1994: 192), but stumbling or blundering along in ways which will allow it to survive and prosper, most particularly through mobilizing a culture which will produce traditions of learning (collective memories which will act both to keep the organization constantly alert and as a reservoir of innovation (Lundvall, 1992)) and extensive intra- and inter-firm social networks (which will act both as conduits of knowledge and as a means of generating trust). In turn, the manager has to become a kind of charismatic itinerant, a 'cultural diplomat' (Hofstede, 1991), constantly imbuing the business organization's values and goals, constantly on a mission to explain and motivate. Management is no longer, therefore, seen as a science. Rather it becomes an art form, dedicated to 'the proposition that a political economy of information is in fact coexistent with a theory of culture' (Boisot, 1995: 7). Capitalism is increasingly a part of the 'humanities' as well as the 'sciences'.

In other words, the rational company man of the 1950s and 1960s, skilled in the highways and byways of bureaucracy, becomes the reflexive corporate social persona of the 1990s, skilled in the arts of social presentation and 'change management'. And the giant multidivisional corporation of the 1950s and 1960s now becomes a 'leaner', 'networked', 'post-bureaucratic', 'virtual' or even 'poststructuralist' organization, a looser form of business which can act like a net floating on an ocean, able to ride the swell and still go forward (Drucker, 1988; Heckscher and Donellon, 1994; Eccles and Nohria, 1990) because it too is perpetually in transition.

This is a discourse that is aware of its own limits, and it is therefore no surprise that, as each year comes and goes, new management ideas appear, as both a means of meeting uncertainty and the never-ending appetite of this form of capitalism for new ideas and practices. New managerial magic 'spells' – like quality circles, the paperless office, the factory of the future, intrapreneurship, brands, strategic alliances, globalization, and business-process re-engineering – usually last for only a limited time but, since capitalist firms have the power to realize their fantasies, these spells are often responsible for quite extraordinary rounds of industrial restructuring with serious consequences for workforces. In a sense, then, capitalism has returned to its magical past.

How has this new 'alchemical' business discourse been produced and disseminated? Mainly through a primarily urban pedagogic infrastructure, which, in effect, dates only from the 1960s and which has now grown into a powerful, specialized system for the production and distribution of business knowledge, which elsewhere I have called the 'capitalist circuit of cultural capital' (Thrift, 1997). The institutions chiefly responsible for producing business knowledge include the institutions of formal business education, and especially the MBA course which has produced a large number of academics and students who have acted both to generate and transport the new knowledge. (In the US, admittedly the most extreme example, almost one in four students now majors in business, while the number of business schools has grown fivefold since 1957). Then there are a further set of institutions which have been responsible for dissemination. These include management consultants and management gurus. Consultants have acted as a key link between theory and practice and they have often had a vital interest in disseminating the new discourse because of their attempts at authorship. 'Management gurus' like Peter Drucker, Charles Handy, John Kay, Theodore Levitt, Gareth Morgan, John Naisbitt, Tom Peters, Rosabeth Moss Kanter, Kenichi Ohmae, and the like, have become increasingly important as embodiments of new managerialist arguments. These gurus have been responsible for the diffusion of a whole host of the 'business fads' taught on management courses which, jointly and singly, have promoted a new managerial worldview (Huczynski, 1993). Distribution of the discourse takes place via a selection of different media, some specialized, some not. The discourse thereby reaches various audiences (though we know remarkably little about these audiences).

In sum, what can be said is that cities are now the sites of a *knowledgeable capitalism* which, through its institutional infrastructure, is intent on producing and disseminating a specific discourse of adaptability and learning. This discourse is not passive for two reasons. Firstly, it is a discourse of constant change and adaptation, Secondly, it is intent on providing new managerial and worker identities, based on a concept of a person which is broader than before but still radically attenuated. Right now, for example, at numerous sites around the world, seminars and workshops are being held which are attempts to engrain this 'enterprising' identity by harnessing *belief*. In these seminars and workshops:

> the worker is represented as an individual in search of meaning in work, and working to achieve fulfilment through work. Excellent organisations are those that 'make money for people' by encouraging them to believe that they have control of their own destinies; that, no matter what position they may hold in an organisation, their contribution is vital, not only to the success of the company for which they work but also to the enterprise of their own lives. Peters and Waterman (1982, 81, 45) for example, quote approvingly Nietzsche's axiom 'that he who has a *why* to live for can bear most any *how*'. They argue that 'the fact that ... we think we have a bit of discretion leads to much greater commitment' and that 'we desperately need meaning in our lives, and will sacrifice a great deal to institutions that will provide meaning for us. We simultaneously need independence to feel as though we are in charge of our own destinies, and to have the ability to stand out.' In this vision work is a sphere within which the individual constitutes and confirms his or her identity. Excellent organisations get the most out of everyone, not by manipulating group human relations to secure a sense of 'belonging', but by harnessing the psychological stirring of individuals for autonomy and creativity and channelling them into the search for excellence and success. (Du Gay, 1996: 60)

Thus, in some senses, it is *belief itself*, rather than what is believed, that counts as important. This means that all manner of cues can be regarded as legitimate. For example, that recent staple of New Age thought, *The Celestine Prophecy* (Redfield, 1994) is used as a key text in a number of management seminars. Again the galactic outings of the Starship Enterprise form the material for lessons of management in other seminars. The magic has not gone away: if anything, it is coming back!

The unjust city (unequal relations of power, new forms of injustice and alternative financial institutions)

I want to end this chapter by alighting on what might seem familiar terrain: the unjust city of Engels, Marx and Harvey, of Dickens, Doré and Campbell, and, currently, of a burgeoning 'social exclusion' industry. In much of this work, resistance to established orders is assumed to take particular forms: it is driven by a cognitivist agenda, it is tightly organized, it is heroic, and only certain groups are therefore deemed to be truly resistant. The corollary is that social groups without this profile, which are *ad hoc*, mundane and only tangentially resistant to the established order, are seen as second-class citizens. A modest sense of disparity is disparaged. Thus, for example, in the cultural studies literature, only certain social groups are usually deemed to be resistant to the established order. Indeed, there is often an implied hierarchy, usually with subcultures engaged in direct action occupying the apex (McKay, 1996). The corollary is that the concerns of more mundane movements often appear to be written out.[7] In turn, the function that such social groups have of disclosing new *skills* and *styles of activity* is lost (Spinosa, Flores and Dreyfus, 1997).

In the final part of this chapter I want to underline this point by directing attention to an important set of alternatives to the established order which are currently being constituted and which, because they seem so mundane and everyday, have received relatively little attention but which have a clear potential to help in the transformation of the lives of those living in low-income urban communities. The problem is this. In British cities, and indeed in numerous other cities around the world, many low-income people have little or no access to the formal financial system. In Britain, the degree of access has expanded because of the juxtaposition of free banking and an extensive bank and building society branch network, although it would be romantic to suggest that access to the financial system ever reached the lowliest of low-income communities. But, for a period, more people have had access to the formal financial system and with it not only to normal banking facilities, but, more importantly, credit. In the 1990s, as in other countries around the world, this access is in danger of being stripped away. Free banking has disappeared in the United States and will soon disappear in the UK. Physical access is disappearing as branch networks close down. (Further, the closure of branch networks is concentrated in the inner areas of the larger British cities).

In theory, there is an alternative. This is the application of information technology, from automatic telling machines (ATMs) to the telephone to the personal computer, which can at least solve some of the problems of physical access. But, in fact, information technology is not a panacea. ATMs can only be used to obtain cash. Telephone banking depends on access to a telephone, and, in any case, is specifically aimed at the relatively well off; indeed it is being used as a way of 'cherry picking' customers. The personal computer can only be used by the small (and affluent) percentage of the population with a PC and a modem. Then again, banks, building societies, and insurance companies are drawing on more and more sophisticated information systems, operated by fewer and fewer companies (in fact, in Britain there are now only two major information companies), to obtain financial information and to serve customers. People are therefore increasingly provided with access to the financial system not only on the basis of monetary worth but also what might be called their 'information worth' etched in computer memories.

In other words, we are in the midst of a process of very rapid uneven development. On one side of the coin, there is a process of 'telematic super-inclusion' taking place, as many more financial institutions fight to sell an ever-greater range of services and products to a group of relatively affluent people using the new information technologies. But the other side of the coin is *financial exclusion* as lower-income communities are deprived of whatever basic financial services they had, through the withdrawal of financial infrastructure like bank and building society branches, through pricing out (for example, charging unmanageable insurance premiums) or through credit scoring policies. Thus many low-income areas are being *de facto* 'redlined' without any explicit policy of redlining.[8]

What can be done to halt this slide towards financial desertification? Barring legislation such as EU social banking directives, of which there is only a very distant prospect, the answer would seem to be to draw on the tradition of *alternative financial institutions*, from the early friendly societies on, which have provided both a material and symbolic resistance to prevailing views of money as a natural medium of exchange. These institutions are mundane. But they contain within them the seed of different means of imagining money, different everyday moralities which can refigure the everyday life of the city.

Firstly, there are *community development banks*, which attempt to substitute for the absence of development funds in mostly run-down and economically distressed communities, which have normally been abandoned by large, more formal financial institutions, and which seek to provide finance for urban redevelopment and business financing. Secondly, there are *credit unions*, which seek to substitute for the absence of formal retail financial institutions, by providing relatively small loans to individuals and households. Thirdly, there are *rotating savings and credit associations* (ROSCAs). These institutions have long been the subject of anthropological and sociological investigation in developing countries, and are defined by Ardener (1995: 1) as associations 'formed upon a core of participants who make regular contributions to a fund which is given in whole or in part to each contributor in turn'. These informal pools of money have had important developmental effects in many parts of the world lacking more formal financial infrastructure. One interpretation of ROSCAs was to see them as intermediate financial institutions, which would disappear as traditional rural societies give way in the face of the advance of industrialization and modernity, and which would be replaced by more formal, bureaucratic financial organizations. However, the limited, transitional role ascribed to ROSCAs is challenged by an alternative view which sees them as durable and longlasting institutions, which can have a development role beyond the pre-industrial societies from which they emerged. One reason for this has been the general failure of development in much of the Third World, which means that the need for ROSCAs remains pressing. Another reason has been the success of organizations such as the Grameen Bank of Bangladesh, in essence an institutionalized form of ROSCA, which has gained international recognition for the economic gains it has delivered to many poor and disadvantaged women. Yet another reason is that in the past thirty years or so ROSCAs have migrated from the developing world to the industrialized world, brought along by various diasporic communities. As a result, these institutions now provide much-needed supplies of credit within a constellation of minority communities in the West, ranging from Eritrean and South Asian women in Oxford (Almedon, 1995; Srinivon, 1995) to Koreans in Los Angeles (Light and Deng, 1995).

The fourth and final type of alternative institution of accumulation is *LETS* (Local Exchange and Trading Systems) (Lee, 1996; Thorne, 1996; Williams, 1996). The idea of creating what is in effect an alternative financial system with its own unique currency is hardly a new one. For example, in 1832 Robert Owen attempted to set up an alternative financial system in London where the value of the currency was based upon labour time (Angell, 1930). Similar schemes grew up elsewhere in Britain and in the United States, as communities drew inspiration from Owen's attempt to 'revalue' money, although, as Angell reveals, all were relatively short-lived.

LETS are in many ways merely the latest incarnation of such schemes, being developed as part of a community-based response to economic recession in Vancouver in the 1980s (Thorne, 1996). From there, LETS have spread to the United States, Australia, Britain and elsewhere. In doing, they have created an emerging global 'necklace' of alternative financial systems (cf. Storper, 1992), which are linked to one another both by institutions which serve to promote the LETS movement, and by a dense information network which makes extensive use of e-mail and the World Wide Web (Internet). In all this, as Thorne (1996:1365) points out, the LETS movement is infused with a desire to remake social relations at a local scale, so that 'the search for material benefit *emerges* in the context of a primarily social motivation which then *re-emerges* through the exchange of goods and services ... In this way, social networking is institutionalized into a permanent system through a money created locally' (original emphasis).

Of course, care has to be taken with each of these different manifestations of a wish for there to be something more to money than just money. Firstly, it cannot be assumed that these developments are inherently radical. As the example of many ROSCAs shows, they can be means of self-help but, equally, they can also be a means of concentrating and recycling resources amongst those who are already relatively well-off. Secondly, nor should it be assumed that just because these institutions are alternative they operate on more straightforward principles than more formal financial institutions. Many alternative financial institutions are of Byzantine complexity and require their participants to possess a high level of financial literacy. Thirdly, the problems of operating beyond the mainstream economy should not be underestimated and, as Thorne (1996) reveals, many LETS in the UK are encountering difficulties as the initial rush of enthusiasm wears off and as members become more aware of the limitations as well as the possibilities.

Nevertheless, this should not cloud the fact that such institutions do possess considerable possibilities. To begin with, they are a means, however modest, of fighting back against a 'Gradgrind' mindset which assumes that money is only ever about harvesting profits. Then again, most alternative financial institutions have begun as 'bottom-up' and largely practical initiatives, often as examples of institutional 'templates' moving from the Third World (as in the case of the micro-credit movement inspired by the example of the Grameen Bank). As such, they are a means of once more underlining the dangers of an academic gaze which tends to ignore social developments that are practical in nature, and therefore often not written down.

Alternative financial institutions are also a means of promoting, in conjunction with other institutions, the kind of democratic associationism which is now much in vogue, by pointing out that money and finance is as much a part of civil society as it is of the economy (for example, Cohen and Rogers, 1995; Amin and Thrift, 1995). And, finally, their emergence is symbolic of a wider movement which is re-evaluating the role of money and finance (for example, Dodd, 1994; Pollin, 1995), and which is intent on bringing it back to the social and cultural realm from which it has only ever partially escaped. These attempts to rework the meaning of money can give the city back to lower-income families who are currently trapped by monetary practices that do not include them.

Conclusions

I have tried to paint a picture of cities as a set of diverse and interacting practical orders in which the interaction is more important than the order. In this sense of how things are, new urban spaces are constantly being disclosed through the hubbub of cultural production. This sense therefore refuses the naturalizing epistemological account that assumes there is a common urban order we can all access. (Indeed, in some cases, e.g. Newman & Holzman, 1997, it rejects the epistemological account altogether.) Instead it is interested in an ontological politics which searches after good ways of making partial connections between worlds. In doing so, it gives us back the city as a means of performing difference (Hetherington & Munro, 1997) and validates many different performances.

References

Acuna, R. F. (1996) *Anything But Mexican. Chicanos in Contemporary Los Angeles*, Verso, London.

Ahearne, J. (1995) *Michel de Certeau. Interpretation and its Other*, Polity Press, Cambridge.

Allan, G. (1996) *Kinship and Friendship in Modern Britain*, Oxford University Press, Oxford.

Almedon, A. M. (1995) A note on ROSCAs among Ethiopian women in Addis Ababa and Eritrean women in Oxford. In Ardener, S. and Burman, S. (eds) *Money-Go-Rounds: The Importance of Rotating Savings and Credit Associations for Women*, Berg, Oxford, 71–6.

Alvarez, A. (1995) *Night. An Exploration of Night Life, Night Language, Sleep and Dreams*, Cape, London.

Amin, A. and Thrift, N. (1995) Institutional issues for the European regions: from markets and plans to socioeconomics and powers of association. *Economy and Society*, **24**, 41–66.

Angell, N. (1930) *The Story of Money*, Cassell, London.

Ardener, S. (1995) Women making money go round: ROSCAs revisited. In Ardener, S. and Burman, S. (eds) *Money-Go-Rounds: The Importance of Rotating Savings and Credit Associations for Women*, Berg, Oxford, 1–19.

Armstrong, I. (1996) Transparency: towards a poetics of glass in the nineteenth century. In Spufford, F. and Uglow, J. (eds) *Cultural Babbage. Technology, Time and Invention*, Faber and Faber, London, 123–48.

Bachelard, G. (1964) *The Poetics of Space*, Beacon Press, Boston.

Bateson, G. (1973) *Steps to an Ecology of Mind*, Picador, London.

Bauman, Z. (1993) *Postmodern Ethics*, Blackwell, Oxford.

Bell, D. and Valentine, G. (eds) (1995) *Mapping Desire*, Routledge, London.

Benjamin, W. (1969) *Illuminations*, Schocken Books, New York.

Benjamin, W. (1979) *One Way Street*, New Left Books, London.

Bernstein, J. (1994) *Foregone Conclusions*, University of California Press, Berkeley.

Boden, D. (1994) *The Business of Talk*, Polity Press, Cambridge.

Botwinick, A. (1986) *Participation and Tacit Knowledge in Plato, Machiavelli and Hobbes*, University of American Press, Lanham.

Bruno, G. (1993) *Streetwalking on a Ruined Map. Cultural Theory and the Films of Elvira Notari*, Princeton University Press; Princeton.

Buck-Morss, S. (1995) Envisioning capital: political economy on display. In Cooke, L. and Wollen, P. (eds) *Visual Display. Culture Beyond Appearances*, Bay Press, Seattle pp.110–41.

Callon, M. and Law, J. (1995) Agency and the hybrid collectif. *South Atlantic Quarterly*, **94**, 481–507.

Casey, E. W. (1993) *Getting Back into Place. Towards a Renewed Understanding of the Place-World*, University of Indiana Press, Bloomington.

Castells, M. (1989) *The Informational City*, Blackwell, Oxford.

Clark, J. and Newman J. (1993) The right to manage: a second managerial revolution? *Cultural Studies* **7**, 427–41.

Cohen, J. and Rogers, R. (1995) *Associate Democracy*, Verso, London.

Collins, J. (1995) *Architectures of Excess. Cultural Life in the Information Age*, Routledge, London.

Crossley, N. (1996) *Intersubjectivity. The Fabric of Becoming*, Sage, London.

Davis, E. (1994) Techgnosis, magic, memory and the angels of information. *South Atlantic Quarterly*, **92**, 518–617.

Davis, M. (1990) *City of Quartz. Excavating Los Angeles*, Verso, London.

Davis, M. (1996) Cosmic dancers on history's stage? The permanent revolution in the earth sciences. *New Left Review*, **217**, 48–84.

de Certeau, M. (1984) *The Practice of Everyday Life*, University of California Press, Berkeley.

Debray, R. (1996) *Media Manifestations. On the Technological Transmission of Cultural Forms*, Verso, London.

Dodd, N. (1994) *The Sociology of Money: Economics, Reason and Contemporary Society*, Polity Press, Cambridge.

Drucker, P. (1988) The coming of the new organisation. *Harvard Business Review*, **88**, 45–53.

Du Gay, P. (1996) Organising identity. Entrepreneurial governance and public management. In Hall, S. and Du Gay, P. (eds) *Questions of Identity*, Sage, London, 151–69.

Eccles, R. and Nohria, N. (1990) *The Post-structuralist Organisation*, Harvard Business School, Cambridge (Working Paper 92–003).

Edwards, M. (1996) *Sunday Times*, Books Section (10 March).

Finch, J. (1989) *Family Obligations and Social Change*, Polity Press, Cambridge.

Finch, J. and Mason, J. (1993) *Negotiating Family Responsibilities*, Routledge, London.

Finnegan, R. (1989) *The Hidden Musicians*, Cambridge University Press, Cambridge.

Fisher, C. S. (1992) *Calling America. The Social History of the Telephone*, University of California Press, Berkeley.

Frow, J. (1997) *Time and Commodity Culture. Essays in Cultural Theory and Postmodernity*, Oxford University Press, Oxford.

Geertz, C. (1972) Deep play: notes on the Balinese cock fight. *Dadalus*, **101**, 1–37.

Gibbons, M., Limoges, C., Nowotny, H., Schwartzman, S., Scott, P. and Trow, S. (1994) *The New Production of Knowledge*, Sage, London.

Giddens, A. (1992) *The Transformation of Intimacy*, Polity Press, Cambridge.

Glennie, P. and Thrift, N. J. (1996) Consumers, identities, and consumption spaces in early-modern England. *Environment and Planning A*, **28**, 35–46.

Grabher, G. and Stark, D. (1996) Organising diversity: evolutionary theory, network analysis and postsocialist transformations. In Grabher, G. and Stark, D. (eds) *Restructuring Networks: Legacies, Linkages and Localities in Postsocialism*, Oxford University Press, Oxford.

Grosz, E. (1996) *Space, Time and Perversions*, Routledge, London.

Harraway, D. (1995) The promises of monsters: a regenerative politics for inappropriate/d others. In Grossberg, L., Nelson, C. and Treichler, P. (eds) *Cultural Studies*, Routledge, New York, 295–337.

Harvey, D. (1989) *The Condition of Postmodernity*, Blackwell, Oxford.

Heckscher, C. and Donellan, A. (1994) *The Post-Bureaucratic Organisation: New Perspectives and Organisational Change*, Sage, London.

Heelas, P. (1996) *The New Age Movement. The Celebration of the Self and the Sacralization of Modernity*, Blackwell, Oxford.

Hetherington, K. and Munro, R. (eds) (1997) *Ideas of Difference. The Labour of Division*, Oxford, Blackwell.

Hofstede, G. (1991) *Cultures and Organisations*, McGraw Hill, New York.

Holmes, R. (1993) *Dr Johnson and Mr Savage*, Flamingo, London.

Huczynski, A. (1993) *Management Gurus: What Makes Them and How to Become One*, Routledge, London.

Huizinga, J. (1949) *Homo Ludens: A Study of the Play Element in Culture*, Routledge and Kegan Paul, London.

Humphrey, N. (1995) *Soul Searching. Human Nature and Supernatural Belief*, Chatto and Windus, London.

Janelle, D. G. (1969) Spatial reorganization: a model and a concept. *Annals of the Association of American Geographers*, **59**, 348–64.

Kay, J. (1993) *Foundations of Corporate Success*, Oxford University Press, Oxford.

King, A. D. (ed.) (1996) *Re-Presenting the City. Ethnicity, Capital and Culture in the 21st-Century Metropolis*, Macmillan, Basingstoke.

Latour, B. (1993) *We Have Never Been Modern*, Harvester-Wheatsheaf, Hemel Hempstead.

Law, J. and Mol, A. (1996) Decision/s. Unpublished paper.

Lee, R. (1996) Moral money? LETS and the social construction of local economic geographies in southeast England. *Environment and Planning A*, **28** (8), 1377–94.

Lefebvre, H. (1995) *Writings on Cities*, Blackwell, Oxford.

Light, I. and Deng, Z. (1995) Gender differences in ROSCA participation within Korean business households in Los Angeles. In Ardener, S. and Burman, S. (eds) *Money-Go-Rounds: The Importance of Rotating Savings and Credit Associations for Women*, Berg, Oxford, 217–40.

Lingis, A. (1994) *Foreign Bodies*, Routledge, New York.

Lundvall, B.A. (1992) *National Systems of Innovation: Towards a Theory of Innovation and Interactive Learning*, Pinter, London.

Martin, E. (1994) *Flexible Bodies*, Beacon Press, Boston.

McKay, G. (1996) *Cultures of Resistance*, Verso, London.

McNeill, W. H. (1995) *Keeping Together in Time: Dance and Drill in Human History*, Harvard University Press, Cambridge, Mass.

McRobbie, A. (1984) Dance and social fantasy. In McRobbie, A. and Nava, M. (eds) *Gender and Generation*, Macmillan, London.

Morson, G. S. (1994) *Narrative and Freedom. The Shadows of Time*, Yale University Press, New Haven.

Mort, F. (1996) *Cultures of Consumption*, Routledge, London.

Newman, F. and Holzman, L. (1997) *The End of Knowing. A New Developmental Way of Learning*, Routledge, New York.

Nye, D. (1990) *Electrifying America*, MIT Press, Cambridge, MA.

Nye, D. (1995) *The Technological Sublime*, MIT Press, Cambridge, MA.

Pahl, R. (1995) *After Success. Fin-de-Siècle Anxiety and Identity*, Polity Press, Cambridge.

Pini, M. (1996) Dance classes – dancing between classifications. *Feminism and Psychology*, **6**, 411–26.

Polanyi, M. (1958) *Personal Knowledge: Towards a Postcritical Philosophy*, Routledge and Kegan Paul, London.

Polanyi, M. (1967) *The Tacit Dimension*, Routledge and Kegan Paul, London.

Pollin, R. (1995) Financial structures and egalitarian economic policy. *New Left Review*, **214**, 26–61.

Pollock, G. (1994) Territories of desire: reconsiderations of an African childhood. In Robertson, G., Mash, M., Tickner, L., Bird J., Curtis, B. and Putnam, J. (eds) *Travellers' Tales: Narratives of Home and Displacement*, Routledge, London, 63–89.

Porter, R. (1993) Baudrillard, hystory, hysteria and consumption. In Rojek, C. and Turner, B. (eds) *Forget Baudrillard?*, Routledge, London, 1–21.

Prendergast, C. (1992) *Paris and the Nineteenth Century*, Blackwell, Oxford.

Probyn, E. (1996) *Outside Belongings*, Routledge, London.

Radley, A. (1995) The elusory body and social constructionist theory. *Body and Society*, **1**, 3–24.

Redfield, J. (1994) *The Celestine Prophecy*, Bantam, London.

Reed, D. (ed.) *Not at Home. The Suppression of Domesticity in Modern Art and Architecture*, Thames and Hudson, London.

Relph, E. (1981) *Place and Placelessness*, Pion, London.

Rose, N. (1996) *Inventing Our Selves*, Cambridge University Press, Cambridge.

Samuel, R. (1994) *Theatres of Memory*, Verso, London.

Savage, M. (1995) Walter Benjamin's urban thought: a critical analysis. *Environment and Planning D: Society and Space*, **13**, pp.201–16.

Savage, M. and Warde, A. (1993) *Urban Sociology, Capitalism and Modernity*, Macmillan, London.

Scott, A. and Soja, E. (1986) Los Angeles: the capital of the twentieth century. *Environment and Planning D: Society and Space*, **4**, 249–54.

Sennett, R. (1995) *Flesh and Stone. The Body and the City in Western Civilization*, Allen Lane, London.

Serres, M. (1995) *Genesis*, University of Michigan Press, Ann Arbor.

Serres, M. and Latour, B. (1995) *Conversations on Science, Culture and Time*, University of Michigan Press, Ann Arbor.

Sheringham, M. (ed.) (1996) *Parisian Fields*, Reaktion Books, London.

Social and Community Planning Research (1996) *British Social Attitudes*, HMSO, London.

Soja, E. (1989) *Postmodern Geographies*, Verso, London.

Spinosa, C., Flores, F. and Dreyfus, H. L. (1997) *Disclosing New Worlds. Entrepreneurship, Democratic Action and the Cultivation of Solidarity*, MIT Press, Cambridge, MA.

Spufford, F. (1996) The difference engine and *The Difference Engine*. In Spufford, F. and Uglow, J. (eds) *Cultural Babbage. Technology, Time and Invention*, Faber and Faber, London, 266–90.

Srinivon, S. (1995) ROSCAs among South Asians in Oxford. In Ardener, S. and Burman, S. (eds) *Money-Go-Rounds: The Importance of Rotating Savings and Credit Associations for Women*, Berg, Oxford, 199–208.

Stebbins, R. A. (1979) *Amateurs. On the Margin Between Work and Leisure*, Sage, Beverly Hills.

Storper, M. (1992) The limits of globalisation: technology districts and international trade. *Economic Geography*, **68**, 60–93.

Strathern, M. (1995) *Shifting Contexts*, Routledge, London.

Taussig, M. (1992) *The Nervous System*, Routledge, London.

Thomas, H. (1995) *Dance, Modernity and Culture*, Routledge, London.

Thomas, H. (ed.) (1996) *Dance in the City,* Macmillan, London

Thorne, L. (1996) Local exchange trading systems in the United Kingdom: a case of reembedding? *Environment and Planning A*, **28** (8), 1361–76.

Thornton, S. (1995) *Club Cultures*, Polity Press, Cambridge.

Thrift, N. J. (1996) *Spatial Formations*, Sage, London.

Thrift, N. J. (1996) Old technological fears and new urban eras: reconfiguring the goodwill of electronic things. *Urban Studies*, **33**, 1463–93.

Thrift, N. J. (1996) The place of complexity: towards a metaphorical geography. Paper presented to the Centre for the Study of Science and Technology Conference on Complexity, Keele University, November.

Thrift, N. J. (1997) The rise of soft capitalism. In Herod, A., Roberts, S. and Toal, G. (eds) *Globalising Worlds*, Routledge, London.

Tickell, A. and Peck, J. (1995) Social regulation after Fordism: regulation theory, neo-liberalism and the global-local nexus. *Economy and Society*, **24**, 357–86.

Virilio, P. (1993) The third interval: a critical transition. In Conley, V. A. (ed.) *Rethinking Technologies*, University of Minnesota Press, Minneapolis, 3–12.

Walker, R. A. (1994) The playground of US capitalism? The political economy of the San Francisco Bay Area in the 1980s. In Davis, M. (ed.) *Fire in the Hearth*, Verso, London, 43–82.

Walker, R. A. (1995) Landscape and city life: four ecologies of resistence in San Francisco. *Ecumene*, **2**, 33–64.

Walker, R. A. (1996) Another round of globalisation in San Francisco, *Urban Geography*, **17**, 60–94.

Wellman, B. and Berkowitz, S. D. (eds) (1988) *Social Structures. A Network Approach*, Cambridge University Press, Cambridge.

Weston, K. (1991) *Families We Choose: Lesbians, Gays, Kinship*, Columbia University Press, New York.

Williams, C. C. (1996) Local exchange and trading systems: a new source of work and credit for the poor and unemployed. *Environment and Planning A*, **28** (8), 1395–416.

Williams, R. H. (1991) *Notes on the Underground. An Essay on Technology, Society and the Imagination*, MIT Press, Cambridge, MA.

Winnicott, D. (1971) *Playing and Reality*, Penguin, Harmondsworth.

Wittgenstein, L. (1998) *Culture and Value*, revised edition, Blackwell, Oxford.

Wollen, P. (1995) Tales of total art and dreams of the total museum. In Cooke, L. and Wollen, P. (eds) *Visual Display: Culture Beyond Appearances*, Bay Press, Seattle, 154–77.

Woods, T. (1995) 'Looking for signs in the air': urban space and the postmodern in *In the Country of Last Things*. In Barone, D. (ed.) *Beyond the Red Notebook. Essays on Paul Auster*, University of Pennsylvania Press, Philadelphia, 107–28.

Wooldridge, A. (1995) Big is back: a survey of multinationals. *The Economist* (24 June), 1–22.

Zajonc, A. (1993) *Catching the Light. The Entwined History of Light and Mind*, Oxford University Press, Oxford.

Notes

* This paper is an extended version of N. J. Thrift (1997) 'Cities without modernity, cities with magic', *Scottish Geographical Magazine*, **113**, 138–49. It is reprinted here with permission.

1   I am aware that I can be accused of setting up straw persons by formulating these four myths, but their currency is now so wide – even when ringed with caveats – that I feel little compunction in making this move.

2   The telephone is a particularly good example of this process. See Fisher, 1992; Thrift, 1996.

3   This is to ignore other media, for example the camera, and especially the impact of the flash camera.

4   Vidler's (1992) account of modern architecture as, in part, an attempt to make buildings more like bodies, is a fascinating addendum to this discussion.

5 The work of Deleuze is an obvious catalyst here.

6 Again, rather like music, dance is partially invisible because so much of its activity is 'amateur' (Stebbins, 1979).

7 Giovanna di Chiro (1995) makes a similar point in her work on the environmental movement in the United States which has found it difficult to come to terms with the more mundane concerns of many urban low-income communities.

8 Of course, it is important not to paint lower-income households as simply financial dupes; they use numerous clever financial strategies to get by (Kempson, 1996). But, most particularly in terms of access to credit and insurance, they are being more and more seriously disadvantaged.

Index*

* *Information in notes is indexed as 89(n13), ie. note 13 on page 89*

Printed and bound by CPI Group (UK) Ltd, Croydon, CR0 4YY

01/11/2024

01782614-0006